HERE BE DRAGONS

Science, Technology
and the Future of Humanity

OLLE HÄGGSTRÖM

OXFORD
UNIVERSITY PRESS

OXFORD
UNIVERSITY PRESS

Great Clarendon Street, Oxford, OX2 6DP,
United Kingdom

Oxford University Press is a department of the University of Oxford.
It furthers the University's objective of excellence in research, scholarship,
and education by publishing worldwide. Oxford is a registered trade mark of
Oxford University Press in the UK and in certain other countries

Published in the United States of America by Oxford University Press
198 Madison Avenue, New York, NY 10016, United States of America

British Library Cataloguing in Publication Data

Data available

Library of Congress Control Number: 2015937609

ISBN 978-0-19-872354-7

Printed in Great Britain by
Clays Ltd, St Ives plc

PREFACE

There is a widely held conception that progress in science and technology is our salvation, and the more of it, the better. This seems to be a default assumption not only among the general public but also in the research community, from university administration and research funding agencies all the way up to government ministries. I believe the assumption to be wrong, and very dangerous. There is no denying that advances in science and technology have brought us prosperity and improved our lives tremendously, or that further advances have the potential to bring glorious further benefits, but there is a flip side: some of the advances that may lie ahead of us can actually make us worse off, *a lot* worse, and, in the extreme case, cause the extinction of the human race.

We urgently need to find ways to push scientific and technological progress in directions that are likely to bring us good, and away from those directions that spell doom. This cannot be done if we stick to the erroneous view that all such progress is good for us. The first thing we need is to be able to distinguish those advances whose potential is mostly in the direction of prosperity and human flourishing from those whose potential is more in the direction of destruction and doom, and we need to find safe ways to handle those technologies that come with elements of both. Our ability to do so today is very limited, and my ambition with this book is to draw attention to the problem, so that we can work together to improve the situation, and avoid running blindfolded and at full speed into a dangerous future.

This is a book on science, technology and their impact on the future of humanity. The future is an exceedingly difficult topic to say anything certain about. I have tried to approach it with the appropriate level of epistemic humility – perhaps to the extent that some readers may get tired of all the "might possibly" and "could potentially" and "if this theory is correct, then . . . " that permeate the text, but my ambition of intellectual honesty compels me to include them. This humility will, however, not prevent me from stating bluntly and clearly when I find someone's argument to be dead wrong, a case in point being my treatment in Section 4.7 of philosopher John Searle's confused argument for why we have no reason to fear a *Terminator*-like scenario where robots take over the world and set out to destroy us.

I am a professor of mathematical statistics, and the reader will notice that, apart from a few sections in Chapters 6 and 7, I treat mostly topics that lie distinctly outside my professional area of expertise. To do so may seem immodest and reckless, but I have two things to say in my defense. First, I have tried very hard to respect the expertise of those who know more than I do about a particular topic.[1] Second, the subject of the book is so multifaceted and cross-disciplinary that *anyone* who takes it on will find themselves to be a non-expert on most parts of it; yet, this is a book that needs to be written, so *someone* needs to write it, and it might as well be me.

The first text I tried writing on the subject of emerging technologies and their future impact on humanity was a chapter named *Brytningstid* (which is Swedish, meaning "transition period" but with a somewhat more poetic ring to it) in my 2008 book *Riktig vetenskap och dåliga imitationer* ("Real science and bad imitations").[2] I have since then read much and thought hard about the topic, and my views on it, although still tentative, have matured considerably. Still, the present book can be seen as having grown out of that chapter, which was inspired by the late James Martin's book *The Meaning of the 21st Century: A Vital Blueprint for Ensuring our Future*.[3] That book, which did much to open my eyes to this topic, is driven by a similar sentiment to the present book, which, however, is in no way meant to replace Martin's, or even compete with it: his style is inimitable, and the subject common to our books is sufficiently broad that there is ample scope to distribute the emphasis differently. Anyone who decides to read Martin's book and mine in tandem is likely to find that they complement each other at least as much as they overlap.

James Martin passed away in 2013 at the age of 79, but fortunately there is a younger generation (within which I count myself) of scientists, writers and philosophers who share his passionate ambition to think about the future rationally and without fear of overstepping the narrow boundaries of the box of mainstream discourse. Among these, there are many brilliant thinkers, and many whose ideas and writings have influenced this book – too many to be listed here, except for the following four, whose influence on this book has been particularly profound: Nick Bostrom, Robin Hanson, Anders Sandberg and Eliezer Yudkowsky. A quick look at the reference list will give a hint of the extent of their influence; every chapter draws on work by at least one of them, and most chapters by more than one.

*

I am grateful to my colleagues and bosses at the Department of Mathematical Sciences at Chalmers, and even more so to my dear wife Marita Olsson, for

---

[1] There is an obvious tension between this statement and passages such as the aforementioned treatment of philosophical arguments by renowned philosopher John Searle. What can I say? One has to strike a balance. When someone is wrong, they're wrong, but I have refrained from delivering such verdicts without first having seriously entertained the possibility that they're right.

[2] Häggström (2008).

[3] Martin (2006).

their charitable acceptance and support of my rather offbeat project of writing this book. My gratitude furthermore extends to Björn Bengtsson, Bengt Brülde, Joar Guterstam, Robin Hanson, Karim Jebari, Johan Jonasson, Patrik Lindenfors, Keith Mansfield, Klas Markström, Luke Muehlhauser, Peter Olofsson, Ulf Persson, Vilhelm Verendel, Thomas Weibull and Peter Winkler; their influence on the book, through reading and criticizing drafts of various parts, has been considerable, although of course I am solely to blame for the errors and other shortcomings that undoubtedly remain, and for the opinions I have decided to stick to despite their protests.

CONTENTS

CHAPTER 1

Science for good and science for bad

1.1 A horrible discovery

About half a millennium into the future, humanity encounters, for the very first time, an extraterrestrial intelligent species. That is how Eliezer Yudkowsky's fascinating and provocative short story *Three Worlds Collide* (2009) begins. The following subplot can be revealed without spoiling the overall plot too much.

After some preliminary exchanges, the human space travelers and their alien counterparts compare notes regarding the scientific findings of their respective species. Allowing for major cultural differences in how the findings are expressed, it turns out that for the most part, the physics, chemistry and so on found by the two species are basically the same. Perhaps this is not so surprising, since after all we inhabit the same universe. In particular, the aliens have, like us, the notion of Planck's constant (although of course with a different name), and a value for it that agrees with ours to great precision. The same goes for all the other fundamental constants of nature, with one glaring exception. Namely, the aliens' value for the so-called Alderson constant, which plays a fundamental role in the part of physics that was discovered some centuries earlier and that allows us to travel through wormholes in space, differs from our value by almost ten orders of magnitude. This is a huge surprise. How can the aliens, who appear otherwise to be so scientifically literate and who have developed the same kind of wormhole traveling technology that we have, be so deluded on this one single issue?

The human protagonists of Yudkowsky's story decide to consult their spaceship's extensive library in order to dig deeper into Alderson physics and try to figure out how and why the aliens have gone so wrong. What they find is an even bigger surprise. The aliens' value for the Alderson constant is right. It is humanity that has been badly wrong, for centuries, about the fundamentals of that part of physics. But how can that be? The space travelers dig further into the archives, and eventually they manage to reconstruct what has happened.

Here Be Dragons. First Edition. Olle Häggström.
© Olle Häggström 2016. Published in 2016 by Oxford University Press.

What the human Alderson physics pioneers had discovered – to their great horror – was a simple way to turn the Sun into a supernova using no equipment beyond what can be picked up at the local hardware store. Faced with that discovery, they had judged that making it public would spell doom for the human race, and decided to keep it secret.[4] This had turned out to require a cover-up involving a fake value of the Alderson constant.

1.2 The ethical dilemma of hiding research findings

I will not reveal any more of Yudkowsky's story. Instead, let us imagine ourselves in the shoes of those early Alderson physicists at the moment when they realize what a terrible doomsday weapon they have discovered. What is the ethically correct way to proceed at that point?

Having spent more than two decades in the scientific community, I have picked up a good deal on the topic of research ethics along the way. If there is one thing that seems to be carved in stone, it is this: *Never ever fake your research findings!* Follow the evidence wherever it leads you, regardless of whether or not it is in agreement with whatever you had (perhaps secretly) hoped to find. Under no circumstances is it OK to refrain from going public with your research findings just because you would have preferred a different outcome. Science is about the endless disinterested pursuit of Truth, whatever Truth happens to be.

I have heard these proclamations many times, and spoken them almost as often myself.[5] Taken at face value, they dictate that the Alderson physicists go public with their sensational findings. On the other hand, their papers and press releases would then, most likely, be quickly followed by the end of the world.[6] What to do? As for myself, I would have done as the physicists in Yudkowsky's story, overruling all the beautiful talk about the endless disinterested pursuit of truth, and trying my best to find a cover-up. (Hopefully the reader would do the same.) Still, those ideas about the impartiality of science carry great weight and should not be abandoned

---

[4] Why, you may ask, had the alien civilization survived, despite their scientists' openness about their corresponding scientific discovery? This is because they have a radically different psychological makeup compared to us humans, so that for them, the situation where any one of several billion citizens has the option of single-handedly causing immediate Armageddon is not a cause for alarm.

[5] I have even authored an essay whose title, in English translation, would read "In defense of the Harvard president and the uninhibited search for truth" (Häggström, 2005). For the context, see *The Harvard Crimson* (2005).

[6] This assumes that the psychological and sociological composition of humanity during the golden era of Alderson physics – several hundred years hence – will be roughly similar to today's, with about the same frequency of suicide bombers and other desperados (with and without the various psychoses that tend to show up in school shootings and similar tragedies), and of intelligent but bored teenagers and 20-somethings of the kind who enjoy constructing and releasing malign computer viruses. On the other hand, as we will see in Chapter 3, human nature being modified into something radically different from how we know it today is a definite possibility.

lightly, but in the critical situation sketched in Yudkowsky's story the right thing to do seems to be to override them, for the greater good of not destroying the Earth and exterminating mankind.

1.3 Some real-world examples

But *Three Worlds Collide* is just fiction. Does such a far-fetched scenario really have any bearing on how we should act in real life?

While the story lacks somewhat in realism, it does make the point that even if we like the ideal of an impartial science marching forward uninhibitedly, we might not want to defend it dogmatically in every situation and at any cost. Furthermore, there are more realistic scenarios that share relevant features with Yudkowsky's. In the coming chapters, several such scenarios that we may or may not encounter during the remainder of the 21st century will be sketched. In fact, during the 20th century and the early years of the 21st we have already encountered some cases where the ethical considerations are of a similar kind.

A major example is the Manhattan project, in which an extraordinarily gifted group of physicists during the 1940s developed the atomic bomb. The success of the project led immediately to the bombings of Hiroshima and Nagasaki in August 1945, with around 200,000 victims killed. It also led to a nuclear arms race. Exactly how lucky we were in getting through the cold war avoiding global nuclear war is difficult to assess (see Sections 8.1 and 8.3 for more on this), but considering incidents such as the Cuban missile crisis, and the one in 1983 when (in the tense political situation just weeks after Korean Air Lines Flight 007 had been shot down) Soviet air force officer Stanislav Petrov judged, in a close call, that a suspected US missile attack was in fact a false alarm,[7] it seems clear that the risk was non-negligible. And of course, the danger of nuclear Holocaust has not yet been averted.

The above is not meant to take a stand on whether the physicists of the Manhattan project were morally right or morally wrong in deciding to participate. The issue is highly complex, and in particular there were reasons to fear for what might happen if Germany were to beat the US in the race to complete an atomic bomb. Much has been written on the ethical issues, especially by the physicists themselves.[8] Particularly interesting is Richard Feynman's reflection that while he had plenty of qualms before and after the project, he had little or no such thoughts during it, when he was too absorbed by the problem-solving to even notice that when Germany surrendered in May 1945, his original motivation for participating in the project vanished.

Why, almost 70 years after Hiroshima and Nagasaki, is the cumulative number of individuals killed in nuclear war still a six-digit number rather than a

---

[7] Hoffman (1999).

[8] Dyson (1979), Feynman (1985), Bethe (1991), Ottaviani and Myrick (2011). See also Sotala (2013) for some further references.

ten-digit number? There are many reasons, including luck, but a particularly crucial fact is that nuclear weapons technology has (unlike the doomsday weapon in Yudkowsky's story) turned out to be very slow in becoming easy: it is still very difficult for single individuals, for terrorist groups, and even for small states, to attain it. Not every technology has this property. It is hard to imagine an alternate history where the technology of, say, aeroplanes or electronic computers remained the exclusive property of a few of the world's nations many decades after their invention.

We did not know in advance that nuclear weapons technology – our first technology for weapons of mass destruction – would have this slowness property, but we were lucky that it did. We cannot count on being equally lucky next time.

Creation (or recreation) of deadly viruses immediately springs to mind as a case for concern. In 2005, a group of US scientists announced that they had reconstructed the Spanish Flu virus that in 1918 and 1919 had killed maybe 50 million people. Their publication generated significant controversy about safety issues concerning both the laboratory work and the publication of their findings.[9] A recent similar controversy concerned the release of details about the laboratory-based mutation of a bird flu virus to make it transmissible between mammals.[10] Here, a US government biosecurity advisory board initially advised that the papers be censored before publication, but later changed their mind on the grounds that "it had heard new evidence that sharing information about the mutations would help in guarding against a pandemic" and that "the data didn't appear to pose any immediate terrorism threat."[11] The revised papers were then published in full. At the time of the controversy, one microbiologist interviewed in a TV news broadcast[12] defended the publication by pointing out that terrorists lack the technological infrastructure to exploit the research findings on the bird flu mutation. Be that as it may, the argument is flawed, because once the findings are published they are irreversibly out there, and we need to be worried not only about whatever infrastructure terrorists have now, but also what they may possess 10 years from now, or 20, or 50. (It may of course be that the microbiologist had access to a careful study showing that no such threat was forthcoming, but that seems unlikely).[13]

1.4 The need for informed research policy

The focus above has been on the terrible risks that can be associated with research breakthroughs. There will be more to come in later chapters. But of course the

---

[9] Bubnoff (2005), van Aken (2007).

[10] Yong (2012), Ritter (2012).

[11] Ritter (2012).

[12] Unfortunately, I have not been able to retrieve a more specific reference.

[13] Breaking news: Just as I'm putting the finishing touches to this chapter, I'm reached by the news that, prompted by this controversy, the White House has announced their intention to "temporarily halt all new funding for experiments that seek to study certain infectious agents by making them more dangerous" (McNeil, 2014).

discussion so far has been terribly lopsided, ignoring entirely the (much less controversial) fact that science has brought humanity enormous benefits, and has even more good to offer in the future.[14] Hardly anyone doubts that medical research has huge potential to continue to cure diseases, improve our health and prolong our lives. It is widely understood that emerging green technologies for energy and transportation hold an important key to maintaining prosperity while simultaneously avoiding man-made climate catastrophe. And so on and so forth. We can even add the radical visions of futurists like Eric Drexler in his book *Radical Abundance: How a Revolution in Nanotechnology Will Change Civilization* (2013), summarizing his view on the prospects and consequences of the emerging technology of atomically precise manufacturing:[15]

Imagine what the world might be like if we were *really* good at making things – better things – cleanly, inexpensively, and on a global scale. What if ultra-efficient solar arrays cost no more to make than cardboard and aluminum foil and laptop computers cost about the same? Now add ultra-efficient vehicles, lighting, and the entire behind-the-scenes infrastructure of an industrial civilization, all made at low cost and delivered and operated with a zero carbon footprint.

If we were *that* good at making things, the global prospect would be, not scarcity, but unprecedented abundance – radical, transformative, and sustainable abundance. We would be able to produce radically more of what people want and at a radically lower cost – in every sense of the word, both economic and environmental.

. . .

Imagine a world where the gadgets and goods that run our society are produced not in a far-flung supply chain of industrial facilities, but in compact, even desktop-scale machines. Imagine replacing an enormous automobile factory and all of its multi-million dollar equipment with a garage-sized facility that can assemble cars from inexpensive, microscopic parts, with production times measured in minutes. Then imagine that the technologies that can make these visions real are emerging – under many names, behind the scenes, with a long road still ahead, yet moving surprisingly fast. (pp ix–x)

This Drexlerian vision is speculative, but we still know enough at this point to warrant the conclusion that

$$\text{scientific progress has the potential both to cause} \atop \text{humanity great harm and to bring it great benefit.} \qquad (1)$$

---

[14] Still, some writers have doubted the overall benefit of science. Here is Dummett (1981):

It seems to me indisputable, with hindsight, that we should be, on balance, far better off than we are if, in 1900 or in 1920, all scientific research had come to a permanent stop. With the experience of what happened, we have little reason to doubt that the net practical result of future research will be increasingly disastrous.

See also Bergström (1994) – which is where I found the Dummett quote – for a strongly pessimistic view on the value of science.

[15] More will be said about Drexler's visions in Chapter 5.

Of course I do not claim priority for this observation, which has been made many times before. Here, for instance, is Albert Einstein:[16]

Penetrating research and keen scientific work have often had tragic implications for mankind, producing, on the one hand, inventions which liberated man from exhausting physical labour, making his life easier and richer; but on the other hand, introducing a grave restlessness into his life, making him slave to his technological environment, and – most catastrophic of all – creating the means for his own mass destruction. (p 147)

Still, it is the observation (1) that motivates this book. If science has this double-edged potential, then we'd better do what we can to steer towards the benefits and away from the harm. In order to improve the odds of a successful journey, it helps to have a good map of the territories we are entering. One of my aims here will be to put together some bits and pieces of what we do understand about them. But it must be admitted that our present knowledge is sadly reminiscent of how medieval mapmakers liked to decorate uncharted territories with dragons and other mythological creatures in order to warn travelers about unknown dangers – *Here Be Dragons!* Such a vague warning alone is perhaps not entirely useless, but we do need better maps. In fact, with nothing less than the fate of humanity at stake, our need to improve our understanding of the landscapes we are about to travel is urgent and desperate. Drawing attention to this need (which is mostly overlooked in mainstream public debate) is my main ambition with this book.

Mapping these uncharted territories is important to many areas of public policy, ranging from industrial and environmental policy to military issues and education. There is one area, however, in which I have personally been involved on various levels of Swedish academia and governance, where the lack of knowledge – and even the near-total absence of discussion – of the territories lying ahead, leads to a more or less absurd situation. Namely, research funding policy. My involvement in decision-making in this area is part of what led me to write this book, so spending a few pages on the background will help the reader see where I am coming from. To this end, the rest of this section will focus mainly on research funding and research policy in Sweden, but it is relevant also in a broader international perspective, as the Swedish case seems to be reasonably representative of how these things are handled elsewhere in the West.

For several years I have served on the natural and engineering sciences council of the Swedish Research Council (VR[17]), which is the largest agency for allocation of research funding in Sweden. It is by far the dominant one when it comes to funding so-called basic science, i.e., research that is not immediately aimed at generating patents, new industrial products and so on. Thousands of research proposals are submitted to us each year. Our total budget prevents us from granting more than a minority of these, so we need to choose between them. By what criteria?

---

[16] Einstein (1954).

[17] This is short for the agency's name in Swedish, *Vetenskapsrådet.*

The official policy of VR is to make the selection purely based on the quality of the proposed research. High quality research here means, roughly, research capable of expanding the horizons of human knowledge in directions that are interesting but which do not necessarily have any obvious or immediate practical applicability. I take part in the selection process politely, despite the policy being, in my opinion, mad. There are two reasons why I judge the policy so harshly.

The first reason is that it is in general impossible to make true sense of ranking the quality of research projects across different areas and disciplines. Geologist A wishes to try out a new method for analyzing lake sediments in order to reconstruct variations in how warm summers were in the lake's vicinity thousands of years ago. Computer scientist B aspires to develop new and faster algorithms for searching a database using parallelization techniques. Microbiologist C wants to work out what kinds of mutations it takes to make a bird flu virus airborne. Political scientist D intends to identify the causal factors behind varying levels of corruption in developing countries. How are we to judge how good A's research is compared to that of B or C or D without invoking our personal and subjective ideas about what kind of research we find interesting? In general, it can't be done. We are comparing apples and oranges, in terms both of methodology and of the kinds of results that can be expected. In order to judge whether A's work is better than B's, we need in particular to judge how interesting ancient climate variations are compared to complexity bounds for search algorithms. This is a matter of taste, not objective quality.

The only major attempt to avoid this subjectivity is bibliometry, i.e., to rank the applications based on counting the applicants' scientific papers or how many times they are cited in the scientific literature (or variations or combinations thereof). But this begs the question of quality, because counting papers measures not quality but quantity, whereas counting citations is just a popularity contest. Different academic fields have different publishing traditions, and to try to correct for the resulting "unfairness" between researchers in different fields, the publication and citation counts obtained are sometimes divided by subject-specific averages. However, such a correction does not address the fundamental criticism that popularity is not the same thing as quality. Furthermore, if we consider basing our funding system on bibliometry, we must also consider the incentives and pathological behaviors that bibliometry risks creating in the scientific community.[18]

VR, like most other research funding agencies, tries to avoid the subjectivity issue by a combination of weighing in bibliometry and having a broad representation of researchers from different areas in its panels and committees. This does not solve the problem either. For instance, suppose the representation of subject areas among panel members is taken to mimic the composition of the research community as a whole (as is, to a large extent, the case with VR). The likely effect of this on the distribution of research funding will be more towards conservation of existing proportions, rather than towards rewarding this mysterious thing we call "quality."

[18] Parnas (2007), Collini (2012).

The second reason why I judge the VR policy of using quality as the single criterion for distributing research funding so harshly is as follows. For the sake of argument, let us disregard my first criticism and instead assume that we do know how to define quality and to identify those research projects that have much of it and those that have less. Since the notion of quality, as understood by VR, is supposed to ignore practical applicability, quality as the sole selection criterion means that we value the production of new knowledge in its own right, rather than just as a means towards attaining other goals. I have long been – and still am – highly sympathetic to this romantic view of knowledge and intellectual achievements. To improve our understanding of the world we live in really is one of the most magnificent and worthy goals of a human activity one can think of. And yet, it is not the *only* worthy goal. A bright future for humanity, where everyone has the best possible prospects of leading a happy and prosperous life, and where such things as poverty, pain and misery are reduced to a minimum, seems like another goal worth striving for, at least as important as the quest for ever-increasing knowledge.

Now, as suggested in this chapter (and as will be argued at greater length in coming chapters), there is a huge potential for upcoming scientific achievements to impact the future of humanity and the well-being of future people – just as 20th-century science has had such a big impact on *our* lives, via the discoveries and inventions of things like penicillin, atomic bombs, birth-control pills, computers, cell phones and the Internet. To completely ignore this aspect of science seems like negligence bordering on insanity.

For the specific case of VR, one might argue that since the role assigned by the government involves specifically looking after *basic* science, and since there are other research funding agencies in Sweden with a more applied focus,[19] it makes sense that VR focuses on the production of knowledge for its own sake, while other agencies focus on the production of knowledge for more instrumental reasons. Surely we can afford both kinds of research?

I would buy this argument wholesale if it were the case that scientific progress only had the potential to *benefit* humanity, and not to *harm* it. Unfortunately, this is not the case. If research funding agencies are to play a decisive role in avoiding new technologies leading to the destruction of the human race, it really does not suffice to have some of them take into account such existential risks, while others ignore them. It is not enough that *some* researchers refrain from creating the ultimate doomsday weapon – what we need is that *everyone* does.

My attempts to draw the VR's attention to these issues have so far not been crowned with success. One idea was to launch a delegation, either under VR's auspices or (preferably) directly under the Swedish government's Ministry of Education and Research, with the task of putting together an extensive report on what is known and what can plausibly be conjectured concerning those uncharted territories marked *Here Be Dragons*. Such a report could then serve as input for VR's

---

[19] These include VINNOVA (Swedish Government Agency for Innovation Systems) and the Swedish Foundation for Strategic Research.

and other research funding agencies' funding priorities, as well as for government policy and legislation. This might lead to a better informed research funding policy in Sweden. Of course, for the reasons outlined in the previous paragraph, obtaining such a thing in Sweden would not be even nearly enough, but the process must start somewhere, and perhaps a Swedish initiative might inspire similar measures in other countries. Eventually we might hope to see the United Nations launch an advisory expert panel on the pros, cons and lethal dangers across the spectrum of radical emergent technologies, similarly to what the IPCC (Intergovernmental Panel on Climate Change) does today on the topic of climate change.

Those were my thoughts, but the delegation and the report that I suggested may not materialize anytime soon, or at all. The book the reader is holding in her hand right now cannot replace it, but is for the moment the best I can offer.

1.5 A hopeless task?

Sometimes when I speak of these matters, and of the need for a systematic study of the dangers associated with future technologies, I get responses along the lines of "Yes, the dangers seem to be enormous, but the project is naive and futile," backed up with one (or both) of the following reasons for deeming the project useless.

The first reason is that mapping those dangerous uncharted territories that lie ahead of us is extremely difficult. Predicting the future is generally very difficult, as is immediately evident if we imagine asking a futurologist 50 years ago to outline the most important likely developments that will take place before 2015. Would she mention the breakups of the Soviet Union and Yugoslavia? 9/11 and the War on Terror? The European Monetary Union? The ozone depletion crisis? The AIDS epidemic? Cell phones, iPads and the Internet? The enormous rise of the Chinese economy from the 1990s and onwards? She'd be very lucky to get any one of these things right. Predicting future scientific and technological breakthroughs is especially difficult. Taleb (2007) phrased it well:

If you are a Stone Age historical thinker called on to predict the future in a comprehensive report for your chief tribal planner, you must project the invention of the wheel or you will miss pretty much all of the action. Now, if you can prophesy the invention of the wheel, you already know what a wheel looks like, and thus you already know how to build a wheel, so you are already on your way. (p 172)

The second reason why my hoped-for project of systematically mapping the possible and likely consequences of future technologies might be considered a naive way to try to save the world is this: Assume, for the sake of argument, that we *are* able to predict, with some degree of confidence, that some emerging technology X – say, whole-brain emulation or nanorobotics with self-replication – holds out great promise for economic gains in a short perspective (10 or 20 years) but is likely, if allowed to flourish unimpeded, to trigger a global catastrophe involving the extinction of the human race in a slightly longer time perspective. What in the world can we do with this information? Surely you cannot be so naive as to expect to be able to halt the development of area X? We live in a complex society, with large parts of the world dominated by market economy and free enterprise ideas, and surely there will be researchers and entrepreneurs looking for those short-term gains from technology X regardless of what some doomsday prophet is saying. And note that in case you're considering a legislative ban on X, it is not enough to convince 205 out of the world's 206 sovereign countries to legislate such a ban and to enforce it – you need all 206. It's hopeless!

Granted, both objections have valid points. It *is* very difficult to make accurate predictions about future technological developments, and even if we should be able to identify some particularly risky line of technological development that for the sake of humanity's survival ought to be avoided, it might *still* be very difficult to arrange a worldwide moratorium. But there is a huge difference between, on one hand, admitting that there are severe difficulties, and, on the other, throwing our hands in the air and fatalistically declaring the problem to be unsolvable. We don't know that they are unsolvable until we have *tried*, and tried really hard. Given the magnitude of what's at stake, just giving up on the problem is in my opinion unacceptable. The extent to which we are currently neglecting the problem is shocking. Nick Bostrom, in a recent paper, illustrates this with a diagram showing how the number of academic publications on snowboarding outnumbers those on risks of human extinction by a factor 20 or so, while those on dung beetles beat those on snowboarding by another factor 2.[20] This should not be taken as a suggestion that too much effort is spent on academic studies of snowboarding and dung beetles, but rather as an indication that current efforts into the study of existential risks to humanity could easily be significantly scaled up without major disruption to the current academic landscape as a whole.

Since the central case I'm making is limited to pointing out the urgent need to map the uncharted territories spanned by the pros and cons of technological developments that may be feasible within the next few decades or centuries, I do not want to overemphasize the issue of what we might subsequently do to steer technological development away from the most malign areas and towards the more benign ones. Still, it is worth pointing out that, although our experience of attempts to negotiate global cuts in greenhouse gas emissions is disheartening, a much more encouraging case is the relative success of the 1987 Montreal Protocol for phasing out our emissions of CFCs (chlorofluorocarbons) and related chemicals in order to prevent the destruction of the ozone layer. One reason that CFC regulation has turned out to be easier than greenhouse gas regulation is that the role that CFCs played in the economy was nowhere near as large as that played by greenhouse gases (in particular by carbon dioxide), and so the phasing out was, from a global perspective, relatively painless.

Another important aspect of the ozone layer crisis (which it shares with the global warming crisis) is that the detrimental effects of emissions are gradual and

---

[20] Bostrom (2013).

cumulative, so that it need not be such a big deal if one or a few small countries refuse to join the agreement, provided their contribution to the problem as a whole is small enough to begin with. This may not be the case with some dangerous future technologies such as the aforementioned technology X. Two hundred and five countries enforcing bans on technology X will not suffice if the 206th country – say, North Korea – proceeds with X and causes global destruction of the entire biosphere. This clearly highlights international political structures for arriving at and enforcing such bans as an urgent area to look into. But let's say we fail, and North Korea eventually destroys the world – does that mean that the efforts of the other 205 countries to limit technology X have been a total failure? A failure, yes – but not a *total* failure. It seems likely that North Korea, on its own, would not be able to lift technology X to the lethal point quite as fast as a race (and/or interactions) between research groups in the US, in Europe and in China might have done. Perhaps the 205-country agreement to avoid X bought humanity a full decade of flourishing compared to what might otherwise have happened. A total destruction of humanity and the biosphere would of course be very, very bad, but even given that such an event eventually happens there are things that are still very much worth striving for, and a delay of the catastrophe is surely such a thing.[21]

1.6 Preview

In the following chapters, I will take the reader on a tour through a wide range of areas of science and technology that strike me as relevant to the issue of what lies ahead. I will offer sketches of a few of the things that we do know already, stressing specific questions that we need to answer in order to be in a position to make informed decisions. It goes without saying that the presentation and the choice of material to cover will be colored by my various personal biases and hangups. Equally obvious is the fact that no single-authored and modestly sized book like this one can provide anything close to an exhaustive treatment of the general problem or even of the chosen specific subtopics. But if even one reader is inspired to roll up her sleeves and start working towards a better understanding of what the future may have in store for us and what we might want to do about it, then the book will have served its purpose.

Here is a brief outline of the chapters to come:

Chapter 2. Our planet and its biosphere. Until we emigrate to outer space (a potential possibility to be discussed in Chapter 9) or radically alter our physical appearance (Chapter 3) we need planet Earth to remain habitable to humans and able to support agriculture and other sectors vital

---

[21] The argument here is that the extra time we get before Armageddon is valuable *in its own right*, but a perhaps even stronger argument is that we might *use* that extra time to figure out a way to avert disaster.

to civilization as we know it today. The planet has gone through drastic changes in the past, and it seems that we are presently pushing it, at record speed, towards further such changes, with risky outcomes. Several environmental issues are crucial, but here the focus will mainly be on the much-discussed global warming and various ways to handle it, including some recent (and highly controversial) suggestions for technological quick fixes.

Chapter 3. Engineering better humans? We may be on the verge of radically transforming ourselves using pharmaceutic, genetic, electronic and other means, leading to huge enhancements of our physical, cognitive and perhaps also moral abilities. The possible gains seem enormous, but so do the risks. So should we go down this road? That is a highly controversial question riddled with unfamiliar moral issues. Particularly radical futures may be in store for us if we learn how to upload our minds onto computer hardware.

Chapter 4. Computer revolution. The ongoing computer revolution is changing our lives. After a very brief sketch of its history up to now, the rest of the chapter will focus on where it might be heading. An issue of particular interest is what might happen if and when the holy grail of artificial intelligence is achieved: a machine matching or exceeding humans in terms of general intelligence. This might turn out to be a computer revolution of much greater magnitude than anything we have seen so far, and there are interesting theoretical discussions on whether it might snowball into a very fast development towards ever more advanced superhuman intelligences. Such a snowballing effect has often been called the Singularity, but adhereing to the more precise terminology used by Yudkowsky (2013a) and Bostrom (2014), it will here be called an **intelligence explosion.**

Chapter 5. Going nano. Our ability to manipulate matter is growing increasingly refined, and we may be heading towards general-purpose atomically precise manufacturing. While today's 3D printers are expensive and specific to a limited range of materials, overcoming these limitations may potentially lead the way to the kind of we-can-have-whatever-we-want paradise sketched in the Eric Drexler quote in Section 1.4. But there are also risks.

Chapter 6. What is science? For some questions regarding where science may be leading us, it may help to understand what science actually *is*. In this chapter a number of key issues in the theory of science will be discussed, with particular emphasis on the central role played by the tools and theory of statistics. This will also lay the ground for some (mathematically simple, but sometimes philosophically quite subtle) statistical arguments to be discussed in later chapters. The chapter ends with a discussion of the relation between science and engineering.

Chapter 7. The fallacious Doomsday Argument and **Chapter 8. Doomsday nevertheless?** Among the various possible future paths taken by humanity, some end in extinction within a time span of perhaps as little as centuries or even decades. The possibility of human extinction is the topic of both Chapter 7 and Chapter 8. Many thinkers worth taking seriously, such as the astrophysicist and former president of the Royal Society Martin Rees in his book with the alarming title *Our Final Century* (2003), attach substantial probability to the event of extinction before the end of the 21st century. Some of the most plausible or likely causes that may trigger such an apocalypse will be discussed. Following the anthology *Global Catastrophic Risks* by Bostrom and Ćirković (2008), these can roughly be organized into three categories: (a) risks from nature, (b) risks from unintended consequences of human actions, and (c) risks from hostile human acts. Although such a classification turns out to be partly problematic (because several of the most plausible apocalypse scenarios can be argued to belong in more than one category), an examination of the various risks nevertheless tends to indicate that, at least on the time scales of decades to centuries, (a) is dwarfed by (b) and (c), i.e., the greatest risks emanate from ourselves rather than from natural disasters. All this is the topic of Chapter 8. Before that, Chapter 7 will offer a critical assessment of an oft-discussed abstract argument, known as **the Doomsday Argument**, that purports to show that the end of humanity is most likely near. I view that argument as mostly a distraction, and Chapter 7 is meant get it out of the way before we get down to serious business about existential risk.

Chapter 9. Space colonization and the Fermi Paradox. The Fermi Paradox is the question of why it is that we have not, so far, seen any signs of extraterrestrial civilizations. There are many candidate answers to this still unresolved problem, which turns out to be related, via a very simple mathematical relation known as **the Great Filter**, to the issue of whether or not humanity is likely to be able, eventually, to colonize the stars. Some radical ideas for how that might be doable will be presented.

Chapter 10. What do we want and what should we do? In this final chapter, the various topics from earlier chapters will be pulled together and put in the perspective of ethics. No science and no analysis of the future consequences of various actions taken today can in itself tell us what to do. We need, in addition, to factor in what kind of future we value, and to what extent we care at all about the future compared to more immediate concerns here and now. The latter aspect is usually modeled in economics by the so-called **discount rate**, which has played a prominent role in discussions of climate change on a decadal to centennial time scale, but hardly at all in the context of longer time perspectives or the various radical technologies discussed in Chapters 3, 4, 5, 8 and 9. We are less used to thinking about ethical issues on long time scales, so our intuitions tend to fail us and lead to paradoxes. These issues need to be resolved, because dodging the bullet would in my opinion be unacceptably irresponsible.

CHAPTER 2

Our planet and its biosphere

2.1 A note to the reader

This chapter is mainly about climate change and about various radical (but risky) high-tech approaches to solving the climate crisis, in case we decide at some point that the more conventional approach of cutting down on fossil fuel burning and other sources of anthropogenic greenhouse gas emissions is undoable or insufficient. One such radical idea is to deliberately pump large amounts of sulfur into the stratosphere, for the purpose of blocking out part of the electromagnetic radiation reaching us from the Sun.

Climate science is a subject that (despite remaining uncertainties) we understand a lot better than many other fields to be discussed, in a more speculative mode, later in this book. In order to take advantage of this and start on relatively solid ground, the first four sections of this chapter will be devoted to digging into the fascinating scientific background of today's dire warnings about how we are changing the climate in dangerous ways. Readers already familiar with this background or willing to accept the scientific consensus without looking into the details, and who are eager to get to the core topic of the book – emerging technologies and their potential consequences for the future of humanity – are advised to skip ahead directly to Section 2.6. But for the rest of you, I offer Sections 2.2–2.5.

2.2 Dramatic changes in past climate

Living conditions on planet Earth have varied enormously during prehistoric times. For those of us who were taught about the ice age in elementary school, it is hard to fathom how strange and revolutionary the idea must at first have seemed: that until around 10,000 years ago there were huge ice sheets, mostly several kilometers thick, covering most of northern Europe, almost all of Canada and parts of northern United States. Several European naturalists and geologists put forth ideas during the 18th and early 19th centuries on how features of today's landscapes might be explained by past glaciations. Spencer Weart, in his wonderful

Here Be Dragons. First Edition. Olle Häggström.
© Olle Häggström 2016. Published in 2016 by Oxford University Press.

book *The Discovery of Global Warming* (2003), tells us about what the scientific state-of-the-art was in the mid-19th century, when Irish-born British physicist John Tyndall tried to understand the mechanisms that control the Earth's climate:

He [Tyndall] hoped to solve a puzzle that was exciting great controversy among the scientists of his day: the prehistoric Ice Age. The claims were incredible, yet the evidence was eloquent. The scraped-down rock beds, the bizarre deposits of gravel found all around northern Europe and Northern United States, these looked exactly like the effects of Alpine glaciers, only immensely larger. Amid fierce debate, scientists were coming to accept the incredible. Long ago – although not so long as geological time went, for Stone Age humans had lived through it – northern regions had been buried a mile deep in continental sheets of ice. What could have caused this? (p 4)

Today, Tyndall is known, together with French mathematician and physicist Joseph Fourier and Swedish chemist Svante Arrhenius, as one of the pioneers in understanding the role of greenhouse gases in shaping the climate. The general acceptance of the reality of past ice ages is approximately marked by the publication of Scottish scientist James Croll's book *Climate and Time, in Their Geological Relations* in 1875.

In Section 2.4, we will return to Croll and the issue of what caused these enormous past climate variations, but for now let us put the ice age into further perspective. The ice age that we are used to talking about lasted from about 110,000 to 10,000 years ago, and is merely the latest in a repeated pattern of glaciations. Ice core data from Antarctica dating back 740,000 years indicate climate variations amounting to no less than eight such glacial cycles,[22] and these form part of a longer ice age called the **Quaternary glaciation**, defined by the uninterrupted existence of a major ice sheet on Antarctica. In this sense of the term "ice age" (which is the one preferred by professional geologists), we are still in it!

The further back in time we go, the sketchier our knowledge gets about what conditions on our planet were like, but we do know that before the Quarternary glaciation, the Earth went through both earlier glaciations and much warmer periods. For instance, during the era of the dinosaurs, lasting from about 230 million years ago until the so-called Cretaceous–Paleogene extinction event 66 million years ago,[23] the climate seems to have been generally warm, with average temperatures mostly several degrees above today's. Some of the glaciations, going back 650 million years or more, are conjectured – although the hypothesis is still highly controversial – to have been of the extreme kind where the entire Earth's surface is frozen; this is the so-called Snowball Earth hypothesis.[24]

---

[22] Augustin et al. (2004).

[23] This extinction event is believed to have been caused by a large asteroid impact on the Yucatan peninsula in Mexico and the aerosol injection into the atmosphere that followed and shielded a substantial fraction of the sunlight for several years; see Schulte et al. (2010). For more on this and other asteroid impacts, see Section 8.2.

[24] See, e.g., Sankaran (2003).

2.3 Greenhouse warming

Whether or not the Snowball Earth hypothesis is true, we can still conclude that life on Earth has survived through highly varying conditions. In contrast, the period from around 10,000 years ago until recently – roughly what geologists call the Holocene – has been unusually stable. It may well not be a coincidence that this is the period when human civilizations developed far beyond anything ever seen before. Since the industrial revolution, this relative stability of the environment has been interrupted to an extent that has led some thinkers to suggest that we are no longer living in the Holocene but in a new era called the Anthropocene.[25] While leaving the stable zone of the Holocene is most likely not an immediate threat to life on Earth as a whole, it may mean an end to the favorable conditions under which humanity has thrived, and perhaps therefore an end to (or even a reversal of) the progress our civilization has been making. In a paper entitled "A safe operating space for humanity" published in *Nature* a few years ago, a Stockholm-based team of scientists led by Johan Rockström attempted to identify the relevant quantities defining the safe and stable zone, and to pin down its boundaries.[26]

Rockström's project is important. Until the day comes – if it ever does – when truly radical technological developments allow us to break free, e.g., by migrating into outer space (a possibility discussed in Chapter 9) or onto computer hardware (Chapter 3), we will remain dependent on the Earth's biosphere for agriculture and other crucial needs. In their paper they propose, tentatively, nine aspects of our global environment that they consider crucial for identifying what they call planetary boundaries – thresholds with the property that crossing them risks leading to environmental instabilities and away from the benign conditions of the Holocene. These aspects are (i) climate change, (ii) rate of terrestrial and marine biodiversity loss, (iii) changes in the nitrogen and phosphorus cycles, (iv) depletion of the stratospheric ozone layer, (v) ocean acidification, (vi) depletion of global freshwater resources, (vii) changes in land use, (viii) chemical pollution, and (ix) atmospheric aerosol content. Concerning the last two aspects, Rockström et al. consider our current understanding of the problem to be so poor that it would be premature to even suggest a quantitative threshold, but for the first seven they do propose such levels. Concerning four of these, they judge that we are still within the safety zone, but for three of them – namely climate change, rate of biodiversity loss and the nitrogen cycle – they find, alarmingly, that we seem to have crossed the thresholds, thereby endangering the future hospitality of our environment.

Discussing all of these aspects at the length they deserve could easily fill this whole book and many more.[27] The rest of this chapter will focus on just one of the

---

[25] Crutzen (2002).

[26] Rockström et al. (2009).

[27] In fact, there are already at least two books out there that explicitly take the Rockström et al. study as their starting points: Wijkman and Rockström (2011) and Lynas (2011).

nine aspects: climate change. This choice goes somewhat against one of the overall aims in this book, which is to put on the table important issues related to scientific and technological advances that have so far been mostly ignored in public discourse. Climate change does meet the importance criterion, but, as the reader is probably aware, it fails the other criterion of being "mostly ignored in public discourse." My choice can nevertheless be defended on precisely that same ground. It may help to start out on relatively familiar territory, before trying, in later chapters, to deal with the Dragons out there. Climate change is the one area where we have begun to talk about decision-making with a fairly long (centennial) time perspective, and to develop ethical and economical frameworks for such decisions – frameworks that may be of some use to us in other contexts (see Chapter 10).

Rockström et al. choose two key quantities to define the climate dimension of "safe operating space for humanity" that their paper speaks of, namely (a) the CO_2 (carbon dioxide) concentration in the atmosphere, and (b) the change in incoming radiative forcing as measured in W/m^2 (watts per square meter) as compared to pre-industrial levels. Let me briefly explain the relevance of these quantities to climate change, and in particular to global warming.

The quantity (a) is probably the more familiar one to most readers. Before the industrial revolution, the CO_2 concentration was around 280 ppm (parts per million), having increased very slowly during most of the Holocene from about 260 ppm 8000 years ago.[28] During the 20th century, the concentration began to grow rapidly, with a concentration around 315 ppm in 1960 and hitting 400 ppm in 2013. Following Hansen et al. (2008) and Hansen (2009), Rockström et al. settle for a threshold of 350 ppm, meaning that we need to bring back the CO_2 concentration, not necessarily to pre-industrial levels, but by at least 50 ppm; in Section 2.8 later in this chapter we will see how such a reduction might be feasible. The figure 350 ppm has caught on in public discourse and given rise to the so-called 350.org movement.[29]

As to the quantity (b), the change in radiative forcing, this is strongly related to (a), because changes in the atmospheric CO_2 concentration are a main contributor to changes in the radiative forcing. On the other hand, CO_2 concentration is not the *only* such contributor, so the quantities are not equivalent.

The radiative forcing in (b) is the amount of energy from incoming radiation that the Earth's surface absorbs, per time and area unit – averaged over all of the Earth's surface, all hours of the day and all days of the year. The basic relation between radiative forcing and surface temperature comes from the so-called **Stefan–Boltzmann law**, as follows. Let I_{in} and I_{out} be, respectively, the average power at which the Earth's surface absorbs the incoming radiation, and the average power at which the surface emits radiation. In equilibrium we have

$$I_{in} = I_{out} . \tag{2}$$

---

[28] Indermühle et al. (1999).
[29] McKibben (2009).

The outgoing radiation I_{out} is governed by the Stefan–Boltzmann law, which states that

$$I_{out} = \varepsilon \sigma T^4, \tag{3}$$

where T is the surface temperature (as measured in Kelvin), σ is the universal so-called Stefan–Boltzmann constant, and ε is a number between 0 and 1 describing the surface's emissivity, which describes how well it approximates a perfect blackbody;[30] the quantities ε and σ can be considered fixed, so (3) should be seen as describing how the outgoing radiation I_{out} depends on the surface temperature T. If the equilibrium is disrupted by an increase in I_{in}, then the Earth accumulates heat and the surface temperature goes up,[31] and continues to go up until the right-hand side $\varepsilon \sigma T^4$ of (3) matches the new value of I_{in}, so that equality is restored in the balance equation (2) and a new equilibrium is attained. In short: if the incoming radiation I_{in} goes up, then so does the temperature T. This, and much of what follows, is beautifully explained by David Archer in his 2007 book *Global Warming: Understanding the Forecast*.

The causal link between the CO_2 concentration in (a) and the radiative forcing in (b) involves the greenhouse effect, which works as follows. Greenhouse gases, such as CO_2 and water vapor, have the capacity to absorb and emit electromagnetic radiation in the infrared spectrum. Incoming sunlight is mostly in the visible spectrum, and passes right through these gases. The radiation emitted from the Earth's surface, however, is mostly in the infrared spectrum. Without the greenhouse gases, this radiation would escape straight into space, but as things are, much of it is absorbed upon hitting, e.g., a CO_2 or an H_2O molecule, and later emitted again in a random direction. In this way, the photons perform a kind of random walk in the atmosphere, and a substantial fraction ends up hitting the Earth's surface again, thus contributing to the incoming radiative forcing I_{in}. When the greenhouse gas concentrations go up, then so does this fraction and (consequently) the radiative forcing in (b). This is then turned into a temperature increase via the balance equation (2) and the Stefan–Boltzmann law (3).

It turns out that, at realistic levels of greenhouse gas concentrations, the dependence of temperature on CO_2 concentration is approximately logarithmic.[32]

---

[30] When I_{out} and T are averaged over the Earth's surface and over some time interval – rather than pertaining to some specific time and point on the surface – this is not exactly right, due to the fact that the relation between I_{out} and T is nonlinear. It is good enough, however, for the qualitative reasoning that follows.

[31] This is a bit of a simplification, because the oceans can transport heat in such a way that some of the energy imbalance leads to an accumulation of heat in the deep oceans rather than at the surface (understanding and quantifying this process is a key area of continued research for understanding climate change; for a recent important contribution, see Balmaseda, Trenberth and Källén, 2013). This accumulated heat will, however, resurface sooner or later. At this point the reader might wonder whether something similar is true of the Earth's crust. Qualitatively it is of course true that there is an exchange of heat between the surface and lower layers in the crust, but quantitatively the heat conductivity is small enough that this effect is negligible.

[32] Archer (2007), Chapter 4.

To the extent that this logarithmic dependence is exact, this means that a doubling of the CO_2 concentration yields the same temperature increase regardless of its initial level. It is therefore natural (and a standard choice) to quantify the capacity of CO_2 to impact the global average temperature as the increase in equilibrium temperature that would result from a hypothetical situation where the CO_2 concentration is doubled. We denote this temperature increase by $\Delta T_{2\times CO_2}$ and call it the **climate sensitivity** of CO_2. This is one of the most crucial quantities for understanding climate change.

If we only had to worry about the most direct impact of the CO_2 concentration on the temperature, via the greenhouse effect and the Stefan–Boltzmann law, then the value of $\Delta T_{2\times CO_2}$ would be close to 1 K, or equivalently 1°C. However, the resulting warming triggers a number of other effects in the climate system, some of which form so-called feedbacks that can either increase the warming further (positive feedbacks) or attenuate it (negative feedbacks). These need to be taken into account when specifying the full impact of CO_2. Whereas the greenhouse effect in itself is physically well understood, both qualitatively and quantitatively, the total influence of the various feedback effects is much less so, and for this reason science has not (yet) been able to pinpoint a nice single value of $\Delta T_{2\times CO_2}$. The latest report from the IPCC summarizing the current state of the art gives a likely interval of

$$1.5\,\text{K} \leq \Delta T_{2\times CO_2} \leq 4.5\,\text{K}, \tag{4}$$

while an oft-quoted figure for an approximate best estimate of $\Delta T_{2\times CO_2}$ is around 3 K.[33] This interval, which in particular tells us that the positive feedbacks are (in total) stronger than the negative ones, has been arrived at via a variety of empirical studies in combination with models and theoretical arguments. For a rudimentary understanding of some of this important work, let us turn back to the ice ages and the issue that puzzled researchers back in the days of John Tyndall: what caused them?

2.4 Milankovitch cycles

One of the most important research directions for understanding the ongoing anthropogenic global warming is to look at climate history on various time scales, including the cycles of glacial maxima and minima during the Quaternary glaciation. The climate change mechanisms during those prehistoric times were hardly identical in every detail to what is happening right now (after all, nobody in his right mind would suggest that the CO_2 increase during the transition from the last glacial maximum to the Holocene was caused by coal power plants and automobiles), but we're still talking about the same planet, so it makes sense to expect that many of the particular mechanisms involved are similar today to how they behaved in the past.

[33] IPCC (2013).

According to Weart (2003), in the mid-19th century there were several suggestions for what might have caused those huge climate variations of the past.[34] Might they have been caused by variations in the Sun's intensity, or perhaps by geological events altering the ocean currents that have such a crucial effect on regional climate? Most of these suggestions provided no explanation for the apparent cyclicity of the glacial cycles, but there was one exception, namely the one about climate change due to slow variations in the Earth's orbit around the Sun, pioneered by James Croll and presented in his 1875 book, and further developed by the Serbian astronomer Milutin Milankovitch.

We all learned in school that the Earth moves around the Sun in circular orbit, and later most of us picked up that the asserted circularity is not exact but a fairly crude approximation: the orbit is in fact an ellipse, whose perihelion (closest point to the Sun) is 3.3% closer to the Sun than is its aphelion (furthest point from the Sun), corresponding to a so-called eccentricity[35] of 0.017. But this, too, is just an approximation. Deviations from the elliptic orbit are mainly due to the gravitational influence of other celestial bodies than the Sun. A prime suspect is the Moon: the Moon and the Earth both rotate around their joint center of gravity (which, due to the much greater mass of the Earth, is located about 1700 km below the Earth's surface), causing a slight wobble in the Earth's orbit. And then there are the other planets, mainly Venus (due to its relative vicinity) and Jupiter and Saturn (due to their sizes).

The whole issue suddenly seems incredibly complicated, but Croll saw a way forward, building on earlier contributions to celestial mechanics by French mathematician Urbain Le Verrier, whose studies of anomalies in the orbit of Uranus had led to the hypothesized existence of a hitherto unseen planet, and eventually (in 1846) the discovery of Neptune.[36] Croll applied celestial-mechanical calculations to the Earth's orbit, and tried to derive consequences for the climate. His study predicted (or, rather, retrodicted) a sequence of ice ages with warmer periods between. A crucial difference, however, between Croll's theory and the view that prevails today is that Croll thought that climate in the southern and the northern hemispheres would correlate negatively, so that warm periods in the northern hemisphere would correspond to cold ones in the southern hemisphere, and vice versa. Dating prehistoric geological events was in the 19th century a very uncertain affair, but by the turn of the century the evidence seemed to point away from Croll's theory. The reality of the great glaciations of the past was no longer in doubt, but their causes were still up for grabs.

Enter Milutin Milankovitch. In the 1920s, he carried out careful celestial-mechanical calculations to see how the insolation (amount of solar radiation) during summer at various northern latitudes had varied over the millennia.

[34] See also Imbrie and Imbrie (1979) for a more detailed history of how 19th and 20th century science gradually came to terms with the ice ages.

[35] The eccentricity e of an ellipse is a number between 0 and 1 measuring how much it deviates from a circle ($e = 0$).

[36] See Section 6.3 for more on the discovery of Neptune, from a philosophy of science point of view.

He found that this variation depended on a number of very slow cycles in the Earth's orbit, of which the three most important are the following:

(i) The eccentricity of the elliptical orbit, varying between 0.005 and 0.058, in a complicated pattern with a main period of 413,000 years but with several other cyclic components with periods around 100,000 years.

(ii) The tilt between the Earth's own rotational axis and the axis of its orbit around the Sun, varying between 22.1° and 24.5° and currently at 23.4°, cyclically with a 41,000 year period.

(iii) The axial precession, i.e., the direction of the axial tilt, varying cyclically with a period of 26,000 years.[37]

These have come to be called **Milankovitch cycles**. Milankovitch believed them to have a crucial influence on the Earth's climate variations, but this remained disputable for a long time. It was only in 1976, 18 years after Milankovitch's death, that the issue was firmly settled, in a seminal paper by James Hays, John Imbrie and Nicholas Shackleton who used deep-sea sediment cores to reconstruct climatic variations over the past 450,000 years, and found these to be in beautiful agreement with the Milankovitch cycles.[38] It is, indeed, these variations in the Earth's orbit that have triggered the cycles of glaciation during the past half a million years and more.

2.5 The role of carbon dioxide

But where does CO_2 enter the picture? Some readers may recall, e.g., from Al Gore's Academy Award-winning 2006 movie *An Inconvenient Truth*, diagrams of changes in temperature and in CO_2 concentration during these glacial cycles, used to emphasize the role of CO_2 in global warming.[39] These diagrams exhibit striking agreement between temperature and CO_2 concentration – when one goes up (or down), then so does the other. But given the findings I've discussed above concerning the causes (Milankovitch cycles, not CO_2) of the glacial cycles, it seems, at least on first sight, that the conclusion suggested by Gore and others concerning the causal role of CO_2 is wrong. To make matters worse, a closer inspection of the temperature and CO_2 concentration diagrams shows that when a glaciation ends, the temperature begins to increase several hundred years before there is an increase in CO_2, and similarly at the end of the glaciation. So has the global warming media alarm gotten the causality entirely backwards? Perhaps the 20th century increase of atmospheric CO_2 is the *consequence*, not the *cause*, of the approximately 0.8 K increase in global average temperature we have seen during the same

---

[37] If the Earth had a circular orbit around the Sun, then the precession would be unimportant, but with the nontrivial eccentricity the precession relative to the perihelion has consequences for the seasonal distribution of insolation in the two hemispheres.

[38] Hays, Imbrie and Shackleton (1976).

[39] Gore (2006).

period? In fact, this is what countless climate deniers[40] (but hardly any climate scientists) have claimed in recent years.[41]

To resolve this, let me first stress that our qualitative understanding of the greenhouse effect of CO_2 in no way rests on paleoclimate data such as those exhibited by Gore. The same goes for the rough quantitative result that, without feedback effects, the greenhouse effect on its own would imply $\Delta T_{2 \times CO_2} \approx 1\,K$. These insights predate those data, and are based on basic physics such as the emission spectra for the CO_2 molecule and the Stefan–Boltzmann law. So an increase in atmospheric CO_2 does cause warming – that much we do know without any paleoclimatic data (contrary to the common misunderstanding that this conclusion is derived from the ice age data in Gore's movie).

We furthermore know that there is a causal link in the other direction: warming causes increased atmospheric CO_2 levels. The mechanism is that with global warming, the oceans warm up as well, and the solubility of CO_2 in water decreases with increasing temperature, so CO_2 will leak from the oceans into the atmosphere. Here's a summary of these main causal mechanisms between warming and atmospheric CO_2 level:

$$\begin{array}{ccc} \text{warming} & \Rightarrow & \begin{array}{c}\text{less } CO_2 \text{ solubility} \\ \text{in ocean}\end{array} \\[1em] \Uparrow & & \Downarrow \\[0.5em] \begin{array}{c}\text{more greenhouse} \\ \text{effect}\end{array} & \Leftarrow & \begin{array}{c}\text{more } CO_2 \text{ in} \\ \text{atmosphere}\end{array} \end{array} \tag{5}$$

This is our first example of a feedback mechanism. A bit of initial warming leads to transport of CO_2 to the atmosphere, increasing the greenhouse effect, and further warming. It is called a positive feedback because it *strengthens* the initial warming. If instead it had counteracted the initial warming, we would have spoken of a *negative* feedback. At first sight, a positive feedback like this may look like a terribly dangerous thing, because this new warming triggers another dose of CO_2 into the

---

[40] These are the contrarians who like to call themselves "climate skeptics" – a term that should be avoided, as it carries erroneous connotations of the kind of healthy skepticism that is one of the characteristics and cornerstones of good science (cf. Chapter 6). Such skepticism involves critical scrutiny of models, methods and results – those of others as well as (especially!) one's own. The "skepticism" of the "climate skeptics" is not like that at all – this I know from ample experience since 2008 when I was drawn into (the Swedish branch of) the debate between them and the climate scientists. Rather, they have fixed ideas about what they oppose, in particular the idea of a dangerous anthropogenic global warming, and they tend to be happy to use every argument and every one-liner, no matter how thoroughly refuted, that serves this purpose in the eyes of a non-specialist audience. See Oreskes and Conway (2010) for some of the sad history of this movement, and Häggström (2011a) for my own critical take on it. As to the term "climate denialism," this is not, of course, meant to say that they deny the existence of climate. What they do deny is large parts of the knowledge that climate science has accumulated.

[41] I cannot bring myself to burden this book's bibliography with any of these intellectually lousy references. Suffice it to say that the reader who Googles, e.g, "CO2 lags temperature" will find plenty of such stuff.

atmosphere, and the whole thing keeps going round and round. Doesn't this lead to a runaway global warming?

Not necessarily. It depends on how strong the feedback loop is, i.e., how much extra warming an initial increase of, say, $1°C$ yields in the first iteration of the loop. If this increase is $0.5°C$, then the second iteration of the loop is only half as strong as the first, yielding another $0.25°C$ the second time around, and so on. And since

$$1 + \frac{1}{2} + \frac{1}{4} + \frac{1}{8} + \cdots = 2,$$

the total warming resulting from the initial $1°C$ stops at $2°C$. So a positive feedback doesn't cause the system to go berserk, provided the feedback is not too strong.[42] Note that there are different ways to start the feedback loop in (5). In the passage from the latest glacial maximum, warming came first, caused by the Milankovitch cycles. Today, we're initiating the same feedback loop, but at the other end, through our injection of vast amounts of CO_2 into the atmosphere (mostly from fossil fuel burning).

Another major feedback contributing to the glacial cycles is the changing sizes of the northern hemisphere's ice sheets. With cooling, they increase in size, and since ice reflects more of the incoming sunlight than the ground below it would have done, less of the incoming sunlight is absorbed, leading to further cooling (and conversely, warming tends to melt away the ice sheets, making the surface less reflective and leading to further warming). The asymmetry between the northern and the southern hemispheres is interesting here: in the north there is plenty of land at latitudes susceptible to such fluctuations, whereas in the south the ice sheet is confined to the small continent of Antarctica, which is mostly so far south as to be covered even during interglacials.[43]

These feedbacks are crucial to understanding the glacial cycles. The global changes in average insolation caused by the Milankovitch cycles are way too small to explain those huge climate changes on their own. A more crucial quantity, compared to global average insolation, for triggering those changes is the amount of northern hemisphere summer insolation, which initiates changes in ice cover, with a snowballing effect on temperature and atmospheric CO_2, resulting eventually in major global climate change. So even when the southern hemisphere is out of sync with the northern in terms of Milankovitch-cycle-generated changes

---

[42] Here's the criterion for stability in this kind of positive feedback system: If $1°C$ of initial warming and one lap around the feedback cycle produces an extra $a°C$, then the total warming after an unlimited number of rounds around the cycle remains bounded if and only if $a < 1$. This is due to the mathematical fact, known as the convergence criterion for geometric series, that

$$1 + a + a^2 + a^3 + \cdots = \begin{cases} 1/(1-a) & \text{if } 0 < a < 1 \\ \infty & \text{if } a \geq 1. \end{cases}$$

Note, however, that even if we had $a \geq 1$, the total warming would not literally hit ∞, because eventually the system would hit upon physical barriers, such as the oceans simply boiling away, not included in the simplified model (5).

[43] Due to continental drift, this situation is not fixed forever, but on the time scales we discuss here, continental drift is small enough that it can be ignored.

in insolation, it is led into sync by these strong feedbacks – contrary to what Croll (1875) believed.

The climate system involves other significant feedbacks, such as the release of greenhouse gases from the thawing Arctic tundra. Especially important is the one involving water vapor: with higher temperature, the capacity of air to hold water vapor increases, leading to an increased greenhouse effect and further warming.[44] It is the combined effect of these feedbacks that we cannot yet quantify with great precision, and this is the reason for the greater part of the wide uncertainty range in the IPCC interval (4) for the climate sensitivity $T_{2\times CO_2}$. But studying how temperature has scaled with ice cover, CO_2 concentration and other factors during the glacial cycles has provided important clues and estimates. To return to Al Gore's movie and its message concerning the role of CO_2: it is indeed phrased in overly simplistic terms (of the kind one often has to revert to when presenting popular science), but the basic message is correct. The large fluctuations in atmospheric CO_2 were crucial for producing the dramatic climate change of the glacial cycles – and a large increase in CO_2 today may potentially produce similarly dramatic effects.

Recent decades have witnessed a wide variety of studies producing estimates of $T_{2\times CO_2}$ based on many different kinds of data, including paleoclimatic data, not only pertaining to the glacial cycles of the last million years, but on various other time scales as well. While none of these studies has been able to pinpoint anything like an exact value for $T_{2\times CO_2}$, their combined evidence seems to converge on the IPCC interval (4).[45]

Here two caveats concerning climate sensitivity, emphasized by James Hansen and coauthors, deserve to be mentioned.[46] The first is that $T_{2\times CO_2}$, as standardly defined, *excludes* the very slow feedbacks of ice sheet shrinkage[47] and vegetation changes. This means that $T_{2\times CO_2}$ may be useful to estimate the effect of given emission scenarios on time scales up to a century or so, but perhaps less useful on longer time scales when these slow feedbacks begin to make themselves felt. $T_{2\times CO_2} = 3\,\mathrm{K}$ may well correspond to a climate sensitivity that is twice as large when taking into account the effects of the melting of the Greenland and West Antarctic ice sheets.

---

[44] One sometimes hears (from the same kinds of sources as those discussed in Footnotes 40 and 41) the claim that the role of CO_2 in climate change is overestimated, because H_2O is a more important greenhouse gas than CO_2. This is a misunderstanding. It is true that there is much more water vapor then CO_2 in the atmosphere, and its contribution to the greenhouse effect is larger, so it is certainly important for climate. Yet, as far as climate *change* is concerned, CO_2 plays a more leading role; see Lacis et al. (2010). This has to do with the short-livedness of H_2O in the atmosphere. Anthropogenic emissions of H_2O are of the same order of magnitude as those of CO_2, but lack almost entirely the potential to cause climate change, because all the added H_2O leaves the atmosphere within weeks (ending up, e.g., in oceans and in the ground).

[45] Knutti and Hegerl (2008) offer an excellent survey of this work. For a recent study in which I took part, see Johansson et al. (2015).

[46] Hansen et al. (2008), Hansen (2009), Hansen and Sato (2012).

[47] The other cryospheric feedbacks, involving sea ice and snow cover, are much faster, and are included in $T_{2\times CO_2}$.

The second caveat is that having a single value of $T_{2 \times CO_2}$, independent of the state of the climate system, is a modeling approximation – a linearization of the dependence of temperature upon external forcings. Such linearizations tend to work well in sufficiently narrow ranges, but may be somewhat less valid if one moves far in the state space – such as comparing the state now to that during the last glacial maximum. It also means that as we continue our CO_2 emissions pushing the climate further away from familiar territories, $T_{2 \times CO_2}$ need not stay constant. It has been suggested – see, e.g., Chapter 10 in Hansen (2009) – that climate sensitivity could increase in that direction and, if we go far enough by burning all our extractable fossil fuels, diverge, causing runaway global warming, boiling away the oceans and ending in a Venus-like state with a planet that is no longer inhabitable. The view that the latter scenario is a real risk is not widely held in the climate science community, and Hansen later distanced himself from it.[48] Climate change resulting from fossil fuel burning can take us far away from the current climate, but probably not as far as the Venus syndrome.

2.6 The need for action

What to do, then? The uncertainty concerning climate sensitivity (and of course other key quantities) means that we cannot say with much accuracy where climate will be in, say, 2100, even if we knew what our future emissions would be. What we do know, however, is that no matter where $T_{2 \times CO_2}$ should happen to sit within the IPCC interval (4), a business-as-usual scenario for our greenhouse gas emissions would lead to far-reaching climate change. A recent report from the World Bank, with the title *Turn Down the Heat: Why a 4° Warmer Planet Must be Avoided*,[49] outlines what might happen in the case of a 4°C warming relative to the pre-industrial level[50] – a scenario towards the higher end but not at the extreme high end of what might happen by 2100 under business as usual – and the picture they paint is not comforting. From their foreword:

The 4°C scenarios are devastating: the inundation of coastal cities; increasing risks for food production potentially leading to higher malnutrition rates; many dry regions becoming dryer, wet regions wetter; unprecedented heat waves in many regions, especially in the tropics; substantially exacerbated water scarcity in many regions; increased frequency of high-intensity tropical cyclones; and irreversible loss of biodiversity, including coral reef systems.

And most importantly, a 4°C world is so different from the current one that it comes with high uncertainty and new risks that threaten our ability to anticipate and plan for future adaptation needs. (p ix)

---

[48] Hansen (2013).

[49] Potsdam Institute (2012).

[50] On this scale, which is the customary one in discussions of climate change, we have (as mentioned earlier) already experienced about a 0.8°C warming, so 4°C warming in this sense means about 3.2°C above current climate.

This we might want to avoid. The most obvious way to do so is to restructure major sectors of our economy (energy, transportation, agriculture) in a way that cuts down on our greenhouse gas emissions. In purely technological terms, this is no doubt doable. For instance, it will take merely a few percent of the Sahara desert covered by solar power plants to produce electricity amounting to the world's energy consumption as of today. Plenty has been written about how to go about concretely cutting down on our greenhouse gas emissions.[51] As far as the overall question addressed in this book – which directions of research are likely to benefit humanity and which directions are likely to harm us – it seems highly plausible that research into technologies for green and sustainable energy, transportation and agriculture is likely to end up favorably in this respect. But I will leave it at that, and instead, in the interest of bringing some slightly less familiar material up for discussion, move on to ideas on how to prevent dangerous climate change by means other than cutting down on our greenhouse gas emissions. Some of them are quite drastic, and many (but not all) of them are often lumped together under the general heading **geoengineering**.

A major reason for considering these more radical ideas is that the more conventional idea of cutting down on our greenhouse gas emissions has been very slow in generating any real action among (most of) the world's leaders. It seems fair to say that, at least currently, the most difficult obstacles in switching to green technology are political rather than technological. The switch is associated with a nontrivial cost, and it seems that politicians are currently trapped by their inability to decide on how to distribute this cost – between countries and over time. This paralysis in switching to green technology can be illustrated by the global annual CO_2 emissions from fossil fuel burning and cement production, which have grown from 22.7 gigatonnes in 1990 to 24.4 in 1997 (the year that the Kyoto protocol was signed, with the intention of cutting back the emissions) and 33.0 in 2010 – an increase of 45% in the two decades when we supposedly woke up to the problem of global warming.[52] The main allure of geoengineering is that some of the solutions on offer seem to be vastly cheaper than the conventional approach of switching to green and fossil-free technology.

2.7 A geoengineering proposal: sulfur in the stratosphere

So let's see if there are other ways to solve the climate crisis. One such suggestion is promoted in the 2009 bestseller *SuperFreakonomics: Global Cooling, Patriotic Prostitutes, and Why Suicide Bombers Should Buy Life Insurance* by Steven Levitt

---

[51] Monbiot (2006) is a nice place to start, where the author shows how to cut CO_2 emissions in the UK by 90% by 2030 (without catastrophically damaging the economy). Readers more interested in other countries will still find plenty that generalizes beyond the UK.

[52] PBL Netherlands (2011).

and Stephen Dubner.[53] The "global cooling" of the title is not some crackpot idea that the climate change we're experiencing involves a lowering of the global average temperature, but alludes to an engineering scheme to purposely bring back global temperatures to pre-industrial levels (or whatever level we may decide to aim for).

The scheme advocated by Levitt and Dubner is inspired by volcanic eruptions, which eject large amounts of sulfur into the atmosphere; this sulfur turns into aerosols (tiny particles) that tend to shield away some of the solar radiation as long as they stay in the atmosphere. Those particles that only reach the troposphere, which is the lowest part of the atmosphere extending about 10 km up, are washed away in days or weeks. Some, however, are blasted by the volcano all the way up to the stratosphere, which is the layer spanning altitudes from about 10 km to about 50 km. Once in the stratosphere, the aerosols remain there much longer, with a typical half-life of a couple of years. Large volcanic eruptions have in the past caused temporary cooling for about that amount of time. There has also been an anthropogenic contribution to the atmosphere's aerosol content, cooling the climate and therefore offsetting (hiding) much of the warming caused by our greenhouse gas emissions. So why not put some more of that stuff into the stratosphere, to offset as much of global warming as we would like? It's a strikingly simple solution, Levitt and Dubner explain, based on discussions with physicists Nathan Myhrvold (of Footnote 53 fame) and Lowell Wood. We can use a very long hose to inject the stratosphere with SO_2 (sulfur dioxide), which then creates aerosols:

At a base station, sulfur would be burned into sulfur dioxide and then liquified.... The hose, stretching from the base station into the stratosphere, would be about eighteen miles long but extremely light. "The diameter is just a couple of inches . . . " says Myhrvold. "It's literally a specialized fire hose."

The hose would be suspended from a series of high-strength, helium-filled balloons fastened to the hose at 100- to 300-yard intervals . . . , ranging in diameter from 25 feet near the ground to 100 feet near the top.

---

[53] After the authors' even better selling and quite interesting *Freakonomics* from 2005, this follow-up was a huge disappointment. The authors continue to demonstrate their *faiblesse* for applying economic reasoning in contexts where this is not usually done and where it leads to counterintuitive conclusions, but the conclusions are way too often based on sloppy reasoning and poor knowledge of the topic at hand. A typical example is their quick dismissal (attributed to Nathan Myhrvold) of solar power:

The problem with solar cells is that they're black, because they are designed to absorb light from the Sun. But only about 12 percent gets turned into electricity, and the rest is reradiated as heat – which contributes to global warming. (p 187)

The physical mechanism here is of course a reality, but to judge whether the heat dissipation is a serious problem in the context of global warming requires further quantitative analysis, and Levitt and Dubner offer no such analysis. Fortunately, Pierrehumbert (2009) does the necessary estimates in a very instructive rebuttal, and the argument put forth by Levitt and Dubner is shown to be moot. Their treatment of geoengineering (to be discussed in what follows) has similar serious shortcomings, and further examples are discussed in my review of the book (Häggström, 2010a).

The liquified sulfur dioxide would be sent skyward by a series of pumps, affixed to the hose at every 100 yards. These too would be relatively light, about forty-five pounds each – "smaller than the pumps in my swimming pool," Myhrvold says. . . .

At the end of the hose, a cluster of nozzles would spritz the stratosphere with a fine mist of colorless liquid sulfur dioxide. Thanks to stratospheric winds that typically reach one hundred miles per hour, the spritz would wrap around the world in ten days' time. (p 194)

Five base stations with three hoses each, we learn, would be enough to entirely cancel global warming. And here's the best thing: the almost negligible cost. This method could "effectively reverse global warming at a total cost of \$250 million" (p 196). This means that even a small nation (or even a single rich person like Bill Gates) could easily afford to cancel out global warming on its own – no need for tedious and painfully difficult international negotiations, or for giving up on cheap and convenient fossil fuels!

Such an incredibly simple solution to the climate change problem sounds like it's too good to be true. And, alas, too good to be true is exactly what it is. The proposed solution has (at least) the following serious drawbacks.

(a) **The sword of Damocles.** The influence on climate of our CO_2 emissions will last for thousands of years, whereas the aerosols put into the stratosphere will go away in a couple of years. So if we settle for the SO_2-in-stratosphere solution, future generations will have the choice of either continuing our SO_2 injection, or experiencing a very sudden catastrophic climate change. This is sometimes known as putting them under the sword of Damocles.[54] What moral right do we have to impose such a burden on future generations?[55] And how do we even know that future generations will have the necessary infrastructure for keeping the SO_2-in-stratosphere project running? If at some point their society is hit by social collapse, then the project will most likely be abandoned, and they will find themselves in two major catastophes simultaneously: collapse of society, and sudden drastic climate change.[56]

(b) **Global warming versus climate change.** The SO_2-in-stratosphere solution has the potential to bring global average temperatures down to whatever level we decide to be optimal. It will not stop climate change,

---

[54] According to Greek legend, Damocles was envious of the king's power and prosperity. King Dionysos II offered to switch places, and Damocles quickly accepted. He sat down in Dionysos' throne, surrounded by luxury, but Dionysos had also arranged for a sword to hang by a single hair above Damocles' head, so as to convey to Damocles a more nuanced view of what it is like to be king. Damocles eventually begged to switch back.

[55] This is somewhat reminiscent of how today we generate nuclear waste that will remain lethally radioactive for thousands of years. One difference, however, is that in the case of nuclear waste our intention is to hide the waste away deep underground in such a way that no effort is needed from future people except for keeping out of those sites, whereas the SO_2 geoengineering scheme requires their continued active participation.

[56] Baum, Maher and Haqq-Misra (2013), Baum (2014).

however. When we increase global average temperature by one mechanism (increased greenhouse gas concentrations), and reduce it by a different mechanism (aerosols in the stratosphere), we cannot expect regional climate to remain the same. Average temperatures will go up in some regions and down in others, and precipitation patterns will change. Simulations by Caldeira and Wood (2008) suggest that regional climate after a doubling of CO_2 concentration and an offsetting SO_2-in-stratosphere injection will mostly be closer to the climate we had before these changes than in the case of only a doubling of CO_2 concentration, but still with substantial variations. Global average precipitation will likely increase compared to the undisturbed climate system, and geographic redistribution of precipitation patterns will be substantial.

(c) **Ocean acidification.** The contribution of our CO_2 emissions to global warming is only their most well-known effect. There are others, including the fact that an increased CO_2 concentration in the atmosphere leads to an increase in the oceanic uptake of CO_2, which in turn leads to acidification of the oceans. This is item (v) on the Rockström et al. (2009) list of planetary boundaries discussed in Section 2.3. With their proposed boundary we are still in their "safe operating space" as far as this aspect is concerned, but well on our way to crossing the boundary.[57,58]

(d) **The ozone layer.** It is highly unclear how large an effect this tampering with the stratosphere would have on stratospheric ozone, but there are results suggesting that its contribution to ozone depletion would be substantial.[59]

Any one of these problems should be enough to make us reluctant to view the SO_2-in-stratosphere idea as the wonderful solution to the climate crisis that Levitt and Dubner (2009) suggest it to be. However, it is just one out of a collection of proposals, under the common heading geoengineering, for radical engineering solutions to the climate change problem.

2.8 Other forms of geoengineering

In the excellent survey *Geoengineering the Climate* by the Royal Society (2009), geoengineering is defined as "deliberate large-scale manipulation of the planetary

---

[57] See Royal Society (2005) for an in-depth survey of the ocean acidification problem.

[58] There is a related issue of acidification that is sometimes raised in discussion of the SO_2-in-stratosphere injection idea, namely that SO_2 also leads to environmental acidification. However, this concern seems to be of relatively less importance compared to the idea's other downsides, because, due to the relatively long residence time of particles in the stratosphere (as opposed to in the troposphere), the amount of SO_2 that we need to inject in order to counteract greenhouse-gas-driven climate change is not more than a percent or so compared to what we already emit today into the troposphere (Levitt and Dubner, 2009, p 196).

[59] Tilmes, Müller and Salawitch (2008).

environment to counteract anthropogenic climate change." Various ways to cut down our greenhouse gas emissions do *not* count as geoengineering, because they are attempts to *avoid* affecting the climate, rather than actively trying to alter it. On the other hand, should we at some point consider purposely changing the climate towards a target climate that is not simply the reversal of anthropogenic climate change,[60] then it would be natural to extend the term geoengineering to include such actions. It has sometimes been suggested (polemically) that our on-going greenhouse gas emissions should count as a geoengineering project, but that terminology is not advisable, because altering the climate is not the *purpose* of our greenhouse gas emissions, but rather a (highly unfortunate) *side effect.*

I will not attempt to repeat the Royal Society's (2009) systematic survey of the various proposed methods for geoengineering, but will mention a few of them. It is useful, as the Royal Society report does, to split the methods in two basic categories, namely **solar radiation management** techniques that seek to reflect away (or redistribute) some of the incoming sunlight, and CO_2 **removal** techniques that seek to decrease the atmospheric CO_2 concentration.

The solar radiation management techniques can be split into two further sub-categories, depending on whether they seek to limit the amount of sunlight reaching the Earth's surface, or if they seek to increase the reflectivity of that surface. The SO_2-in-stratoshere proposal discussed above falls in the first category, as does the one inspired by Saturn's rings in Footnote 60. Another proposal is the creation of artificial clouds.[61] Yet another one is placing a huge collection of reflectors in space, right at the so-called Lagrange point L1, which is the point located about 1.5 million kilometers from Earth in the direction of the Sun, where the Earth's and the Sun's gravity balance each other.[62]

In the other subcategory – that of increasing the surface reflectivity – we have the method of painting roads and rooftops white. This one sounds really simple, but it turns out that it is less cost efficient than other methods, and in fact the amount of settled area in the world is insufficient for fully offsetting anthropogenic global warming in this way. A more ambitious proposal would be to cover deserts with highly reflective material, but this is still not very cost effective, and carries ecological and other risks.

Moving on to CO_2 removal techniques, these too are of many different kinds. A process which is part of the natural carbon cycle and plays a role in climate change on geological time scales is weathering, where silicate minerals react with atmospheric CO_2 to form carbonates, thereby removing CO_2 from the atmosphere. We could accelerate this process by seeking to expose more of these minerals directly to the atmosphere. This method has, in principle, high potential, but is considered

---

[60] In fact, such ideas have already been seriously entertained and are not even new. Around 1960, there were proposals in the Soviet Union to place Saturn-style rings, made of aerosols, in orbit around the Earth, with the twin aims of warming Siberia to make it amenable to agriculture and shadowing equatorial regions to yield a more temperate climate. See Keith (2000).

[61] Royal Society (2009).

[62] Angel (2006).

costly, because it requires mining, processing and transportation of vast amounts of silicate minerals.[63]

It is also possible to remove CO_2 from the atmosphere using industrial processes – so-called air capture. This is likely to eventually become feasible on a large scale, but a lot of research and development is still needed to make it affordable.[64] Just like for so-called CCS (carbon capture and storage, which is viewed by many to be a possible future way to keep burning fossil coal without contributing much to global warming),[65] it involves the still partly unsolved problem of safe underground storage of CO_2. Air capture has the advantage of being more flexible than CCS in terms of geographic locations, which can be decoupled from power plants, whereas the chemical engineering involved in CCS is much easier than for air capture, because it is applied to the exhaust fumes of power plants, where the CO_2 concentration is orders of magnitude larger than in ambient air. CCS as applied to fossil fuel burning does not count as geoengineering, because the combination of the fossil fuel power plant and the CCS cannot yield a net negative effect on atmospheric CO_2. Applying CCS to bioenergy is a different story: it has the capacity to yield a net removal of CO_2 from the atmosphere. As with bioenergy in general, there are other downsides, however, including the issue of to what extent it can be implemented without causing famine by taking land away from the world's food production.

Yet another CO_2 removal technique is ocean fertilization. The idea is that fertilization will increase the uptake of CO_2 by microorganisms in the ocean. Much of this biological material sinks to the deep sea and is offset on the ocean bed. Various schemes for how to do this in more detail have been proposed, but with our current state of knowledge they are ecologically extremely risky.[66]

2.9 No miracle solution

There is obviously a lot more to say about the pros and cons of the various geoengineering techniques. Generalizing somewhat, it can be said that CO_2 removal techniques are to be preferred to the solar radiation management techniques. This is because they attack the problem at its root (increased greenhouse gas concentrations) rather than the symptoms (global warming). All the solar radiation management techniques will suffer from at least the downsides (b) and (c) – and to varying extents also (a) – discussed above for the SO_2-in-stratosphere idea, whereas the CO_2 removal techniques generally do not suffer from these.

David Keith, who is one of the world's foremost experts on geoengineering, spoke on these matters at my home university, Chalmers, in December 2009.

He emphasized that if we should ever go about implementing the SO_2-in-stratosphere idea or other solar radiation management techniques on large scales, it should only be as a temporary contribution to solving a very urgent climate crisis. On the other hand, he pointed out one general advantage to solar radiation management techniques over CO_2 removal techniques, namely their timeliness: they can change the Earth's energy balance at full power almost immediately, and can cool climate substantially within a year. The climate effects of CO_2 removal will typically take decades to become visible. Therefore, according to Keith, it may make sense to leave open the possibility at some point to combine slower methods (CO_2 emission cuts and/or CO_2 removal techniques) with the use of, e.g., SO_2-in-stratosphere in order to cut down a temporary peak in global mean temperature that would otherwise be unavoidable and is judged to risk pushing climate over some tipping point (such as dieback of the Amazon rain forest or collapse of the Greenland ice sheet). Such a scenario might well come about in the second half of the 21st centrury even if we should make those ambitious cuts in global greenhouse gas emissions that politicians today seem unable to agree upon. This is due, first, to the momentum in the climate system that can cause temperature to keep rising for many decades after we've been able to turn atmospheric CO_2 content downwards, and, second, to the significant uncertainty (4) in our estimate of climate sensitivity – an uncertainty that we cannot trust to be resolved anytime soon.

Another important lesson emphasized in David Keith's talk is that none of the geoengineering methods that have been proposed (including the air capture technology that his own research group is developing) offers a miracle solution to the climate crisis or an excuse for continuing with unconstrained greenhouse gas emissions. Further research into the various methods does seem highly warranted, however, to get a better grip, e.g., on their cost effectiveness and the environmental risks involved. But not *too* much – perhaps surprisingly, Keith says in an interview in *The Atlantic* that "the most disastrous thing that could happen would be for Barack Obama to stand up tomorrow and announce the creation of a geo-engineering task force with hundreds of millions in funds".[67,68]

---

[67] Wood (2009).

[68] In later work, Keith seems to have shifted towards a somewhat more favorable view of SO_2-in-stratosphere geoengineering. He still feels that we are far from a level of knowledge where we can responsibly initiate such geoengineering, and he is still far from certain that it will ever be a good idea, but he no longer has the same reservations about launching a high-profile research program on the topic. The main reason for his shift seems to stem from recent work that indicates that solar radiation management techniques, properly implemented, may suffer less than previously thought from downside (b) discussed above. While earlier simulation studies of the effects of solar radiation management focused on the case of a dosage chosen to entirely cancel the man-made rise in global average temperature, more recent studies have considered more modest dosages, canceling only part of man-made global warming. In particular, this more recent work suggests (tentatively) that if the dosage is chosen so as to cancel about *half* of the average temperature rise (which corresponds approximately to restoring global average precipitation), then, in contrast to what happens with full dosage, almost

Why, then, does he express this modest or even negative view towards funding of his own research area? The answer is to be found in the concept of **moral hazard**. This is the phenomenon where, e.g., someone who buys insurance might feel tempted towards riskier behavior than otherwise. In the present case, the moral hazard is that the more attention geoengineering gets (as, for instance, in a high-profile announcement from the American president), the greater will the temptation be for decision-makers and voters to think that there is no big hurry to cut down on greenhouse gas emissions, because if worst comes to worst we will surely be able to work things out with one geoengineering scheme or another.[69,70]

2.10 Searching for solutions further outside the box

The geoengineering proposals discussed above all have a certain flavor of outside-the-box thinking compared to the more mainstream discourse about curbing our greenhouse gas emissions. But is it possible to think even further outside the box in the search for solutions to the climate crisis? It turns out that the answer is yes, and this final section of the chapter will be devoted to some truly radical proposals that have been put forth by unconventional and brave thinkers in recent years. Many readers will no doubt immediately find the proposals extreme or even outrageous, but some of these readers, when looking back on this section after having read the book from beginning to end, may find these proposals relatively less extreme.

all regions will get a climate that is closer to its pre-industrial climate than would have been the case without the geoengineering. The effects of the geoengineering intervention at this more modest level can thus be argued to be more uniformly beneficial than with full dosage. See, e.g., Moreno-Cruz, Ricke and Keith (2012), Keith (2013) and Häggström (2014a).

[69] Keith (2013) expresses some regret over having introduced, in earlier work, the term moral hazard in the present context, and prefers instead **risk compensation**. While moral hazard in general refers to situations (such as insurance) that encourage riskier behavior because someone else bears the costs, risk compensation refers to situations where the riskier behavior is triggered by a true *reduction* (rather than the transfer to someone else) of risk, an example being how safety belt regulations for automobiles resulted in less reduction in traffic casualties and injuries than expected, because seat belts instilled a feeling of safety in drivers that made them more careless.

[70] This raises the issue of whether it might be wrong of me to draw attention in this book to geoengineering – might it be that by doing so I am contributing to a more dangerous world? Analogous questions can be asked in connection with a number of other issues brought up elsewhere in this book. These questions are very difficult to answer, and I feel that being overly worried about them (and, more generally, being overly strategic about what to say) may result in paralysis. In writing this book, I am simply assuming – right or wrong – that an open discussion about the pros and cons of upcoming emerging technologies is necessary for informed decision-making, and that it thereby may contribute to a safer and better world. See Bostrom (2011) for an attempt at systematic treatment of cases where the value of knowledge is negative.

Moral philosophers Julian Savulescu and Ingmar Persson view the climate crisis as a difficult ethical problem, and propose a radical new approach based on the idea that an ethical problem that appears difficult or even unsurmountable need not be so on an absolute scale, but perhaps only relative to the limited moral capabilities of the agents confronting it. That the climate crisis is ethically difficult to us mere mortal human beings is undoubtedly correct. Another moral philosopher, Stephen Gardiner, calls it "a perfect moral storm,"[71] based on the juxtaposition of a number of aspects, each contributing heavily to its difficulty. One is that the harm caused by greenhouse gas emissions does not hit mainly upon the individual emitter, but it is dispersed over the entire population on Earth. This means that an individual looking after his personal well-being has little incentive to contribute to the common good by cutting his emissions. Conditional on what everyone else does, then, no matter what their emission levels are, my own emissions make little difference, so if it is more convenient for me to take the car than to take the bus, that will be the best action, from a prudential perspective (i.e., when optimizing what is best for me without regard to what is good for others). In other words, the problem is an instance of the social dilemma known as **the tragedy of the commons**.[72] To this spatial dispersion of causes (greenhouse gas emissions) and effects (climate change), we must also add the temporal aspect: our emissions today affect not only people living today but also future generations. This makes the problem even more difficult, due to the asymmetry that our actions influence the well-being of future generations whereas theirs do not affect us. A further contribution to the perfect storm is our lack of adequate theoretical and institutional tools to handle issues involving the far future, intergenerational justice, scientific uncertainty and our actions' effects on nature.

Savulescu and Persson (2012) realize all this, and conclude that it is unlikely, given our current level of moral competence, that we will be able to solve the problem. Hence we need to improve our moral competence. More precisely, we need to change in the direction away from our tendency to do what is egoistically prudentially good and towards doing what is morally good, in the sense that each person does what is best for everyone on the whole.[73] To some extent we have done so via education, and while progress has not been zero, it is still insufficient according

---

[71] Gardiner (2006).

[72] Hardin (1968). The prototypical case giving rise to its name is as follows. The farmers of a village share a common pasture that each farmer can use at will for his sheep. If all the farmers use the common pasture fully, it will collapse. Nevertheless, the incentive for each farmer is always to use it, even in the situation where it is already overexploited, because individual constraint would just be to let the neighbors get all the benefits from the pasture before its eventual collapse. So the pasture seems doomed.

[73] This, of course, is Savulescu's and Persson's *opinion* about what is morally good. I tend to agree with their view, and will take it as a given in the following discussion, despite my skepticism about the existence of objective moral truth. This last issue is slightly thorny, and further discussion on it is deferred to Section 10.1.

to Savulescu and Persson, who argue in favor of another path that is beginning to open up, namely enhancement of our moral capacities by biomedical means:[74]

Our knowledge of human biology – in particular of genetics and neurobiology – is beginning to enable us to directly affect the biological or physiological bases of human motivation, either through drugs, or through genetic selection or engineering, or by using external devices that affect the brain or the learning process. We could use these techniques to overcome the moral and psychological shortcomings that imperil the human species. We are at the early stages of such research, but there are few cogent philosophical or moral objections to the use of specifically *biomedical* moral enhancement – or *moral bioenhancement*. In fact, the risks we face are so serious that it is imperative we explore every possibility of developing moral bioenhancement technologies – not to replace traditional moral education, but to complement it.

Another group of authors discussing moral bioenhancement as a response to the climate crisis, including some of the possible biochemistry that might be involved in an implementation (more oxytocin, less testosterone), is Liao, Sandberg and Roache (2012). They discuss moral bioenhancement in the broader context of human enhancement to counteract climate change, and brainstorm the following further examples of possible modifications of humans or humanity:

Meat intolerance. Citing work that demonstrates that a large fraction of the global warming triggered by greenhouse gas emissions comes from our meat production,[75] Liao et al. observe that, although reducing our meat consumption would not solve the climate crisis on its own, it might nevertheless play a nontrivial role. As a means to cutting down on meat consumption, they suggest pharmacologically induced meat intolerance, for instance by stimulating our immune system against some characteristic bovine proteins.

Smaller humans. If we were smaller, we would eat less, require smaller cars, etc., and would in general be less of a burden on the environment and on climate. Techniques for making us smaller need not be particularly difficult, and may, e.g., involve pre-implantation genetic profiling, or of the growth-hormone-inhibiting hormone somatostatin.

Cognitive enhancement. Enhancement of our cognitive capacities by pharmacological, genetic and other means will be a major topic in Chapter 3. In the present context, Liao et al. note that it is reasonable to

---

[74] Savulescu and Persson (2012). See also Persson and Savulescu (2008) for a more detailed discussion of their moral bioenhancement proposal, and a motivation that focuses less on global warming than on the kind of situation described in Section 1.1 in connection with Eliezer Yudkowsky's science fiction story *Three Worlds Collide*, where individuals have the capacity to single-handedly destroy civilization. What Persson and Savulescu hope for is nothing less that a modification of human nature to become more like the aliens discussed in Footnote 4. Given where synthetic biology and other areas are heading, their concern can hardly be dismissed as irrelevant. See Section 8.3 for more on such existential threats.

[75] Steinfeld et al. (2006).

expect this to have a side effect of reducing nativity, so that eventually there will be fewer of us to burden the climate system. It is well established that young women's access to education reduces birth rates, and cognitive ability has also been observed to correlate negatively with birth rate.[76]

To avoid misunderstandings at this point, let me emphasize that – here, as elsewhere in the book – my mere mentioning of a proposal does not imply that I endorse it. In fact, not even Liao et al. claim that the suggested modifications of human nature ought to be implemented. What they do defend, however, is, to quote the final sentence of their paper, the idea that proposals such as these "deserve to be considered and explored further in this debate," and in this I agree.[77] Suggestions such as those discussed here tend nevertheless to generate considerable controversy and heat. For instance, critiques of Persson and Savulescu are offered by Fenton (2010) and Gunson and McLachlan (2013); these, however, are fairly technical discussions of risk and probability, and I wish to defer discussion of those matters to later chapters. Let me instead discuss a couple of the critiques of Liao et al., namely Bognar (2012) and Trachtenberg (2012). Both of them are agressively dismissive, but in being so, Trachtenberg (2012) deserves kudos for the witty passage where, alluding to Swift (1729), he says that Liao et al.

perhaps risk misunderstanding by adopting a rhetorical technique typically associated with satire. They are not, after all, the first to propose a dietary solution to a pressing social problem, and theirs is more modest than others that come to mind.

On a more serious note, Trachtenberg suggests that it is unlikely that "people who are not willing to stop eating meat would be willing to undergo a treatment to make them stop eating meat." While such a notion may seem absurd on first sight, it is actually consistent with basic human psychology. Due to the psychological phenomenon known as **hyperbolic discounting**, which will be treated briefly in Section 10.2 and has been dealt with beautifully and at book length by Ainslie (2001), we are not always in agreement with our future selves regarding what these should do. This is why we lock away our money in pension funds and install time locks on our refrigerators. Similarly, I may well be in a position where I, OH_{today}, cannot resist the beef for tonight's dinner, but feel that $OH_{nextweek}$ really ought to switch to vegetarianism, while realizing that he (just like me) will most likely be too weak to resist the beef offered on the menu. (As a description of my actual position on vegetarianism, this would not be totally off the mark.) In such a situation, taking that meat intolerance injection does sound like an excellent way for me to help $OH_{nextweek}$ overcome his unfortunate weakness for red meat.

---

[76] Shearer et al. (2002).

[77] But even this last and rather cautious position is not obviously unassailable; cf. Footnote 70.

Moving on to Bognar (2012), who expresses his disdain for Liao et al. already in his note's title "When philosophers shoot themselves in the leg," we find that he goes on to explain what he means by this, namely

philosophers [who propose] policies with little regard to their feasibility and real costs. They do themselves – and their profession – a disservice. They are unlikely to be taken seriously if they give the impression that philosophers do not understand what makes a policy proposal worthy of consideration.

Bognar is most certainly right that politicians are highly unlikely to jump on the human enhancement ideas of Liao et al. anytime soon, but in his general conclusion that philosophers ought to avoid bringing up policy ideas for public discussion unless they enclose full and favorable cost–benefit analyses and have made sure that a parliamentary majority in favor of the proposed policy will be within reach some time in the next few years, I could not agree less. Suppose everyone, even including *philosophers* (whose subject area Collini (2012, p 138) describes as "that form of enquiry in which no question can be ruled out as inappropriate in advance"), exhibited that kind of restraint and never allowed themselves to publicly think outside the box or present half-baked ideas. That would surely make Bognar happy, but it would also mean that in humanity's collective journey through the uncharted territories discussed in Section 1.4 we will suffer from a self-inflicted myopia that makes us unable to see even a hint of any dragons until the moment when we stand face to face with them. I will make sure that Bognar receives a review copy of this book, which he will no doubt brand as a disgrace for all of academia, and a threat to our reputation as serious thinkers and to our future prospects of being able to influence decision-makers on important issues such as whether public funding of dental insurance should cover 75% or just 60% of the costs.

What we have touched upon in this section are a few examples of the broader issues of human enhancement and transhumanism, which is the subject of the next chapter, where we will see plenty of further examples of possible modifications of the flesh-and-blood packages that are us, and the controversies concerning whether we should go down these paths.

CHAPTER 3

Engineering better humans?

3.1 Human enhancement

When discussing possible future scenarios involving human enhancement and radical biotechnological changes in human nature, we should first realize that we are already well under way. The human population is changing, in various striking ways, at rates far faster than can be explained by biological evolution and changes in the gene pool. Take, for instance, height: as a consequence of improved nutrition, and perhaps also improvements in hygiene, public health and other socioeconomic factors, average height among young men in Western Europe has increased by about 11 cm in a century.[78] Related trends in developed countries are those towards increased weight,[79] earlier onset of puberty,[80] and others.

None of these physical changes are obvious improvements, but how about the case of intelligence? Consider the so-called **Flynn effect**: the increase of average IQ in large parts of the world, sustained for many decades by about 0.3 IQ points per year.[81] It seems likely that there are several causal factors behind this increase, but the jury is still out concerning what the main ones are. That the whole thing might be an artifact of increased test familiarity is an obvious candidate explanation that needs to be checked, but it turns out that this can account for no more than a small part of the increase. Improvements in nutrition and education are two other candidates, and the latter is supported by substantial correlations between IQ scores and literacy.[82] Due to its seemingly non-biomedical character,[83] education is sometimes overlooked in discussions of human enhancement, yet it seems reasonable to view the ambitious efforts by many governments from the

---

[78] Hatton and Bray (2010).

[79] Komlos and Baur (2004).

[80] Buck Louis et al. (2008). In this case, causal factors are not well understood, but exposure to toxic chemicals is suspected to play a role.

[81] Neisser (1997), Flynn (1999), Wai and Putallaz (2011).

[82] Marks (2010).

[83] I write "seemingly" because on one level education can be seen as a class of interventions that cause biomedical changes in the brain.

Here Be Dragons. First Edition. Olle Häggström.
© Olle Häggström 2016. Published in 2016 by Oxford University Press.

19th century onwards to provide education for the general public as the greatest human enhancement project to date.[84]

It has become increasingly common to improve (or so they say) appearance and performance with Botox and Viagra. And Prozac. There is a strong tendency today to categorize ever more psychological states as disorders that can be treated with psychopharmaceuticals, such as how much of what used to be thought of simply as shyness is nowadays increasingly diagnosed and treated as **social anxiety disorder**; see Crews (2007) for a critical discussion of this development. And what about pharmacological treatments of the "disorder" of becoming exhausted after hours of intensive study? James Miller (2012), who teaches economics at a women's college in the US Northeast, offers an inside view of his students' widespread use of the amphetamine-based drug Adderall for that purpose – as well as his own use.

The very notion of human enhancement is usually defined in contrast to therapy or intervention against disorders: while therapy aims to cure disorder and bring bodily and/or mental functionality up to (or towards) normal levels, enhancement aims to lift functionality *beyond* such levels. But "normal" is a fuzzy concept, and some of the above examples indicate the difficulty of drawing a clear line in the sand. Some writers argue that we can and must do so,[85] whereas others are more skeptical about the distinction, such as Wolpe (2002) who characterizes it relativistically as "what medicine chooses to treat is defined as disease, while altering what it does not treat is enhancement."[86]

The list of enhancements or enhancement-like developments goes on. In vitro fertilization, which already accounts for more than 1% of births in the United States,[87] is beginning to change our reproductive behavior. Under particular circumstances, such as when there is a family history of inherited disease, this is combined with pre-implantation genetic profiling to decide whether a given embryo should be rejected or implanted. Such profiling is often viewed as a first step on a slippery slope towards **designer babies** and **human cloning** – possible future practices whose desirability have been hotly debated and will be discussed in Sections 3.2, 3.3 and 3.5.

Glasses and hearing aids give many of us the chance to go on with whatever occupations or pet projects we have for many decades longer than would otherwise have been possible. Smartphones and similar electronic devices enhance our cognitive and communicative skills considerably. The **extended mind** idea of Andy Clark and David Chalmers[88] holds (analogously to Richard Dawkins' **extended**

---

[84] Closely related to the education hypothesis is the more general idea, originally suggested by Flynn himself, that IQ tests measure abilities in (certain kinds of) abstract reasoning, and that we today practice abstract thinking far more than earlier generations did; see, e.g., Chivers (2014).

[85] Sandel (2004), Colleton (2008).

[86] See also Harris (2009) and several other contributions to Bostrom and Savulescu (2009) for further discussion.

[87] Nair (2008).

[88] Clark and Chalmers (1998), Clark (2008).

phenotype)[89] that our minds do not reside solely within our skulls or even our bodies, but extend further. Even if we should reject that idea, it is still not so easy to distinguish between what is truly part of ourselves and what is merely an external device, and it becomes especially difficult as soon as the devices migrate into our heads. Michael Chorost, who was born with severe loss of hearing and became completely deaf in 2001, had a so-called cochlear implant – an electronic device surgically implanted into his head to let him hear again – and now argues that we should in similar fashion hook up Internet connections directly into our brains.[90] Perhaps the implant should be viewed as truly a part of Chorost himself, but why, then, should not his smartphone be granted the same status?

Given this plethora of ways of improving our bodies, minds and capabilities, one may ask whether it makes sense to treat them all as varieties of the same thing. Nick Bostrom and Julian Savulescu, in the introduction to their anthology *Human Enhancement*,[91] write:

The degree to which human enhancements constitute a distinctive cluster of phenomena for which it would be appropriate to have a (multidisciplinary) academic subfield is debatable ... One common argumentative strategy, used predominantly to buttress pro-enhancement positions, is to highlight the continuities between new controversial enhancement methods and old accepted ways of enhancing human capacities. How is taking modafinil fundamentally different from imbibing a good cup of tea? How is either morally different from getting a full night's sleep? Are not shoes a kind of foot enhancement, clothes an enhancement of our skin? A notepad, similarly, can be viewed as a memory enhancement – it being far from obvious how the fact that a phone number is stored in our pocket instead of our brain is supposed to matter once we abstract from contingent factors such as cost and convenience. In one sense, *all* technology can be viewed as an enhancement of our native human capacities, enabling us to achieve certain effects that would otherwise require more effort or be altogether beyond our power. (p 2)

Terminological issues aside, it is clear that there is much scope for further enhancement. Perhaps more than most of us imagine. Bostrom, in his paper "Why I want to be a posthuman when I grow up," which is an unusual hybrid between a scholarly paper and a manifesto, tries to give us an idea of just how far it might go.[92] As a warmup, he asks the reader to imagine being subjected to a series of improvements that are within (or at least bordering on) normal human limits, such as improved health, making you feel physically "stronger, more energetic, and more balanced," and improved cognition, so that you "can concentrate on difficult material more easily and it begins making sense to you," you "can follow lines of thinking and intricate argumentation farther without losing your foothold," and when listening to music you feel great joy in "perceiv[ing] layers of structure and a kind of musical logic to which you were previously oblivious," and so on and so forth. You also "begin to treasure almost every moment of life," and

---

[89] Dawkins (1982).
[90] Chorost (2006, 2011). For more on this, see Section 3.6.
[91] Bostrom and Savulescu (2009).
[92] Bostrom (2008a).

"feel a deeper warmth and affection for those you love, but you can still be upset and even angry on occasions where upset or anger is truly justified and constructive." All this is just the beginning, and here is how he imagines how life will seem when you have reached a little beyond our current limits:

You have just celebrated your 170th birthday and you feel stronger than ever. Each day is a joy. You have invented entirely new art forms, which exploit the new kinds of cognitive capacities and sensibilities you have developed. . . . You are communicating with your contemporaries using a language that has grown out of English over the past century and that has a vocabulary and expressive power that enables you to share and discuss thoughts and feelings that unaugmented humans could not even think or experience . . . You are always ready to feel with those who suffer misfortunes, and to work hard to help them get back on their feet. You are also involved in a large voluntary organization that works to reduce suffering of animals in their natural environment in ways that permit ecologies to continue to function in traditional ways; this involves political efforts combined with advanced science and information processing services. Things are getting better, but already each day is fantastic.

All of this (or some variant, similarly advanced and similarly good) is probably possible,[93] and there is no reason that this sketch comes even close to the limits of what is possible, but, as Bostrom points out, "as we seek to peer farther into posthumanity, our ability to concretely imagine what it might be like trails off."

Nevertheless, the kind of life that Bostrom sketches is what many, or perhaps most, members of the so-called **transhumanist** movement hope for. The concept of transhumanism is hard to avoid in a chapter on human enhancement. Before sorting out exactly what it means, it is worth pinning down the distinction between the notion of a **posthuman** and that of a **transhuman**. A posthuman is the kind of hypothetical future being who in one way or another descends from humans, but whose physical, cognitive or other properties are enhanced to the extent that it becomes questionable whether he or she (or whatever) can still be called a human. A transhuman is a being who is somewhere on the evolutionary path from unenhanced humans to posthumans.[94] It may, as hinted above, be argued that we are all transhumans, although drawing exact lines distinguishing unenhanced humans from transhumans and posthumans seems to me more of a terminological issue than one of substantial importance.

Not everyone agrees with me on this last point, however. For instance, Francis Fukuyama considers it terribly important to insist on the existence of some human essence – some holistic property that he, unable to specify concretely, calls Factor X – that we must not overstep, for reasons having to do with human dignity, which will be the topic of Section 3.2.[95]

---

[93] Notice also that towards the end of the quote, Bostrom implicitly suggests that the improvements to our minds may come to include not only *cognitive* enhancement, but also *moral* enhancement, which I touched upon in Section 2.10. In the present chapter, most of my focus will be on the former, but we should not forget that the latter is possible (and perhaps also desirable).

[94] Bostrom (2003b).

[95] Fukuyama (2002).

In his *Transhumanist FAQ*, Bostrom gives a definition that ascribes the word transhumanism two meanings:[96]

Transhumanism is a way of thinking about the future that is based on the premise that the human species in its current form does not represent the end of our development but rather a comparatively early phase. We formally define it as follows.

(1) The intellectual and cultural movement that affirms the possibility and desirability of fundamentally improving the human condition through applied reason, especially by developing and making widely available technologies to eliminate ageing and to greatly enhance human intellectual, physical, and psychological capacities.

(2) The study of the ramifications, promises, and potential dangers of technologies that will enable us to overcome fundamental human limitations, and the related study of the ethical matters involved in developing and using such technologies.

A transhumanist is a proponent of transhumanism in this sense, but note that the definition is a bit ambiguous, because it is clearly possible to be a proponent of (1) but not of (2), or vice versa. Bostrom himself is a transhumanist in both senses of the word (and has been, for the better part of two decades, one of the leading representatives of the movement). As for myself, I am wholeheartedly in favor of (2). As to (1), I am more hesitant, because while the overall project of human enhancement seems to have the potential to bring us enormous benefit, there are also dangers and downsides (some of which will be discussed below), so it is hard to unreservedly affirm the "desirability" part of (1). Particularly hard to endorse is the lighthearted techno-optimism that permeates much of the transhumanist movement and that can be exemplified by the so-called **proactionary principle**, formulated by Max More (another leading figure of the transhumanist movement) in opposition to the more well-known precautionary principle. The proactionary principle places great weight on what it calls freedom to innovate, defined as follows.

Freedom to innovate: Our freedom to innovate technologically is valuable to humanity. The burden of proof therefore belongs to those who propose restrictive measures. All proposed measures should be closely scrutinized.[97]

---

[96] Bostrom (2003b).

[97] More (2005), whose full statement admittedly gives a somewhat more nuanced view. Its central passage reads . . .

People's freedom to innovate technologically is highly valuable, even critical, to humanity. This implies a range of responsibilities for those considering whether and how to develop, deploy, or restrict new technologies. Assess risks and opportunities using an objective, open, and comprehensive, yet simple decision process based on science rather than collective emotional reactions. Account for the costs of restrictions and lost opportunities as fully as direct effects. Favor measures that are proportionate to the probability and magnitude of impacts, and that have the highest payoff relative to their costs. Give a high priority to people's freedom to learn, innovate, and advance.

. . . and is followed by careful explications of the concepts involved, and a lengthy discussion of the central conflict between the proactionary and the precautionary principles. This is well worth reading and thinking about, but it still seems just too reckless, as we enter the dangerous uncharted technological

See also Sandberg (2001) for a related view, based on libertarian ideas, promoting what he calls morphological freedom, i.e., the individual right to modify one's own body according to one's own wishes, without government regulation or other interference. It is not at all clear to me that it is in our long-term interest to employ a view so biased in favor of action over precaution when rushing ahead with emerging technologies, with all their known and unknown unknowns (to borrow former US Secretary of Defense Donald Rumsfeld's terminology) and quite possibly with the future survival of humanity at stake.

The transhumanist idea that radical changes to human nature are possible has intellectual roots that can be traced back to various sources, for instance to the ancient Greek myths about Prometheus, Daedalus and Icarus, to Enlightenment thinkers like Marquis de Condorcet,[98] to great novels like Mary Shelley's *Frankenstein* (1818) and Aldous Huxley's *Brave New World* (1932), and to the 20th century British biologist J.B.S. Haldane;[99] see Bostrom (2005a) for a brief history.

The dystopian society envisioned in Huxley's book has sometimes been invoked as an argument against the transhumanistic project of human enhancement. Here is Kass (2002):

If modern life contributes mightily to unhappiness, can we not bring technology to the rescue? Can we not make good on the Cartesian promise to make men stronger and better in mind and in heart, by understanding the material basis of aggression, desire, grief, pain and pleasure? Would this not be the noblest form of mastery, the production of artful self-command, without the need for self-sacrifice and self-restraint? On the contrary. Here the final technical conquest of his own nature would almost certainly leave mankind utterly enfeebled. This form of mastery would be identical with utter dehumanization. Read Huxley's *Brave New*

---

landscapes in which a single group of scientists or engineers may well (accidentally or otherwise) lead humanity off to a lethal path, to put a greater burden of proof on "those who propose restrictive measures" than on those who wish to rush ahead.

[98] Condorcet (1822), which, e.g., contains the following passage:

We feel that the progress of preventive medicine as a preservative, made more effective by the progress of reason and social order, will eventually banish communicable or contagious illnesses and those diseases in general that originate in climate, food, and the nature of work. It would not be difficult to prove that this hope should extend to almost all other diseases, whose more remote causes will eventually be recognized. Would it be absurd now to suppose that the improvement of the human race should be regarded as capable of unlimited progress? That a time will come when death would result only from extraordinary accidents or the more and more gradual wearing out of vitality, and that, finally, the duration of the average interval between birth and wearing out has itself no specific limit whatsoever? No doubt man will not become immortal, but cannot the span constantly increase between the moment he begins to live and the time when naturally, without illness or accident, he finds life a burden?

[99] Haldane's most important work in this respect is his book *Daedalus; or, Science and the Future* (Haldane, 1924), which gives an early vision of human enhancement via directed mutation and in vitro fertilization, but also warnings about how scientific progress might make society and our lives worse rather than better, unless this progress is matched by advances in ethics. The last worry is echoed by Persson and Savulescu (2008), and their suggestion of moral enhancement by biochemical means (as discussed in Section 2.10) can be seen as a response to it. Haldane's book served as a major influence on Huxley's *Brave New World*.

World, read C.S. Lewis's *Abolition of Man*, read Nietzsche's account of the last man, and then read the newspapers. Homogenization, mediocrity, pacification, drug-induced contentment, debasement of taste, souls without loves and longings – these are the inevitable results of making the essence of human nature the last project of technical mastery. In his moment of triumph, Promethean man will become a contented cow. (p 48)

But we should also listen to the response of Bostrom (2005b), claiming Huxley's dystopia to be the very opposite of what transhumanists are arguing for, and suggesting morphological freedom as a way to ensure a benign development:

The fictional inhabitants of *Brave New World*, to pick the best-known of Kass's examples, are admittedly short on dignity (in at least one sense of the word). But the claim that this is the inevitable consequence of our obtaining technological mastery over human nature is exceedingly pessimistic – and unsupported – if understood as a futuristic prediction, and false if construed as a claim about metaphysical necessity.

There are many things wrong with the fictional society that Huxley described. It is static, totalitarian, caste-bound; its culture is a wasteland. The brave new worlders themselves are a dehumanized and undignified lot. Yet posthumans they are not. Their capacities are not superhuman but in many respects substantially inferior to our own.... *Brave New World* is not a tale of human enhancement gone amok but a tragedy of technology and social engineering being used to deliberately cripple moral and intellectual capacities – the exact antithesis of the transhumanist proposal.

Transhumanists argue that the best way to avoid a Brave New World is by vigorously defending morphological and reproductive freedoms against any would-be world controllers.

In the last couple of decades, transhumanism has grown into a versatile and multifaceted (and only to a moderate extent coordinated) intellectual movement. I think it is a very good thing that we have these visionaries, even when their rhetoric borders on the evangelical,[100] but I consider it equally important that the discourse is tempered by critics who appreciate the risks.

3.2 Human dignity

What follows next, in this section and in Sections 3.3 and 3.4, is a discussion of some of the main concerns and objections that have been raised against transhumanism and human enhancement. These concerns tend to come from

---

[100] As is the case, e.g., with Kurzweil (2005) and Goertzel (2010). And, of course, the long quote above from Bostrom (2008a), although in his case it is clear from other writings such as Bostrom (2013, 2014) that he takes the risks very seriously.

bioconservative quarters, and they do so almost by definition if we *define* bioconservatism as the opposition to transhumanism (sometimes, however, bioconservatism is construed more broadly, and taken to include opposition also to, e.g., genetic modification of food crops and livestock).

Before getting down to serious business, let me first mention one type of concern that I will *not* bother to discuss, namely those that are based on the reading of Genesis and other holy scriptures as offering literal truths about what has taken place and infallible declarations from some supernatural being about how we ought to live. Some readers may wonder what the point is of even mentioning this class of invalid arguments, but the fact of the matter is that they have been and continue to be influential. In the collection *Human Dignity and Bioethics: Essays Commissioned by the President's Council on Bioethics*[101] addressed to US President George W. Bush in 2008 at his request, more than one of the essays explicitly took such a stance (and the volume as a whole is dominated by arguments stemming from religious sentiments). For more detail on why such concerns are invalid arguments, see Dennett's (2008) commentary in the same volume, or Pinker (2008).[102]

A more serious class of oft-discussed concerns (although it does have some overlap with that disposed of in the previous paragraph) are those that center around the notion of human dignity, and whether it survives a transition to a posthuman existence. The word "dignity" has several different connotations, causing endless ambiguity and confusion in the bioethics debate (including the *President's Council on Bioethics* volume). It has even been suggested that the debate would benefit from giving up the concept altogether. Pinker (2008) calls it a "squishy, subjective notion, hardly up to the heavyweight moral demands assigned to it," and Macklin (2003) notes that "a close inspection of leading examples shows that appeals to dignity are either vague restatements of other, more precise, notions or mere slogans that add nothing to an understanding of the topic," suggesting in particular that once we recognize a principle of respect for personal autonomy, the word "dignity" offers no further clarification. I am sympathetic to that view, but since "dignity" already features prominently in the debate, it may be worthwile to to try to work out its intended meaning, and understand the arguments invoked by it. Following Jebari (2012), we may distinguish three main meanings:

(i) The Kantian notion of dignity as being a moral subject, something that requires a rational nature and entitles a person to the right not to be used as a "mere means" to further the objectives of others.

---

[101] Pellegrino et al. (2008).

[102] After having read the entire volume, it seems to me that Dennett, as well as Bostrom (2008b), appear in it as a kind of hostage, in order for the Council to be able to claim having included multiple perspectives.

(ii) The identification of dignity and the status of being a moral subject with membership of the species *Homo sapiens*: dignity is possessed by all humans and by no others.

(iii) The "aristocratic" notion of dignity, which is the quality of being honorable, noble and worthy, something that can be taught and that can vary over an individual's life.

The meanings (i) and (ii) are often conflated, but to do so requires an argument, and becomes especially difficult if we insist on including, e.g., fetuses or severely demented individuals in the class of humans. The identification hardly becomes any easier if we replace the "rational nature" requirement in the Kantian definition by "capacity to suffer," because it would then entail the highly questionable position that non-human animals lack such capacity.

It is hard to see why trans- or posthumans should automatically be considered lacking in dignity either in sense (i) or in sense (iii): why would enhancing a person's physical, cognitive or moral capabilities infringe on her status as a moral agent or on her nobility or worthiness? The closest thing to what I might recognize as a valid argument here is the idea, emphasized by McKibben (2003) and Habermas (2003) among others, that an individual's knowledge of, say, her genome having been deliberately designed (by her parents or by someone else) would somehow be demeaning and rob her of dignity.[103] But it is still not very convincing. I, personally, know that the person I am and the personality traits that I have are determined by a complex multitude of factors, some of which are best labeled as the blind and merciless operations of chance (such as the particular combination of my parents' genes that were merged in conception) and others that are better described as deliberate design (including various choices that my parents made in the project of raising me). I don't find those elements of deliberate design at all demeaning or robbing me of any status as an individual in my own right. It is not at all clear to me why I should feel any more traumatized by the "deliberate design" part of the process producing the person that I am, than by the "blind chance" part. Imagining myself as a future transhuman whose genome is entirely the product of deliberate design, I notice that there would still be plenty of environmental factors that influence in random fashion who I turned out to be. In such a scenario, the proportion of "deliberate design" over "blind chance" would be different from what is actually the case today, but that is a difference in degree rather than in kind,[104] and it remains unclear why I would be at all psychologically

---

[103] Or of freedom and autonomy. Here, however, is Jebari (2014a), commenting on Habermas (2003):

> On [Habermas'] view, a person whose genetic predispositions are the result of a random process is more autonomous than one whose genetic makeup is partly the result of deliberate choices. [This argument] fails to convince. Imagine that a person is born with hemophilia as a result of a random process. Would that person be less autonomous than if he or she had received a genetic intervention that had prevented this disorder?

[104] That is, unless we accept the superstition that "who we are" is entirely determined by genetic factors.

damaged by such a shift in proportion.[105,106] The argument about the trauma of being deliberately designed seems just very weak.[107]

This leaves item (ii) in Jebari's (2012) list of meanings assigned to the word "dignity" – dignity as a unique feature of human beings, or Fukuyama's (2002) mysterious Factor X. It seems to me utterly implausible that there should exist such a Factor X with the property that beings who have it objectively deserve the status of being moral subjects, whereas those who do not have it lack such status, be they transhumans or extraterrestrials, and regardless of whether their capacities for suffering and for moral reasoning as well as their cognitive capacities in general should be on par with ours.[108] That, however, still leaves the possibility that we might need, for pragmatic reasons, to *invent* or *postulate* such a Factor X.[109] The argument for such a position that most deserves our attention is one that suggests that human civilization can only withstand so much diversity and inequality, and that a transhumanistic development is likely to overstep those limits. That would result in disaster and needs to be avoided. This view was fervently put forth by Annas, Andrews and Isasit (2002) in their call for a ban on cloning and inheritable genetic modification of humans:

Cloning will inevitably lead to attempts to modify the somatic cell nucleus not to create duplicates of existing people, but "better" children . . . If it succeeds, . . . a new species or subspecies of humans will emerge. The new species, or "posthuman," will likely view the old "normal" humans as inferior, even savages, and fit for slavery or slaughter. The normals, on the other

---

[105] The bioethics literature is abundant with (unfounded) suggestions about such psychological damage. For instance, on the topic of human cloning, Annas, Andrews and Isasit (2002) write that "whether cloned children could ever overcome the psychological problems associated with their origins is unknown and perhaps unknowable". To demonstrate the feebleness of such an argument, we might respond, borrowing an analogy from Bostrom (2005b), that "whether children having undergone deliberate prenatal exposure to classical music could ever overcome the psychological problems associated with such treatments is unknown and perhaps unknowable."

[106] One might also wonder why we so rarely (if ever) hear of this psychological trauma in the (still today not uncommon) case where it is experienced by people who actually believe themselves to be entirely the product of "deliberate design" by God Almighty.

[107] McKibben (2003) also makes a related argument about how our having our genomes deliberately designed will make sports uninteresting. Sports are about testing the limits of our abilities, and with our genomes designed and known, these limits are also known. But this seems just wrong: the limits of what I can achieve in the sports arena are of course influenced by my genome, but not *solely* by my genome, as all sorts of environmental factors might contribute. In fact, mightn't future competitions between clones be *more* interesting than those between participants of different genetic makeups? The outcome of such future competitions would be determined more purely by choices made by the participants (either strategic ones made in the course of the competition, or choices of training methods), uncontaminated by genetic factors for which they cannot plausibly claim any credit. In any case, I've never heard anyone suggest that a race between, say, Swedish track and field stars Jenny and Susanna Kallur would be uninteresting as a result of them being clones (they are identical twins).

[108] No more plausible, anyway, than the idea that some particular ethnic group is "God's Chosen People."

[109] Here is Fukuyama (2002):

Denial of the concept of human dignity – that is, of the idea that there is something unique about the human race that entitles every member of the species to a higher moral status than the rest of the world – leads us down a very perilous path. (p 160)

hand, may see the posthumans as a threat and if they can, may engage in a preemptive strike by killing the posthumans before they themselves are killed or enslaved by them. It is ultimately this predictable potential for genocide that makes species-altering experiments potential weapons of mass destruction, and makes the unaccountable genetic engineer a potential bioterrorist. It is also why cloning and genetic modification is of species-wide concern and why an international treaty to address it is appropriate.

This is a highly legitimate concern,[110] and we do need to think about whether genetic modification of humans (or other developments towards the emergence of posthumans) can lead to social tensions or ultimately global war. However, Annas et al., as well as the bioconservative camp as a whole, have failed to demonstrate that these scenarios are inevitable or likely consequences of the creation of posthumans. Looking back at our history, it is undeniable that humans have often ganged up in some group that views another group as "fit for slavery or slaughter." This still happens, and we have little reason to expect that (with or without radical human enhancement) this phenomenon will automatically disappear. It may even be the case, as Annas et al. suggest, that an increase in diversity and inequality resulting from a transhumanistic development raises the risks of social unrest and war. But we should not be fatalistic about such a development. Our propensity to ingroup thinking and outgroup derogation is not fixed, but depends on various institutional and cultural factors, and as Steven Pinker shows at length in his book *The Better Angels of Our Nature: Why Violence Has Declined*,[111] speaking of, for instance, how the postwar period has witnessed

a growing revulsion against aggression on smaller scales, including violence against ethnic minorities, women, children, homosexuals, and animals. These spin-offs from the concept of human rights – civil rights, womens rights, childrens rights, gay rights, and animal rights – were asserted in a cascade of movements from the late 1950s to the present day. (pp xxiv–xxv)

Might we not be able to continue our civilization's development towards greater moral maturity, and to increase the radius of our "expanding moral circle" to the extent that it encompasses, say, everyone with a capacity to think and to suffer – regardless of whether or not they are human? We should of course not take for granted that this will happen, but it is unhelpful to take for granted that it will *not* happen. The best attitude here seems to be to do what we can to expand our moral circle, and simultaneously to take the nightmare scenario of Annas et al. seriously – not as a decisive argument in favor of the embargo on new technologies that they advocate, but as one of many arguments, risks and prospects to be carefully weighed against each other when thinking about which technologies we wish to promote, and which ones we may want to hold back or even ban.

---

[110] The language about "weapons of mass destruction" and "bioterrorist" is, however, not legitimate rhetoric. Whereas it is conceivable that human cloning starts a chain of events that eventually leads to terrorism, that does not make the genetic engineer a terrorist. To see how inappropriate such terminology is, compare to the case of a surgeon saving the life of a troubled young man. The young man might then go on to join a terrorist organization and take part in horrendous terrorist attacks, but this does not make the surgeon a terrorist.

[111] Pinker (2011).

3.3 The wisdom of repugnance?

In 1997, Leon Kass, an American biochemist and bioethicist who later served as chairman of the aforementioned President's Council on Bioethics, published the essay "The Wisdom of Repugnance," which in my opinion is one of the most important contributions ever to the transhumanism versus bioconservatism debate.[112] It is important not because its arguments are right (I don't think they are), or even because they are new (they are not),[113] but because it draws attention to a view that has been (and remains) highly influential in the debate but which is usually left implicit. Kass focuses on human cloning, and begins by eloquently describing how repugnant and revolting such practice appears from many points of view, such as "the prospect of mass production of human beings, with large clones of look-alikes, compromised in their individuality," and "the bizarre prospects of a woman giving birth to and rearing a genetic copy of herself, her spouse or even her deceased father or mother," and "the narcissism of those who would clone themselves," and the list goes on and on, ending with "man playing God." From there, he immediately arrives at his general point, which is relevant not only to cloning but to a variety of human enhancements and to the transhumanism versus bioconservatism debate as a whole:

Revulsion is not an argument; and some of yesterday's repugnances are today calmly accepted – though, one must add, not always for the better. In crucial cases, however, repugnance is the emotional expression of deep wisdom, beyond reason's power fully to articulate it. Can anyone really give an argument fully adequate to the horror which is father-daughter incest (even with consent), or having sex with animals, or mutilating a corpse, or eating human flesh, or even just (just!) raping or murdering another human being?

Would anybody's failure to give full rational justification for his or her revulsion at these practices make that revulsion ethically suspect? Not at all. On the contrary, we are suspicious of those who think that they can rationalize away our horror, say, by trying to explain the enormity of incest with arguments only about the genetic risks of inbreeding.

Kass asks us to give up the search for a "full rational justification" – a search that is bound to fail, and that is "suspicious." Instead, we should accept the "deep wisdom" that our repugnance expresses, as an argument[114] from which we can conclude that cloning (or whatever it might be that we felt repugnance for) is morally impermissible. This is the **wisdom of repugnance**[115] principle indicated

---

[112] Kass (1997).

[113] Here is Devlin (1965):

I do not think one can ignore disgust if it is deeply felt and not manufactured. Its presence is a good indication that the bounds of toleration are being reached. Not everything is to be tolerated. No society can do without intolerance, indignation and disgust; they are the forces behind the moral law. (p 17)

[114] There seems to be a bit of subtlety going on here concerning the meaning of "argument." "Revulsion is not an argument," says Kass in the above quote, but *pointing out* the revulsion, and appealing to its wisdom, does seem to count as an argument.

[115] Transhumanists tend to use instead the less positively flavored term **yuck factor**; see, e.g., Bostrom (2005a).

in the title of his essay. I will soon explain why this purported wisdom of repugnance is a poor guide to what is morally right or morally wrong, but before that it is worth pointing out two grains of truth that I think the principle captures:

(1) The repugnance that we feel may, in many cases such as for the idea of eating rotten meat, have evolved biologically as a useful defense mechanism against parasites.[116] In such cases, provided that the environmental factors that the disgust reaction evolved to avoid are still relevant today, the reaction can be said to offer a piece of biologically inherited "wisdom," and it is wise of us to take the hint and avoid the rotten meat.

(2) Repugnance can be seen as a kind of unarticulated intuition about something. It is very difficult, perhaps impossible, to argue rationally for a moral position without stopping at some point and appealing to an intuition. Virtually every child discovers at some point the "why game" and finds that it can be used to win discussions over any grown-up: ask "why?," and respond to any answer by asking "why?" again. This can be used against moral philosophers as well: whatever moral position he defends, ask "why?" yet again. This pushes him to formulate ever more fundamental moral positions, but at some point, such as (if he is a hedonistic utilitarian) if he has just declared that "an outcome is better the larger its 'happiness minus suffering' balance, summed over all sentient beings, is," and in response to the next "why?" surrenders with the declaration "because it just seems obvious to me that this is the case."[117] As unsatisfactory as appeals to intuition may seem as philosophical arguments, banning them entirely from moral philosophy seems to pull the rug from under the subject's feet.[118] And why should repugnance count for any less than other intuitions?

---

[116] Curtis, de Barra and Aunger (2011).

[117] See also Section 10.1.

[118] Singer (2005) comes close to suggesting that we do ban the idea of grounding our morals in intuitions. He is aware of the argument against doing so that I give here, as is clear from the following passage:

> Whenever it is suggested that normative ethics should disregard our common moral intuitions, the objection is made that without intuitions, we can go nowhere. There have been many attempts, over the centuries, to find proofs of first principles in ethics, but most philosophers consider that they have all failed. Even a radical ethical theory like utilitarianism must rest on a fundamental intuition about what is good. So we appear to be left with our intuitions, and nothing more. If we reject them all, we must become ethical skeptics or nihilists.
>
> There are many ways in which one might try to respond to this objection, and I do not have the time here to review them all. So let me suggest just one possibility.

What then follows is, however, not convincing – not even to Singer himself, who ends his paper with the suggestion that

> even to specify in what sense a moral judgment can have a rational basis is not easy. Nevertheless, it seems to me worth attempting, for it is the only way to avoid moral skepticism.

The obvious response to (1) is that although it may seem that repugnance *sometimes* carries a piece of wisdom, it is hardly the case that it *always* does. For instance, ever since childhood, I have felt deep disgust every time I so much as contemplate the idea of eating raisins,[119] but surely everyone, including Kass, would agree that I would commit an error if I concluded that eating raisins is morally wrong. Why, then, would it be right for Kass to conclude from the repugnance of cloning that cloning is morally wrong, while it would be wrong for me to draw the corresponding conclusion about eating raisins? This question cannot be answered while honoring Kass' advice against searching for rational justifications of our feelings of repugnance and disgust.[120] We really must look into the matter more seriously than just listening carefully to our feelings of disgust. If Kass and other fans of the wisdom of repugnance had had their way, we would probably still be stuck with ideas about the wrongness of, e.g., homosexuality and interracial marriage. These examples show that the conditions triggering reactions like repugnance and disgust are not necessarily fixed and hard-wired into our brains, but malleable under cultural influences. I propose that we ditch the blind trust in the wisdom of repugnance, and that we instead take the advice of Roache and Clarke (2009) to embrace the wisdom of *reflecting on* repugnance.

3.4 Morphological freedom and the risk of arms races

A concept that appears frequently on the transhumanist side of the debate, and that we have already touched upon, is morphological freedom – the right of the individual to modify her own body however she pleases, without government interference. In Section 3.1, we encountered it in Bostrom's (2005b) response to how Kass invoked Huxley's *Brave New World* as a vision of what would happen if we gave in to the transhumanists' proposal to move ahead with human enhancement. Here is the continuation of Bostrom's response:

Transhumanists argue that the best way to avoid a Brave New World is by vigorously defending morphological and reproductive freedoms against any would-be world controllers. History has shown the dangers in letting governments curtail these freedoms. The last century's government-sponsored coercive eugenics programs, once favored by both the left and

---

[119] Some would find it tempting to point out here that it is not just I who have repugnance intuitions lacking in "wisdom," but that obvious candidates for cases of such false alarms can be found also in the thoughts and writings of Kass himself. Already the "though, one must add, not always for the better" remark in the above quote from his essay does suggest that he has some rather old-fashioned ideas about morality. Pinker (2008) reviewed some of Kass' work and found striking examples, including how Kass is troubled by practices such as "cosmetic surgery, [...] gender reassignment, and [...] women who postpone motherhood or choose to remain single in their twenties." Pinker then goes on to quote a passage where Kass (1994) goes bananas over the practice of licking ice-cream cones in public.

[120] To Kass' credit, his 1997 essay does offer some attempts at arguments for the impermissibility of human cloning – arguments that go (slightly) beyond mere testimonies of feeling repugnance.

the right, have been thoroughly discredited. Because people are likely to differ profoundly in their attitudes towards human enhancement technologies, it is crucial that no one solution be imposed on everyone from above but that individuals get to consult their own consciences as to what is right for themselves and their families. Information, public debate, and education are the appropriate means by which to encourage others to make wise choices, not a global ban on a broad range of potentially beneficial medical and other enhancement options.

Modern society values individual freedom highly, and what freedom could be more basic than my sovereignty over my own body? Yet, things are not quite so simple, due to the familiar liberal dilemma that one person's freedom may hinder that of another: my right to cut my lawn any time I like may infringe upon your right to peace and quiet on a Sunday morning, and our right to burn as much oil as we like may infringe upon Bangladeshi farmers' right not to have their lands and homes flooded by man-made sea level rise. For the case of morphological freedom, the following two concerns come to mind.

The first concern is about genetic enhancement and the scenario where parents make decisions about their future children's genomes. This is a case of one or two individuals' decisions about another individuals' physical makeup, and thus falls outside what we can plausibly mean by individual morphological freedom. Already today, parents have enormous influence on their children, not just by their very choice to mix and pass on their genes, but by dominating the children's environment for many years. This is not really a situation I would suggest that we overthrow – I have enough conservative values not to desire, say, a shift to some sort of kibbutzian socialist society where the nuclear family is banned. On the other hand, we should recognize that the enormous power that one human has over another in a parent–child relationship is not unproblematic, and that there may be reason for society to take various measures to protect the weaker part in such a relationship. Doing so is a delicate balance that deserves continuous discussion, but all in all I am mostly happy with the balance that has been found in a society like modern Sweden.[121] Expanding parents' decision-making capability over their children's genomes (which is a development that has already begun, through the practice of, e.g., prenatal screening for Down's syndrome and the associated selective abortion) represents a shift in this balance, and needs thoughtful and measured discussion.

The second concern is not limited to genetic enhancement, but applies across the full range of techniques for human enhancements, and concerns the risk of arms races. In some situations, an individual who would otherwise not like to

---

[121] Part of the explanation for my having this view is of course that I have been indoctrinated by this very same society into liking it.

At the margin, however, I do have some issues. I do not support parents' right to send their children to denominational schools where the children are protected from encountering religious views that are at odds with those of their parents. This is a case where the children's right to a wider view of society ought to overtrump the parents' right to force their religious views upon their children – especially since, under normal circumstances, parents will still have ample opportunity at home to influence their children on religious matters.

undergo enhancement might – in an environment of others who have chosen to enhance – feel compelled to do so anyway out of fear of falling behind.

An example might be height. As Sandel (2004) points out, this arms race has already to some extent begun. Human growth hormones have been approved in the US for children that have a hormone deficiency leading to a height far below average. Since the treatment works to increase height even if there is no hormone deficiency to begin with, parents of perfectly healthy children who just happened to be short for other reasons began to request the same treatment for their children. This led some doctors to begin prescribing hormone treatment even to such children. By 1996 such treatment made up 40% of human growth hormone prescriptions, and the Food and Drug Administration later approved such use to all healthy children whose projected adult height lands in the lowest percentile (i.e., the lowest 1% of the population), stratified by sex. But why just them, Sandel asks:

This concession raises a large question about the ethics of enhancement: If hormone treatments need not be limited to those with hormone deficiencies, why should they be available only to very short children? Why shouldn't all shorter-than-average children be able to seek treatment? And what about a child of average height who wants to be taller so that he can make the basketball team?

The slope does seem a bit slippery, though proponents of morphological freedom will not be worried by this: let people modify their height as they wish!

A problem here, however, is that height is, at least to a very large extent, a relative asset. The advantage of my 6′4″ height stems almost exclusively from the fact that other people are, on average, a bit shorter. This puts me (if everything else is equal) in a better position to outcompete others in basketball, and it gives me the social advantage of being able to look down on others (or at least avoid being looked down upon) at cocktail parties. Even the advantage of being able to reach the highest shelves in any kitchen cupboard without standing on a chair has to do with other people's average height – if they were all 7′ or more, kitchen cupboards would be designed differently and those highest shelves would be beyond my reach.

This will tend to make hormone treatment for height collectively self-defeating. The fraction of the population in the lowest percentile will always be 1%, so if being in that percentile is a bad thing, then, on a population level, nothing will help.[122] A more widespread practice of height enhancement, perhaps degenerating into an arms race, might even make the population worse off than before: more back problems, and we might end up being a greater burden on the environment by going in the direction diametrically opposite to the

---

[122] Some nuance here may be appropriate. As suggested to me by Karim Jebari (personal communication), hormone treatment given exclusively to short people will tend to decrease the variability of height in the population, so that being in the lowest percentile becomes less of a social handicap. This may benefit the population as a whole, whence it is wrong to say that such treatment is *always* collectively self-defeating.

shrink-humans-for-the-sake-of-the-environment proposal of Liao, Sandberg and Roache (2012) discussed in Section 2.10.

If we go on to consider cognitive enhancement rather than height enhancement the situation is a bit different, because while a statement like "we would all be better off if everyone was taller than they are now" is plain silly, the corresponding statement "'we would all be better off if everyone was smarter than they are now" is not. It is not an obviously true statement (since, e.g., smarter people might be better at hurting others), but it is not obviously false either. The devil may be in the details. But there are other problems. Let us consider what cognitive arms races can do to personal freedom.

James Miller, in his book *Singularity Rising*, writes about the widespread use of the amphetamine-based drug Adderall among students at Smith College, where he teaches.[123] That part of his more general discussion of the use of cognitive-enhancing drugs in academia is anecdotal, but he also quotes a study from 2008 indicating that more than a third of students at another campus had used that or related attention-enhancing drugs.[124] Some of them were diagnosed with ADHD (attention deficit hyperactivity disorder) and had legal prescriptions of the drug, but the majority had used it illegally.[125] Among the latter, common reasons to use the drug were, according to the study, to stay awake, to help concentrate, to help memorize school material, and to make school work seem more interesting. Some quotes from the students, such as "[I'm] so much more productive. I mean, I'm generally productive. It's just a different level on Adderall," suggest that the drug really works to enhance academic performance.[126] It is very easy to imagine a scenario where students who really do not wish to take drugs feel compelled to do so nevertheless upon looking around at the large proportion of classmates who do.

And what happens with students can also happen in the workplace. Miller describes at some length his own use of Adderall, and admits that his book is partly a product of the drug:

While doing research for this book, I've read millions of words. With Adderall, I had a much better ability to recall my readings than I ever remember having before, and bits of text kept

---

[123] Miller (2012).

[124] DeSantis, Webb and Noar (2008).

[125] Miller also highlights the problematic treatment versus enhancement distinction discussed in Section 3.1:

> It is unthinkable that a college such as Smith would notice the learning benefits of Adderall-type drugs and start actively advocating their use. But wait: at exclusive and expensive schools such as Smith, when a student does poorly on several tests she is contacted by a dean. These deans often suggest that the student gets tested to determine whether she has some kind of learning disability, such as ADHD, and doctors routinely prescribe Adderall to those who are diagnosed with ADHD. (p 102)

[126] In case of readers who take these remarks as an inspiration to try amphetamine-based drugs themselves (which is definitely not my intention), please consider that these drugs have a variety of physical and psychological side effects, including addiction; see, e.g., Carvalho et al. (2012). Taking them without being personally advised by a medical doctor to do so (and in what dosages) may be dangerous. Furthermore, a recent meta-study suggests that the positive effects on cognitive abilities are not as strong as the purported anecdotal evidence will have it; see Ilieva, Hook and Farah (2015).

popping into my brain just when I needed them. Adderall also gave me the gift of time. Before taking the drug, to run at peak performance I needed about eight and a half hours of sleep per night. But with Adderall, I would wake up refreshed after being in bed for only seven hours. (p 104)

Now consider my own book project (the fruits of which the reader is holding in her hand). With this book, I am competing in very much the same niche as Miller (and I, too, have read millions of words during research for it) – a competition in which I will be at a disadvantage unless I, too, take drugs to improve my cognitive performance.[127] And of course we can imagine similar arms races in almost any other profession.

All this goes to show that a policy of morphological freedom can infringe on people's ability to go on with their lives without engaging in enhancement. It does not necessarily amount to a decisive argument against such a policy, but it is at the very least a complication that requires careful consideration. And we should keep in mind that if we invoked an uncompromising anti-arms race attitude, we might end up advocating bans on everything that increases productivity at the office, including coffee and word processors.

Leaving that particular issue, we may go on to consider arms races on higher levels than between individuals. Miller treats the case of cognitive arms races between nations, and shows a knack for outlining nightmarish but not implausible future scenarios where the United States and China engage in such competition – a competition that may become so fierce that both feel forced to avoid wasting time on safety precautions and to go on with ruthless experimentation on humans. Let us for simplicity pretend that the world consists of just these two countries, ignoring all others. How will the US react to a Chinese breakthrough in cognitive enhancement (or vice versa)?

There are both economic and military sides to this issue. If we focus on the economic aspect, classical economic theory based on the Ricardian idea of comparative advantage suggests that advances in one country benefit both parties, and from this perspective a Chinese breakthrough ought not to be alarming to

---

[127] Two personal remarks may be in order here:

First, I tell myself that I write this book with the purpose of contributing to making the world a better place. While I believe this is an honest and accurate description of my main drive in the project, it would be delusional if I also told myself that this drive is entirely untainted by lower motives, such as my desire to gain academic prestige, in competition with Miller and others.

Second, the only performance-enhancing drugs I use are coffee and beta blockers, and the latter (for which I have a prescription) I take only on the relatively rare occasions when I am to speak on a controversial subject and expect parts of the audience to react with hostility (here I'm pretending that there is a clear and non-problematic line to be drawn between drugs and nutrition, and I'm excluding alcohol from the list because (a) I only use it recreationally, and (b) it is mostly performance impairing rather than performance enhancing). I really do not want to expand the list. The rationality of this may be questioned, but the wish seems related to a tendency towards conservatism regarding my personality: I want (to a large extent) to remain me, rather than to turn into someone else.

US policy-makers.[128] Miller (2012) points out that "if a Chinese brain-boosting program caused Chinese engineers to innovate in ways that were beyond what Americans were capable of, Americans would almost certainly still benefit economically from Chinese innovation" (p 122) and points at modern Chinese history as a parallel: During the Mao Zhedong era, when the country's economy was kept sealed off from the rest of the world, economic growth was low. His successor Deng Xiaoping opened up the economy, which led to extremely rapid growth, and Miller explains this as China being able to benefit from technologies previously developed in the US and elsewhere.

This is not to suggest that the US would not react to a Chinese cognitive enhancement program. Mao's policy notwithstanding, it is hard to imagine how either the US or China would, now or in the foreseeable future, willingly accept to fall back in what is often perceived as a global competition for economic growth. Governments all over the world have, from the 19th century onwards, spent astronomical amounts on cognitive enhancement of their populations via education. The reasons for doing so have been multifaceted, but, as has been evident from the rhetoric now and in the past, a large part of the motivation comes from the wish to compete well against other countries.[129] Now, cognitive enhancement via pharmaceutics or genetic engineering is not the same thing as cognitive enhancement via education, so perhaps we cannot say with certainty how governments will react to opportunities of the former kind, but it seems at least plausible to think that, out of fear of falling behind other nations, they will be tempted by such technologies.

There is also the military aspect. Here, of course, the "all sides benefit from progress" idea in economics is inapplicable. The military interest in cognitive-enhancing drugs is well established. According to Emonson and Vanderbeek (1995), a majority of US pilots participating in the Iraq war of 1991 used amphetamines, but the practice was already widespread in the Second World War. Such use notwithstanding, the greatest military potential of human

---

[128] This goes back to the early 19th century British economist David Ricardo; see Krugman (1996) for a brief discussion of the core idea. Relatedly, Harford (2006) explains beautifully that restrictions on free trade do not really protect a country (as popularly believed), but rather protects one industry in the country at the cost of another:

> Trade can be thought of as another form of technology. Economist David Friedman observes, for instance, that there are two ways for the United States to produce automobiles: they can build them in Detroit, or they can grow them in Iowa. Growing them in Iowa makes use of a special technology that turns wheat into Toyotas: simply put the wheat onto ships and send them out into the Pacific Ocean. The ships come back a short while later with Toyotas on them. The technology used to turn wheat into Toyotas out in the Pacific is called "Japan," but it could just as easily be a futuristic biofactory floating off the cost of Hawaii. Either way, auto workers in Detroit are in direct competition with farmers in Iowa. (p 213)

[129] Perhaps the most famous example is the so-called Sputnik crisis in the US in the late 1950s, where the success of the Soviet Sputnik program led to great fear of falling behind technologically, triggering a series of responses from the US government, including the launching of education programs to foster new generations of engineers; see, e.g., Mooney and Kirshenbaum (2009). (In this case, however, the fear seems to have had more to do with military than with economic competition.)

enhancement may well lie not in enhanced soldiers, but in cognitively enhanced scientists and engineers who can develop new military technology. It is hard to imagine the US silently sitting still and watching a cognitive enhancement development that can turn China into the world's military overlords. It seems likely that the US would follow suit. Or worse, they might feel compelled to launch a preemptive military strike against China.

3.5 Genetic engineering

After these lengthy discussions about transhumanism, bioconservatism, and the ethics of human enhancement in general, it is time to discuss a few particular enhancement technologies more specifically, beginning with genetic engineering.

Gene therapy is the deliberate modification of the DNA content of cells. We should distinguish between **germline gene therapy** and **somatic gene therapy**. The former is done on so-called germline cells, which are either sperms, eggs or stem cells that can result, via cell division, in sperms or eggs. The latter, in contrast, is done on somatic cells, which are the majority of our bodies' cells and those that are not germline. This means that while genetic changes resulting from germline gene therapy may be inherited by the individual's children and further descendants, those resulting from somatic gene therapy cannot. This makes somatic gene therapy much less ethically controversial, and in the context of the ethics of human enhancement it is in some respects more akin to enhancements by pharmaceutical means than to germline genetic engineering.[130] Somatic gene therapy as a treatment of various genetic defects has been practiced since the 1990s, whereas germline gene therapy has not (yet) been attempted on humans.

Human enhancement via genetic engineering is, however, possible without manipulating the genetic content of a cell, but instead using selection. This has been practiced on plants and non-human animals since prehistory, and the eugenics programs in countries like the US, Germany and Sweden in the 20th century belong in the same category. The latter were barbaric practices that have fallen into well-deserved disrepute,[131] but we do practice prenatal screening for various diseases followed by selective abortion.

It seems very likely that we are currently in a transition period towards increased use of deliberate human genetic selection. A main driver in this development is **in vitro fertilization** (IVF), which involves an egg being fertilized by

[130] This statement becomes less straightforward if in the future we see techniques where genes from cells having undergone somatic gene therapy are inserted into germline cells, thus blurring the distinction between germline and somatic gene therapy – in addition to the blurring that already arises from the Nobel Prize-winning discovery by Takahashi et al. (2006, 2007) of techniques for turning somatic cells into stem cells.

[131] Of course, by inclusion of the word "well-deserved" I go beyond stating mere facts to stating my opinion, which in this case is based on the programs' blatant disrespect for individual autonomy.

sperm outside the body, and then cultured in the laboratory for a few days until it is transferred into the woman's body, resulting (if successful) in pregnancy. Djerassi (2014) quips about the "divorce of coitus from reproduction" that the past half-century has witnessed: first, the Pill gave us sex without reproduction; later, IVF gave us reproduction without sex. The world's first baby conceived by IVF was born in 1978, and the cumulative number of IVF babies now exceeds five million.[132]

IVF opens up a variety of novel developments, including embryo selection. Testing for a standard catalogue of genetic diseases governed by a single gene can now be done at low cost. This is also available for a pregnancy initiated by more traditional means, but in the case of IVF the selection can be done in the test tube, thus avoiding the hassle of early pregnancy and abortion. That shortcut may soon open up possibilities for embryo selection of a different kind: not screening for specific genetic defects, but optimizing for some more complex property.

Shulman and Bostrom (2014) consider the case of cognitive ability. Far from being governed by a single gene, cognitive ability is influenced by many genes, as well as, of course, by environmental factors (nutrition, education and so on). Studies tend to show that a substantial part of the variation in IQ between individuals is caused by genetic factors.[133] Shulman and Bostrom quote recent works[134] suggesting that much of the variation can be attributed to a large number of common genetic variants which, taken one by one, have tiny effects. Detecting these individually, and quantifying the effects, will require very large studies matching genetic data against well-known correlates of IQ such as SAT scores and income, but such studies may be forthcoming.[135] Once the results of such studies are available, IVF may be taken to include a procedure where genomes from some number n of embryos are tested, and the one whose genome predicts the highest IQ is selected. Under certain model assumptions (whose validity are of course open to questioning, so here is where their argument becomes increasingly speculative), Shulman and Bostrom estimate that the average number of IQ points gained would be around 4 for $n = 2$, 11 for $n = 10$ (which they take as the upper limit of what today is practically feasible in a single IVF cycle, but this may well change in the future), 19 for $n = 100$ and 24 for $n = 1000$.[136] They go on to discuss what

---

[132] Brian (2013).

[133] See, e.g., Devlin, Daniels and Roeder (1997).

[134] Such as Benyamin et al. (2014).

[135] They mention the UK Biobank project which "has collected survey data and biological samples from some 500,000 individuals, and has hired a firm to do genetic testing in 2014," and go on to point out that "in the longer term, as DNA testing becomes a routine part of medical care, data sets of tens of millions of individuals may be assembled" (Shulman and Bostrom, 2014). A recent *Nature* news item, however, underlines how difficult the challenge is (Callaway, 2014).

[136] The expected gains will be smaller if parents wish to select also for traits other than IQ: when selecting for a single trait, the one embryo looking most promising as regards this particular trait can be selected, while selecting for several traits may require compromises, as it is unlikely that a single embryo will look best for all of the traits. This is pointed out by Miller (2012) who, true to his taste for shocking statements, suggests that "breast implants will create smarter girls" (p 88), his argument being

the direct societal effects will be for various n and various rates of adoption of the technology, suggesting that for $n = 100$, an adoption rate of a fraction of a percent will still make the "enhanced contingent [form a] noticeable minority in highly cognitively selective positions," while a 10% adoption rate would make them "dominate ranks of scientists, attorneys, physicians [and] engineers," in which case it seems that prospective parents who can afford the embryo selection procedure but have a preference for doing things the natural way face a tough choice.

Cumulative effects over more than one generation will of course be larger, but this development will be relatively slow due to the long maturation period of humans. But embryo selection may possibly be combined with a technology that is not yet available but may be within a decade or two: the derivation of viable sperms and eggs from embryonic stem cells. In combination, these technologies allow for an iterative procedure, where a number of embryos are genotyped and selected based on the desired genetic characteristic, the selected embryos are then used to produce sperms and eggs which are crossed to produce new embryos, and the procedure is repeated a number of times to obtain the desired genetic changes.[137] The Shulman–Bostrom model predicts that with $n = 10$, ten iterations of this procedure gives an expected IQ gain of 130, which, if widely adopted, would take us straight into posthumanity – but also off the scale where IQ has been validated as a meaningful measure, adding doubt to how much predictive value we should assign to the result. Shulman and Bostrom suggest that, in reality, there may be some diminishing returns phenomenon not encompassed by their model, and there are many further uncertainties and complications. For instance, in order to avoid the negative effects of inbreeding, it may turn out to be necessary to start the process with sperms and eggs from (much) more than just two donors, which would diminish the genetic relationship between parents and children, thus perhaps impacting the adoption rate of the technology negatively. Nevertheless, regarding the question asked in the title of the Shulman and Bostrom paper – "Embryo selection for cognitive enhancement: Curiosity or game-changer?" – it seems that the latter is a real possibility.

3.6 Brain–machine interfaces

The next technology up for discussion is brain–machine interfaces. As Michael Chorost describes in his partly autobiographical *Rebuilt: How Becoming Part Computer Made Me More Human*, such technology, connecting our brains to electronic devices without passing through the detour of our ordinary sensory organs,

---

that if breast implant use becomes more widespread, then parents have less reason to select embryos for breast size, and so will be in a better position to optimize IQ. Shulman and Bostrom (2014) suggest "disease risks, height, athleticism or personality" as other traits that may show up in parents' selection criteria.

[137] Sparrow (2013) digs deeper into the ethics of such procedures, and calls them "in vitro eugenics".

already exists and is being used by hundreds of thousands of people.[138] Chorost has a so-called cochlear implant inside his skull, which picks up a radio signal from a small audio receiver sitting on his ear, processes the signal electronically, and transmits the information via the auditory nerves to his brain.[139] Reading his book, I initially felt that it was unclear whether the cochlear implant was in principle so much different from traditional hearing aids, but I changed my mind as soon as Chorost explained that he could turn off the audio receiver and connect his device directly to a telephone or a CD-player, allowing him to experience sound *without the presence of any corresponding sound waves in the room he was sitting in.* This really is different.

Chorost is enthusiastic about what the cochlear implant can do and has done for him and for other hearing-impaired or deaf patients, and he looks favorably upon the development of similar technologies for other disabilities.[140] He tries, however, to draw a firm line between treatment and enhancement, such as in the following passage where he polemicises against Warwick (2004).

These technologies are only for people with serious medical conditions. Warwick consistently neglects this important qualification. Upon having used his implant to pilot an electric wheelchair he comments, "I told everyone that this would ultimately mean, in the future, we should be able to drive a car around by picking up signals directly from the brain, and change direction just by thinking about it, right, left, and so on". For disabled people, this is an exciting prospect. But the rest of us already do drive with signals from our brains, picked up and executed perfectly by our arms and legs.

Warwick might respond that having direct nervous control over an automobile would let us drive it better, since we wouldn't have to move around our heavy meat arms and legs. But then issues of practicality and safety come into play. (Chorost, 2006, pp 176–177)

There is plenty to object to in this passage, such as the almost comically misplaced "perfectly," and the implicit suggestion that the observation "issues of practicality and safety come into play" is a decisive argument against a new technology. It seems, however, that there is no longer any need to explain these things to Chorost, because in his follow-up book *World Wide Mind: The Coming Integration of Humanity, Machines and the Internet* he makes a complete 180, and argues that we should proceed with technologies for hooking up our brains to the Internet and to brain–brain interfaces that allow us direct communication with each other.[141]

---

[138] Chorost (2006).

[139] Another example of an already existing and widely used brain–machine interface technology is **deep brain stimulation**, where a so-called brain pacemaker sends electrical impulses, via implanted electrodes, to specific parts of the brain. It is used (and is approved by the Food and Drug Administration) to treat a variety of disorders like Parkinson's disease and obsessive–compulsive disorder. See, e.g., Kringelbach et al. (2007).

[140] See, e.g., Degnan et al. (2002) for a discussion of implants that help patients suffering from paralysis regain some of their motor function.

[141] Chorost (2011).

He envisions how this technology will bring us closer to each other, eventually dissolving our identities into some sort of collective consciousness. This brings to mind the dystopian Borg society from *Star Trek* – a collection of species whose individuals are no longer individuals but instead serve as drones to the hive mind of the collective. This sounds scary, and Chorost admits that Borg is a relative of his vision, but he prefers the metaphor of a hug: he speaks of the frustrated longing for each other and for closeness that we humans feel (and that he seems to have been haunted by, probably more than the average person, throughout his life), and of the dissolution and merger of our identities as the final answer. This is all very speculative and vague, and whether our eventual merger into a collective consciousness is a desirable outcome seems like a wide-open question, whose answer is likely to depend on details of the outcome.

There are other ethical concerns, however, that will apply to brain–machine interface technologies long before we come anywhere near the utopian (or dystopian) vision of a collective consciousness. Besides the concerns that apply across the board of enhancement technologies – the alleged wisdom of repugnance, morphological freedom, cognitive arms races, and so on – Jebari (2013) mentions two pressing aspects that may be more specific to brain–machine interfaces: privacy and autonomy.

Privacy is a tricky matter. As soon as there are technologies to monitor what goes on in our brains, there will be others who are interested in obtaining this data. Employers may wish to monitor employees in order to maximize their productivity, and business owners may want to monitor consumers' reactions and habits in a similar (but more refined) way to how today they use cookies to monitor our Internet activity. And then there's the old fear of *1984*-style government surveillance. It is possible that at some point we find ourselves in the precarious situation discussed in Section 1.1, where some doomsday technology has become easily available to anyone who (whimsically or otherwise) decides to destroy the world, and what if it turns out that the only way to prevent the impending extinction of the human race is government surveillance on the level of our brain activities? That is a nightmare – but perhaps not an entirely implausible one. To avoid the asymmetry of a government that can monitor us while we cannot monitor the government, Tännsjö (2010) makes the radical suggestion that we give up privacy entirely in favor of an infrastructure that allows everyone to monitor everyone else. To me at least, that is a hard pill to swallow.

Autonomy may be at least as problematic as privacy, and philosophically it is not an entirely straightforward concept, for reasons related to the notorious difficulty of the notion of free will. If we accept a naturalistic worldview, our actions are ultimately caused by factors external to ourselves (the environment we live in, and the genes that were passed to us by our parents), so independence from external causal influence is too much to ask for. According to Jebari (2013), an individual's autonomy is his or her "awareness of and influence over his or her own desires, values, emotions and the formation of these mental states," and moreover "coherence and the lack of internal conflict in one's attitudes is also significant for our autonomy." With this definition, a technology like deep brain stimulation (see Footnote 139) can improve an individual's autonomy by treating depression or other mental illnesses, especially if the treatment is under the direct conscious control of the individual himself. On the other hand, it may under some circumstances be open to abuse by third parties and can then be a serious threat to personal autonomy – a problem that may well become more severe with more advanced brain–machine interface technologies.[142]

What, then, is the cognitive enhancement potential of brain–machine interface technology in a short to medium time frame? At first it may seem very promising, because computers and human brains have abilities that seem to some extent complementary. Computers far outperform human brains in tasks such as accurate arithmetic, fast data transmission and perfect recall, whereas no computer program has so far come anywhere near human-level performance in the kind of flexible and intuitive thinking which we are hard-pressed to specify exactly what it is but which we think of as quintessentially human. If we could devise an interface that unites the computer and a brain into a single functioning unit – wouldn't that be a match made in heaven?[143]

Bostrom, in his recent book *Superintelligence*, is skeptical about the near-term feasibility of such a union.[144] In an overview of the various ways in which superintelligence (meaning, loosely, much-higher-than-human intelligence) may first materialize, he holds brain–machine interface technology to be an unlikely winner.[145] He admits that impressive work has been done on simpler animals, such as by Berger et al. (2012) who worked on rats and implanted a hippocampal prosthesis that partly restored memory-forming abilities lost after damage to the hippocampus. Still, Bostrom argues, to significantly improve human cognitive performance via a brain–machine interface we have to

---

[142] There exists today a small subculture of people (many of whom suffer from paranoid schizophrenia or other mental disorders) who believe themselves to be partly remote controlled, via implants in their heads, by governments or other evil agencies. It seems to me highly unlikely that their (self-)diagnoses are correct, but this is of course not the same as saying that nothing along these lines can happen in the future.

[143] Even in such a seemingly narrow area as chess, the competence profiles of humans and machines diverge substantially. After Garry Kasparov (who was then world chess champion) lost narrowly in 1997 to the machine *Deep Blue*, chess programs have continued to improve, and are by now far stronger players than the best human grandmasters. Yet, the thinking styles of humans and machines (sometimes described, with some simplification, as strategic thinking versus brute force calculation) are sufficiently different that a human grandmaster and a computer in consultation still have a clear edge over a computer on its own. See Kasparov (2010). Might installing a chess computer inside a grandmaster's skull, and working out how to interface it with the brain, be an interesting pilot project for the kind of cognitive enhancement we're discussing here?

[144] Bostrom (2014).

[145] The other main competitors are artificial intelligence (more on which in Chapter 4), whole-brain emulation (more on which in Sections 3.8 and 3.9), genetically engineered (post-)humans (which was discussed in Section 3.5), and the spontaneous emergence of a collective intelligence in the Internet or some other network or organization (an imaginative possibility that I'll mostly ignore).

compete with our brain's natural interfaces to the world, whose performance are in some cases quite impressive:

We do not need to plug a fiber optic cable into our brains in order to access the Internet. Not only can the human retina transmit data at an impressive rate of nearly 10 million bits per second, but it comes pre-packaged with a massive amount of dedicated wetware, the visual cortex, that is highly adapted to extracting meaning from this information torrent and to interfacing with other brain areas for further processing. Even if there were an easy way of pumping more information into our brains, the extra data inflow would do little to increase the rate at which we think and learn unless all the neural machinery necessary for making sense of the data were similarly upgraded. Since this includes almost all of our brain, what would really be needed is a "whole brain prosthesis" – which is just another way of saying artificial general intelligence. Yet if one had a human-level AI, one could dispense with neurosurgery: a computer might as well have a metal casing as one of bone. (pp 45–46)

The question, thus, is why we would opt for a direct brain–machine interface, with all its complications, when we already have such splendid other channels. One answer could be that we might hope for "bypassing words altogether and establishing a connection between two brains that enables concepts, thoughts, or entire areas of expertise to be 'downloaded' from one mind to another" (p 46). This seems like an enormously difficult project – we need to find the precise neural correlates in my brain of concepts like "the meadow right below my summer house" and "the exponential distribution," among countless others. Chorost (2011), citing work of Tsien (2007), suggests that neural clusters in mice have been found whose joint firings represent concepts such as "motion-disturbance" and "wind." These interpretations exhibit a substantial degree of wishful thinking, however. And Bostrom (2014) points out that brains, unlike structured computer programs,

do not use standardized data storage and representation formats ... Which particular neuronal assemblies are recruited to represent a particular concept depends on the unique experiences of the brain in question (along with various genetic factors and stochastic psychological processes). (p 46)

One hope for the dream of the brain–machine union might be to take advantage of the brain's adaptivity and plasticity. Perhaps we could implant an interface device and let the brain learn, over time, a workable mapping between its own internal concept representations and the signals that the device delivers and accepts. But what, Bostrom goes on to ask, is really gained then compared to using our ordinary senses? Assuming that the brain could learn to detect patterns in an input stream from a brain–machine interface, then it seems incomparably simpler to instead just project the same information onto the retina, for the brain's pattern recognition machinery and plasticity to work on.

These are valid questions, that enthusiasts of radical enhancement via brain–machine interface technologies should try to answer.

3.7 Longer lives

Another direction for human enhancement is **life extension**: to extend our life spans. Transhumanists tend to view this as very important, and to wish not just for increasing life expectancy to 90 or 100 or 120 or even 200,[146] but to postpone death indefinitely. They also tend to view research on life extension as vastly underprioritized.[147] When Savulescu, Bostrom and de Grey (2009) complain that only about 0.02% of the money spent by the National Institute of Health (the foremost funder of biomedical and health-related research in the US) addresses the problem, they are in a sense being a bit unfair, because huge parts of the NIH budget are spent on research on some of the most common causes of death, such as cancer and cardiovascular diseases, and any decrease in mortality in these diseases will of course have a positive influence on life expectancy. And we are making good progress: life expectancy in those countries where it is highest has, for the past 150 years, grown at a fairly steady pace of 3 months per year.[148]

What Savulescu et al. mean by their 0.02% is research more narrowly on ageing, defined (roughly) as the accumulative damage to the body's macromolecules, cells and tissues caused by "genomic instability, telomere attrition, epigenetic alterations, loss of proteostasis, deregulated nutrient sensing, mitochondrial dysfunction, cellular senescence, stem cell exhaustion, and altered intercellular communication" (López-Otín et al., 2013). It is a positive sign for proponents of life extension that the journal *Nature* devoted a special *Outlook* section in 2012 to the issues of ageing, life extension and gerontology, but they are likely to be less pleased by the very first sentence of the editorial: "Ageing is inevitable" (Grayson, 2012). The claim is unfounded, fatalistic, and probably false. None of the biophysical mechanisms listed by López-Otín et al. seem impossible in principle to arrest or reverse,[149] and although we are of course miles away from perfecting any technologies that achieve such age halting or age reversing, some early progress is being made; see e.g., the article by Wehrwein (2012) in the *Nature Outlook* section, discussing ways to slow down or halt the ageing of stem cells. And at the other end of the spectrum, compared to Grayson's fatalistic sentiment, we find notorious techno-optimist Ray Kurzweil, writing in his bestselling 2005 book *The Singularity is Near*, that "when nanotechnology is mature, it's going to solve the problems of biology by overcoming biological pathogens, removing toxins, correcting DNA errors, and reversing other sources of aging," and that this can be expected within "about twenty to twenty-five years."[150] This vision will be revisited in Section 5.3.

---

[146] As of 2014, 31 countries including most of Western and Northern Europe have a life expectancy of 80 or more, with Japan topping the list with 84.6. See WHO (2014).

[147] Bostrom (2005c), Savulescu, Bostrom and de Grey (2009).

[148] Savulescu, Bostrom and de Grey (2009), Scully (2012).

[149] Important clues on how this might work may be found, e.g., in a species of jellyfish that seems to have unlimited lifespan; see Dimberu (2011).

[150] Kurzweil (2005), pp 256–257.

There are of course ethical issues in connection with life extension, as with other human enhancements. A particular issue that may superficially sound like an inconsequential matter of terminological convention is that of whether ageing (in López-Otín's sense of the word) should count as a disease. But perhaps the issue goes deeper than that, because it may influence our thinking on whether or not we should try to treat it. Leon Kass insists that we should not, and given his reactions of repugnance and condemnation of human enhancement encountered in Section 3.3, it is perhaps not so surprising to learn that he feels that

the desire to prolong youthfulness is not only a childish desire to eat one's life and keep it; it is also an expression of a childish and narcissistic wish incompatible with devotion to posterity. (Kass, 1985, p 316)

If ageing-as-a-disease became mainstream discourse, then views like Kass' would seem much less defensible. But ageing is not generally viewed as a disease. What sets it apart from conventional diseases is mainly that it hits everyone above a certain age without exception, and that it is 100% lethal (or at least it has so far seemed to have these properties). But transhumanists like Bostrom (2005c) and Savulescu et al. (2009) tend to disagree with the view that ageing is not a disease, and in fact as early as 1954 biologist Robert Perlman coined the expression "ageing syndrome."[151] From the point of view of an individual, the ageing process is something that threatens one's health. My normal reaction when disease strikes so that my health (or perhaps even my life) is in jeopardy is that I wish for a cure to it, and if such a cure is actually on the radar, why should I give up on my wish just because of the hitherto 100% incidence and 100% lethality?

A common negative reaction to radical life extension involves death being somehow essential to the meaning of our lives, and how postponing it to the far future or removing it altogether would have devastating psychological consequences.[152] And although it is important that we discuss the meaning of life, ideas about death being somehow essential to it tend to strike me as far-fetched attempts at rationalization after the fact. And if death is such a wonderful thing, then why do we obsess so much about the absurdity of life in the face of death,[153] and why do we have such a strong tendency to indulge in religious fantasies about an eternal afterlife?[154]

The most important concerns and objections that have been raised regarding the desirability of radical life extension are, in my opinion, issues of demographics. Assuming that we are stuck on planet Earth (which is not a definite given, as we may at some point have the option to migrate to outer space or to virtual realities), we may need, in order to avoid a Malthusian scenario where population

---

[151] Perlman (1954).

[152] See, e.g., Shaw (1921), Jobs (2005) and Midgley (2012).

[153] See, e.g., Camus (1942).

[154] Yes, I am aware of the parallels that have been drawn between such religious fantasies and the topic of this chapter; see, e.g., Smith (2013).

is limited by starvation, to keep the population below some fixed number. This number may depend on various future technologies, e.g., for food production, but just for the sake of argument, let's say we need to keep the number below 10 billion. To do this, we need to reduce fertility, so that fewer people are born.[155] If life expectancy is 100 years, then we can afford 100 million births per year, whereas if life expectancy is 200 years, then we can only afford 50 million births per year. It is a legitimate and difficult question whether a doubling of life expectancy is a good thing if it also means that only half as many people get the privilege of being born. Arrhenius (2008) analyzes precisely this question, and arrives at a "no" answer, but intuitions and opinions differ on these matters, and there seems to be plenty of scope for further discussion.[156]

3.8 Uploading: philosophical issues

The ultimate life extension technology – if it can be made to work – is **uploading**, the possibility of transferring our minds to computer hardware. Uploading is based on **whole-brain emulation**, which I take here to be the idea that (a) if we have a good enough simulation of a human brain, then that simultaneously gives a simulation of the human mind, and that (b) if the simulation is *really* good, then it is not just a simulation but a *copy* of the original mind, faithful in all relevant aspects – an emulation.[157]

An uploaded mind will in principle consist, at any given time, of a finite string of 0's and 1's. Once it is uploaded, copying it is therefore a trivial matter, and here

[155] Here we have to be careful about what we mean by fertility. To illustrate this, let me offer some (extremely idealized) calculations.

Assume that we start from an equilibrium situation where each person has on average two children, and consider the case of increasing life expectancy from 100 to 200 years. If the average number of children people have per time unit is held constant (perhaps because they will want to spend the same proportion as before of their lives raising children), then each person will on average have four children, and we'll see a rapid exponential growth of the total population. So if fertility is taken to be the number of births per adult person per time unit, then a cut in fertility is needed. If instead we mean the average total number of children that a person begets, then maintaining this figure at two is consistent with a stable population size.

But there is then a further complication in that the average age of child bearing matters. Suppose that prior to the enhancement (i.e., when life expectancy is still a mere 100 years) the average age of child bearing is 30 years. Then, at any given time, roughly $100/30 = 3.3$ generations are alive, each with about three billion people. If life expectancy is enhanced to 200 years, and people on average still have two children, then each generation will still consist of three billion people. But look what happens: if the average age of child bearing scales up proportionally to 60 years, then the number of generations alive will remain at roughly 3.3, so that total population stays at ten billion. If instead people keep having children at age 30 as before, then the number of generations alive at a given time will double, and total population will double to 20 billion (but will then remain stable).

Note, however, that this is about a mere doubling of life expectancy. Virtual immortality is a whole different ball game.

[156] See also Jebari (2014b) for a discussion of various societal risks that may be associated with radical life extension.

[157] Of course, the term "relevant" here leaves some room for debate on what exactly it should mean.

is the reason for calling it the ultimate life extension technology: you can create however many backup copies you want, and choose strategic physical locations for them, so that if something lethal happens to you, you can always be restarted from one of the backups – a technology that seems to virtually guarantee your survival for as long as (post-)humanity survives.

Russell Blackford, in his introduction to the excellent recent anthology *Intelligence Unbound: The Future of Uploaded and Machine Minds*, lists some of the features that may make uploading attractive:[158]

It might offer us many advantages: more varied experiences; greatly enhanced speed of thought; and opportunities to extend our cognitive, motor, and perceptual capacities in multiple ways. Most obviously, perhaps, it might be a way to achieve survival beyond the flesh: the computer hardware might be more durable than a human body, and perhaps you could be transferred to successive computational substrates indefinitely into the future, giving you a form of immortality.

He then immediately concedes that complications abound, and that the contributors to the anthology are divided as to whether upoading really is such a good idea. Beyond the technological issue of what will be needed for successful whole-brain emulation and whether this is feasible, the essays in *Intelligence Unbound* (which will be drawn heavily upon in this section and the next) offer a variety of perspectives on uploading, including metaphysical, ethical and sociological. Let us postpone the technological issue to Section 3.9, and focus here on whether perhaps the whole project is a philosophical mistake – might it be that no matter how well we succeed in simulating the brain in every detail, there is no hope of prolonging one's life via uploading, simply because one does not survive the transfer? Following Chalmers (2014) and Danaher (2014a, 2014b), let us split this question in two, and briefly discuss the two halves separately:

(1) Can a digital computer be conscious? In particular, would the uploaded mind be conscious?

(2) If the answer to (1) is "yes," and if I upload my mind, then will the uploaded mind really be *me*, as opposed to being another person who is merely a copy of me?

For an individual to survive in digital form upon uploading, we clearly need "yes" answers to both (1) and (2).

Answering (1) seems to require knowing what consciousness really is, as well as understanding how it fits into and arises in the physical universe. I have found these issues mysterious for as long as I can remember, and after reading many books by contemporary scientists and philosophers on the subject – including but not limited to Hofstadter (1979, 2007), Hofstadter and Dennett (1981), Dennett (1991, 2005), Chalmers (1996), Searle (2004), McGinn (2004) and Koch (2012) – I come out with the impression that they are still fumbling around

---

[158] Blackford (2014), Blackford and Broderick (2014).

for the fundamental nature of consciousness. Neuroscientists are making good progress on finding so-called neural correlates of consciousness,[159] but we still have no answer to the question of what, really, it is that causes some (but, it seems, not all) constellations of neuron firings to translate into conscious experience.

The main idea that supports a "yes" answer to (1) is the so-called **computational theory of mind** (CTOM), which says, roughly, that what matters for consciousness is not the material substance itself but how it is organized,[160] and that the kind of organization that produces consciousness is (the right sort of) information processing or computation.[161] Among the writers listed in the previous paragraph, Hofstadter, Dennett and Koch are strong supporters of (various strands of) CTOM, Chalmers looks agnostically but fairly favorably upon it, Searle is among its most outspoken and fervent opponents, and McGinn holds the so-called mysterian position that it is not within the capabilities of the human cognitive machinery to understand the fundamentals of consciousness.[162]

Many arguments have been suggested for and against CTOM. Personally, I find the theory fairly plausible – at least as plausible as whatever alternative theories have been put forth. One argument in its favor that to me carries some weight is this: It seems clear that it is, at least in principle, possible to write a computer program that has the intelligence of the human mind – provided we define "intelligence" in behavioral terms, i.e., in terms of what the machine is observed from the outside to do, as opposed to "consciousness" which is about the inside view, i.e., what it feels like to be the machine. If nothing else, the program could simulate our brains, plus whatever extras are needed in terms of input/output channels, down to the level of elementary particles. Now, the only consciousness I can directly observe is my own, but I unhesitatingly ascribe consciousness to people I encounter (and even some dogs) based on their intelligent behavior. If one of my friends to whom I ascribe consciousness would open up her skull and reveal its contents, which turn out to consist not of the disgusting-looking thing known as a brain, but of electronic computer hardware – a scenario that may be highly unlikely but in principle seems possible – then I would probably not change my mind about her consciousness. That a computer can produce consciousness may seem strange, but

---

[159] Metzinger (2000).

[160] Philosopher Richard Sharvy found a witty formulation (which in popular culture is not entirely devoid of dirty connotations) when he summarized this view as "it ain't the meat, it's the motion" (Sharvy, 1985).

[161] I am not doing full justice here to CTOM, because it is often construed more broadly as dealing not only with consciousness, but also with other, not necessarily synonymous, notions like "thinking," "understanding," "cognition" and "intelligence." And, of course, "mind."

[162] It would be interesting to hear whether McGinn thinks it might nevertheless be possible for cognitively enhanced posthumans to understand these things. His reply to an interview question in Baker (2002) hints in that direction: "The limits on our understanding are really contingent cognitive biases built into us *at this moment in evolutionary history*" (my emphasis).

it is hardly more strange than a human brain doing the same thing. This commits me to the view that computers can be conscious, and it also goes a long way – since it was what the computer *did*, rather than the material substance it *consisted of*, that convinced me – towards committing me to the CTOM.

As to arguments *against* the CTOM, these tend to be less convincing. A sadly typical example is Pigliucci's repeated assertion, in his contribution to *Intelligence Unbound*, that "consciousness is a biological phenomenon," and his complaint that Chalmers' contribution to the same book "proceeds *as if* we had a decent theory of consciousness, and by that I mean a decent *neurobiological theory*" (emphasis in the original).[163] But this merely presupposes what Pigliucci wants to prove, namely that consciousness is a biological phenomenon. Pigliucci knows of one example of a conscious entity, namely himself, and he has a reasonable case for assuming that most other humans are conscious too, and also that in all these cases consciousness resides (at least to a large extent) in the brain. Since all these brains are neurobiological entities, isn't that just what we need for the claim that a theory of consciousness needs to be neurobiological? Well, no, because being neurobiological entities is not the only thing that these brains have in common. They are also, e.g., material objects, and they are computing devices. Pigliucci thinks that being a neurobiological entity is the relevant property around which a decent theory of consciousness needs to be centered. Other philosophers (panpsychists) think that being a material object is the relevant property, and yet others (supporters of CTOM) think that being a computational device is the relevant property. Any discussion of the pros and cons of CTOM ought to start with the admission that it is an open problem to decide which of these generalities (or some other one, such as being bipedal, or being created in the image of God) is the relevant one for the fundamental nature of consciousness. Instead, postulating from the outset, as Pigliucci does, which of them is the right one, short-circuits the discussion in such a way as to make it trivial.[164]

But there are of course somewhat better arguments than Pigliucci's against CTOM. Possibly the most interesting, and certainly the most widely cited, such argument is John Searle's **Chinese room argument**.[165] It is a beautiful thought experiment, and even its most ardent critic concedes that it has "tremendous appeal and staying power" and dubs it "a classic."[166] It works as a kind of *reductio ad absurdum*: supposing CTOM is correct, Searle gives us a scenario that is so crazy that the theory must be rejected. Here goes:

Suppose CTOM is correct. Then it is possible to write a computer program that *really* thinks, that *really* understands, and that *really* is conscious, as opposed to

---

[163] Pigliucci (2014c), Blackford and Broderick (2014), Chalmers (2014).

[164] There are further arguments against mind uploading in Pigliucci (2014c), and they are all of about the same merit as this one, so I'd side with Danaher's (2014b) judgment that the paper is poorly argued.

[165] Searle (1980).

[166] Dennett (2013), p 320.

merely giving the outwards appearance of such feats. Suppose, for concreteness, that the computer program understands Chinese, and is able to carry out an intelligent conversation in that language, on the level of a typical educated Chinese grown-up. Now, since we're assuming CTOM, the fact that the program is implemented in electronic circuits is incidental, so we are free to choose some other substance or medium on which to implement the structure that the program represents. Let's do it in the following unusual way. Instead of a computer, we have a room, containing John Searle himself – who knows not a word of Chinese – plus several large stacks of paper. One of these stacks contains precise instructions in English for how Searle should act, corresponding to the computer program. The other stacks contain blank paper, corresponding to the computer memory needed to carry out the program. The room also has two windows, one in which people outside can deliver strings of Chinese symbols, and one in which Searle can deliver other such strings, namely the replies he arrives at by following the step-by-step instructions he has at his disposal. Defenders of CTOM are here committed to the view that Searle understands Chinese. But that is crazy, because Searle still does not understand Chinese – he merely knows how to mindlessly follow step-by-step instructions given on bits of paper in the room. Hence (so the argument goes) CTOM must be rejected.

Here I am inclined to go along with Hofstadter and Dennett (1981) in giving what Searle calls **the systems reply**: in the proposed scenario, it is not Searle himself who understands Chinese, but the entire system, consisting of the room, plus the stacks of paper, plus Searle. Searle is just one of many components in the system, and it is wrong to insist that for a system to be conscious we need at least one of its components to be conscious, lest we repeat the mistake of Leibniz (1714):

Moreover, it must be confessed that *perception* and that which depends upon it are *inexplicable on mechanical grounds*, that is to say, by means of figures and motions. And supposing there were a machine, so constructed as to think, feel and have perception, it might be conceived as increased in size, while keeping the same proportions, so that one might go into it as into a mill. That being so, we should, on examining its interior, find only parts which work one upon another, and never anything by which to explain a perception. Thus it is in a simple substance, and not in a compound or in a machine, that perception must be sought for.[167]

The problem with Leibniz' argument is that it compels us either to deny that we humans have perceptions or consciousness, or to give up the view (pretty much universally accepted by scientists, but – it must be admitted – a bit less so among philosophers) that perceptions and consciousness arise from the human brain, which can be broken down into parts (elementary particles, if needed) that lack both perception and consciousness of their own.

---

[167] This passage from an English translation of Leibniz (1714) is borrowed from Dennett (2005), p 3.

Faced with the systems reply, Searle retorts by pointing out that he can internalize the entire system and still not understand a word of Chinese:

Let the individual internalize all of these elements of the system. He memorizes the rules in the ledger and the data banks of Chinese symbols, and he does all the calculations in his head. The individual then incorporates the entire system. There isn't anything at all to the system that he does not encompass. We can even get rid of the room and suppose he works outdoors. All the same, he understands nothing of the Chinese, and a fortiori neither does the system, because there isn't anything in the system that isn't in him. If he doesn't understand, then there is no way the system could understand because the system is just a part of him.[168]

To this, Hofstadter and Dennett offer a lengthy response, beginning with the observation that Searle may have failed to imagine the magnitude of the amount of learning needed to "swallow up the description of another human being" – can it even be done, and if so, can it really be done without thereby achieving true understanding of Chinese (or whatever it is that this other human being understands)?[169] Hofstadter and Dennett point out that "a key part of [Searle's] argument is in glossing over these questions of orders of magnitude" (p 375). The reply by Searle (1982a, 1982b) offers no amendment to this alleged glossing-over, and the reader is left wondering just *why* the individual, under these extreme circumstances, will not understand Chinese.[170] Maybe he will, maybe he won't; Searle merely *claims* he won't. And given what we know about split-brain patients[171] and multiple personality disorder, it is perhaps not a terrible stretch to think that more than one conscious being can inhabit the same skull. In this case, we might have two guys, English-speaking-Searle and Chinese-speaking-Searle, the former claiming (in English) that he doesn't understand a word of Chinese, and the latter insisting (in Chinese) that he does understand Chinese. Why should we trust only the former, and not the latter?

---

[168] Searle (1980), p 419.

[169] Hofstadter and Dennett (1981), pp 375ff. For further discussion on the Chinese room argument and the systems reply, I recommend the exchange by Searle (1982a), Dennett (1982) and Searle (1982b) – especially for those readers who enjoy a good fight and are fascinated by how much heat a philosophical discussion can generate.

[170] So does Searle make Leibniz' mistake of thinking that if none of the components of a system has consciousness, then the system itself also lacks consciousness? What he writes in Searle (1980, 1982a, 1982b) gives very much this impression, until suddenly and a bit confusingly (to this reader at least) he *denies* doing so:

Dennett thinks my only reply to this is to deny the distinction between the supersystem and subsystem and to deny "that a supersystem understands Chinese on the grounds that none of its subsystems do." But that is not and has never been my argument, neither in my published writings nor in my letter to him. My objection to the systems reply is that though there is a quite valid distinction between the level of the supersystem and the level of the subsystem, it is irrelevant to the issue because neither level has any way of attaching any meanings to the Chinese symbols. (Searle, 1982b)

And then Searle goes on to more or less repeat the above-quoted passage from Searle (1980) about the individual internalizing all of the elements of the system. But then, the claim that the new system (the one consisting of the individual internalizing all of the elements of the original Chinese room) does not understand Chinese just stands there as a naked claim, unsupported by argument.

[171] See, e.g., Gazzaniga (1998).

The above is only a scratch of the surface of the philosophical debate surrounding question (1) concerning the possibility of computer consciousness, and the main intellectual framework (CTOM) used to argue for a "yes" answer. This is obviously not the time and place to settle this debate, so let us instead move on to question (2). Given that the answer to (1) is "yes," then, if I attempt to upload my mind to a computer by creating a whole-brain emulation, will the uploaded mind really be *me*, as opposed being another person who is merely a copy of me?

In order to more neatly separate (2) from (1), it will be convenient to modify (2) by considering, instead of uploading, the related procedure of **teleportation**. Here the individual's body is scanned, and the information obtained from the scanning is sent somewhere else, where the body is reassembled (or, if you will, a new identical body is assembled). It will be assumed that teleportation technology is good enough so that the reassembled individual, including its behavior, is indistinguishable from the original one. We will mostly consider the case of **destructive teleportation**, in which the scanning procedure involves destroying the original body. The alternative, **non-destructive teleportation**, where the original body remains intact, is more like copying. A teleportation counterpart of question (2) is this:

(2′) If I undergo destructive teleportation, will I survive – as opposed to dying and being replaced by another person who is merely my copy?

If the answers to (1) and to (2′) are both "yes," then it seems reasonable to hope that the answer to (2) will also be "yes," so that it will be possible for me to transfer my mind to a computer and retain "being me" (whatever than means). Note, however, that this is not a foregone conclusion, because it might be that computer consciousness is possible, and that teleportation works, whereas remaining the same person is impossible if we insist on transferring from one substrate (like flesh and blood) to another (like a digital computer). Still, answering (2′) seems like a good place to start.

So let's have a look at teleportation. As a thought experiment, we fast-forward to 2030, and assume that (destructive) teleportation technology has advanced faster than expected, so that it is by then in common use for long-distance personal transportation.[172] Today is Monday, and I need to be in New York on the other side of the Atlantic for an important meeting on Thursday morning. Wednesday is travel day, and I face the choice between going by air or taking the teleporter. Going by air will be a hassle and take me the better part of a day, while teleportation will take me door to door in less than two hours, at a lower cost. All practical aspects point towards teleportation being the obvious choice, were it not for the issue of my survival. If I take the teleporter, will I survive?

I have never before tried the teleporter myself, so I'm a little anxious. Since teleportation is by now, 2030, such a widely used means of transportation, I can ask people around me who have teleported on one or more occasion. Did they survive? They all say that they did survive, and moreover that they never noticed any

---

[172] The next few paragraphs are loosely based on Häggström (2012a), which in turn owes a great deal to Parfit (1984).

difference between their (immediate) pre- and post-teleportation selves. That reassures me at first, but then it hits me that if teleportation kills the person, only to replace him or her with another person with identical properties, then their answers can be expected to be exactly the same. After all, they come out of the teleporter with exactly the state of mind, psychological characteristics and memories that the originals had.

So the question seems really hard (perhaps impossible?) to answer empirically, so let me try to attack it with a bit of analytic philosophy instead. The key concept is survival, and although we tend to intuitively feel that we know what it means, it is not fully clear how to define it rigorously. Following are two candidate definitions, and to distinguish them let us call the respective concepts they give rise to σurvival and Σurvival.

σ**urvival:** I σurvive until Thursday if, on Thursday, there exists a person who has the same personality traits, the same memories, and so on, as I have today (allowing for some wiggling corresponding to what is considered normal changes during the course of a few days).

Σ**urvival:** I Σurvive until Thursday if (a) I σurvive until Thursday, and (b) whatever extra property that is needed for the guy in New York to really be me (as opposed to just a copy) holds.

The definition of Σurvival may seem fuzzy, but the problem is that I have no good idea what the extra property in (b) would be. We may of course contemplate candidate suggestions like **continuity of my body's trajectory through space-time** (CBTST, for short), which would imply my non-Σurvival in case I teleport, but faced with such a suggestion we ought to ask why this particular property matters. Why is CBTST crucial to my personal identity? If that really is the case, it requires an explanation. But is it? I've asked those friends and acquaintances who have tried teleportation (remember, we're still in the 2030 thought experiment!) whether they are in any way inconvenienced by the lack of CBTST caused by their teleportation, but they categorically deny any such inconvenience. And the same goes for any property that I can think of in place of CBTST, i.e., anything that might serve as replacement for (b) in the definition of Σurvival and that is violated by teleportation. So Occam's razor (the idea that "plurality must never be posited without necessity," or in a more modern formulation that scientific theories ought to be as simple as the evidence admits; see, e.g., Baker, 2010) compels us to reject (at least for the time being) the idea that there is anything to Σurvival beyond σurvival. Now, if σurvival is what matters, then teleportation is unproblematic.[173]

People's intuitions on this matter tend to diverge drastically, and many thinkers insist that teleportation will kill anyone who tries it. A standard argument that it will is to compare it to the case of non-destructive teleportation. Here it is, in

---

[173] On the other hand, it may seem to some readers that if σurvival is all there is to survival, then what point is there to survive at all? I can certainly feel the force of that thought, but have found it is possible to view survival as σurvival while still retaining a feeling that life is not meaningless. Blackmore (2012) explains and defends such a view beautifully.

Eliezer Yudkowsky's words, where the teleportation is taken to be interplanetary rather than merely transatlantic:[174]

Ah, but suppose an improved Scanner were invented, which scanned you non-destructively, but still transmitted the same information to Mars. Now, *clearly*, in this case, *you, the original* have simply stayed on Earth, and the person on Mars is *only a copy*. Therefore [teleportation without the improved scanner] is actually murder and birth, not *travel* at all – it destroys the original, and constructs a copy!

Chalmers (2014) and Pigliucci (2014c) both discuss this argument.[175] Chalmers admits it carries some weight, but finds it nevertheless inconclusive. Pigliucci, on the other hand, finds it conclusive: "if it is possible to do the transporting or uploading in a non-destructive manner, *obviously* we are talking about duplication, not preservation of identity" (italics in original). I agree about "duplication," but if σ urvival is all there is to survival (or to "preservation of identity"), then there is no particular reason to hold that I cannot simultaneously survive in two bodies. So, while Pigliucci doesn't spell out his argument explicitly beyond italicizing "*obviously*" (echoing Yudkowsky's semi-sarcastic italicizing of "*clearly*"), it does seem that he holds some sort of Σ urvival beyond σ urvival to be important. And since, in the case of non-destructive teleportation, he holds an asymmetry between the original body (in which you survive) and the new one (in which you don't), it seems reasonable to speculate that CBTST is the property he thinks is missing in σ urvival but essential for Σ urvival. But that just brings us back to the question about just *why* CBTST is important for Σ urvival. In the absence of such an explanation, we probably ought to reject the relevance of CBTST, not just by appeal to Occam's razor, but also because fundamental physics doesn't seem to support it. Elementary particles do not have identities, making claims like "these electrons come from the original body, but those other ones do not" literally nonsensical,[176] and since our bodies are made up of elementary particles, postulating an asymmetry between the two bodies, where one is the original and the true inheritor of personal identity, while the other one is merely a copy, seems unfounded.[177] And even if physics as we know it did admit elementary particles with individual identities, it would not help much, due to the relentless flux of matter on higher levels. Here is Kurzweil (2005):

The specific set of particles that my body and brain comprise are in fact completely different from the atoms and molecules that I comprised only a short while ago. We know that most of our cells are turned over in a matter of weeks, and even our neurons, which persist as distinct

---

[174] Yudkowsky (2008f). It should be noted that, although his formulation is not unfair, his purpose is not to endorse the argument but to distance himself from it.

[175] They do so mainly in the setting of uploading rather than teleportation, but the arguments carry over from one setting to the other pretty much verbatim.

[176] Bach (1988, 1997).

[177] Yudkowsky (2008f).

cells for a relatively long time, nonetheless change all of their constituent molecules within a month. . . . The half-life of a microtubule is about ten minutes. . . .

So I am a completely different set of stuff than I was a month ago, and all that persists is the pattern of organization of that stuff. The pattern changes also, but slowly and in a continuum. I am rather like the pattern that water makes in a stream as it rushes past the rocks in its path. The actual molecules of water change every millisecond, but the pattern persists for hours or even years. (p 383)

3.9 Uploading: practical issues

The reader will undoubtedly understand from the above discussions that I think there is reasonable hope for "yes" answers to question (1) about the possibility of computer consciousness and question (2) about personal identity surviving uploading. But these are deep philosophical questions about the fundamental nature of reality and issues about which our understanding today is insufficient. Thus, a good deal of epistemic humility is called for: I may well be wrong about (1) and/or (2), and if either of the questions are correctly answered by "no," then all hopes of achieving radical life extension via mind uploading are simply mistaken.

Let's assume, for the sake of argument, that both questions have affirmative answers. Then there are a number of further more practical questions on uploading, including the following:

(3) What will be needed in order to get the required whole-brain emulation technology up and running, and when (if ever) can we expect this to be accomplished?

(4) What will a society be like in which uploading is a widely available technology?

(5) Are there any particular ethical concerns that we ought to keep in mind before we rush ahead with such technology?

Let us begin with (3). Two of the world's most high-profile ongoing scientific projects – the European Union's Human Brain Project[178] and President Obama's BRAIN Initiative[179] – concern advanced computer modeling and neuronal mapping of the human brain, and while neither of them is explicitly concerned with uploading, that seems to be the logical endpoint of such endeavors. The author best known for his unabashedly precise predictions and timelines for future technologies is Ray Kurzweil, who predicts in *The Singularity is Near* that uploading will have its breakthrough in the 2030s.[180] Such a prediction is, however, based on a great many assumptions about which there are a great many uncertainties. More cautious – and preferable – is the approach of Sandberg and Bostrom (2008b), who

---

[178] Yong (2013).
[179] Markoff (2013).
[180] Kurzweil (2005).

admit and systematically explore these uncertainties, giving a wide range of possible scenarios. They break down the capabilities needed into scanning, translation and simulation.

> **Scanning** is the high-resolution microscopy needed for detecting and registering all the relevant properties and details of the brain. What level of resolution will actually be needed is not known, but since techniques are already available for seeing individual atoms, it seems extremely likely that the necessary level has already been attained, but the speed and parallelization that is needed to avoid astronomical scanning times is not there yet. Sandberg and Bostrom focus mainly on techniques involving cutting up the brain in thin slices that are scanned separately. This will obviously destroy the original brain. Non-destructive scanning is even more challenging. A speculative suggestion here, advocated by Kurzweil (2005), is to use nanorobots that enter the brain in large numbers and report back the locations of molecules and atoms (see Section 5.3).
>
> **Translation** is the processing of the data from the scanning that needs to be done before it can be used for simulation. This will involve a huge amount of image analysis: "Cell membranes must be traced, synapses identified, neuron volumes segmented, distribution of synapses, organelles, cell types and other anatomical details (blood vessels, glia) identified" (Sandberg and Bostrom 2008b, p 55).
>
> **Simulation** requires hardware capabilities in terms of memory (for storing the original model and the current state), bandwidth and CPU, probably in a massively parallel architecture. How much is needed depends to a large extent on how detailed the neuron model needs to be. More generally, there's a balance between the amount of high-level scientific understanding of the workings of the brain and the need for computing power. The need for computing power is greatest when little or no such understanding is available, in which case everything needs to be done bottom-up, brute force; the emulated mind will then serve more or less as a black box, where we may understand the inputs and the outputs but little of what goes on inside (although no more of a black box than our brains already are today).

In addition, the emulation needs either to be embedded in a basic virtual reality environment or to be given a robotic embodiment with the appropriate audiovisual and motoric I/O (input/output) channels, but as Bostrom (2014) points out, this seems easy relative to the three main tasks. Minimally adequate I/O is to some extent already available, as exemplified by the cochlear implant technology discussed in Section 3.6.

Complications abound. It is common knowledge today that the brain's function depends crucially on the chemical balance of neurotransmitters and neuromodulators such as dopamine and serotonin. Hundreds of further such substances are known, and there may be even more unknown ones. Knowing the connectivity matrix of all the neurons – knowledge that we have had since the 1980s for the case of the tiny roundworm *Caenorhabditis elegans* with all its 302 hard-wired neurons – is obviously not enough. As Bostrom explains, we also need information about "which synapses are exitatory and which are inhibitory; the strength of the connections; and various dynamical properties of axons, synapses, and dendritic trees" (p 35). Not even for this minuscule model organism do we at present have information at this level of detail.

Yet, all in all, the verdict of Sandberg and Bostrom (2008b) is that the requisite technologies for whole-brain emulation are all attainable by incremental advances, with no in-principle barriers in sight. Bostrom (2014) concludes that "the prerequisite capabilities might be available around mid-century, though with a large uncertainty interval" (p 34).

Once successful whole-brain emulation has been accomplished, it might not be long before it becomes widely available and widely used.[181] This brings us to question (4) – what will society be like when uploading is widely available? Most advocates of an uploaded posthuman existence, such as Kurzweil (2005) and Goertzel (2010), point at the unlimited possibilities for an unimaginably (to us) rich and wonderful life in ditto virtual realities. One researcher that stands out from the rest in actually applying economic theory and social science to attempt to sketch how a society of uploads will turn out is the American economist Robin Hanson, beginning in a 1994 paper, continuing with a series of posts on his extraordinary blog *Overcoming Bias*, and summarizing his findings (so far) in a chapter in *Intelligence Unbound* and in an upcoming book.[182]

Two basic assumptions for Hanson's social theory of uploads are

(i) that whole-brain emulation is achieved mostly by brute force, with relatively little scientific understanding of how thoughts and other high-level phenomena supervene on the lower-level processes that are simulated, and

(ii) that current trends of hardware costs decreasing at a fast exponential rate will continue (if not indefinitely then at least far into the era he describes).

Assumption (i) prevents us from boosting the emulated minds to superhuman intelligence levels, other than in terms of speed, by transferring them to faster hardware. Assumption (ii) opens up the possibility for quickly populating the world with billions and then trillions of uploaded minds, which is in fact what Hanson predicts will happen.

---

[181] This is not a foregone conclusion. Ulf Persson suggested to me (personal communication) that Moon landings serve as a striking example of a technology that, more than four decades after its successful introduction, hasn't become widely available. I suspect, however, that uploading technology is more likely to fall into the category of technologies such as computer hardware and DNA sequencing technologies, that have (up to now) become cheaper and more efficient at a fast exponential rate.

[182] See, e.g., Hanson (1994, 2008a–f, 2012, 2014a).

The consequences for society will be disruptive. Miller (2012) explains pedagogically what can be expected to happen to labor markets:

Think of the ten most productive people in your workplace. Now imagine that your company could cheaply make as many copies of these valuable employees as it wanted. . . . Taking this to the next stage, consider what would happen to your company's profits if it were free to substitute the best minds mankind has to offer for all its current employees. (p 140)

Decreases in hardware costs will push down wages: why would an employer pay you, say, $100,000 per year to work for them, when an upload can do it for a hardware cost of $1 per year? This will send society to the classical Malthusian trap in which population will grow until it is hit by starvation (uploaded minds will not need food, of course, but things like energy, CPU time and disk space). We who live today in the rich part of the world have been able to escape this trap (for the time being), because ever since the industrial revolution we have maintained an innovation rate that has allowed the economy (including food production) to grow much faster than the population. A society of uploads will innovate even faster than ours, but the ease with which uploads admit copying is likely to cause a population increase that the innovation rate cannot match.

There are many exotica in Hanson's future. One is that uploads can run on different hardware and thus at different speeds, depending on circumstances. It may, for instance, make sense to cut middle management levels in a company by letting a single manager run at a thousandfold speed compared to the lower-level workers, giving him time to manage thousands of them individually. Adjustable speed will also make it easier to meet project deadlines, and so on.[183]

Even more exotic is the idea that most work will be done by short-lived so-called spurs, copied from a template upload to work for, say, a few hours and then be terminated (i.e., die). Such a scenario has also been explored at length by David Brin, in his science fiction novel *Kiln People* (Brin, 2002). Among the issues investigated by Brin is to what extent spurs will feel that they live on in their templates. Will they not revolt? The question has been asked, but Hanson maintains that "when life is cheap, death is cheap."[184]

The future outlined in Hanson's theory of uploaded minds may seem dystopian, as emphasized by Nikola Danaylov in a lengthy interview, but Hanson does not accept this label, and his main retorts seem to be twofold.[185] First, population numbers will be huge, which is good if we accept that the value of a future should

---

[183] Hanson (2012).

[184] Shulman (2008), Hanson (2008a). The full passage for the Hanson quote reads:

Taking the long view of human behavior we find that an ordinary range of human personalities have, in a supporting poor culture, accepted genocide, mass slavery, killing of unproductive slaves, killing of unproductive elderly, starvation of the poor, and vast inequalities of wealth and power not obviously justified by raw individual ability. The vast majority of these cultures were not totalitarian. Cultures have found many ways for folks to accept death when "their time has come". When life is cheap, death is cheap as well. Of course that isn't how our culture sees things, but being rich we can afford luxurious attitudes.

[185] Danaylov and Hanson (2013a, b), Hanson (2013).

be measured not by some average level of well-being in a population, but rather by that quantity times population size, i.e., by the total amount of well-being, which in a huge population can be very large even if each individual has only a modest positive level of well-being.[186] Second, the trillions of short-lived uploaded minds working hard for their subsistence right near starvation level can be made to enjoy themselves, e.g., by cheap artificial stimulation of their pleasure centra. Presumably most readers will, like Danaylov and like myself, not find much comfort in those responses.

Then there's the issue of whether and to what extent we should view Hanson's analysis as a trustworthy prediction of what will actually happen. A healthy load of skepticism seems appropriate. His work is very valuable as a first study of the economic and societal implications of a breakthrough in uploading technology, and for suggesting a number of non-obvious possible effects that clearly merit our attention, but it also seems that he works so far outside of the comfort zones where economic theory has been tested empirically, and uses so many explicit and implicit assumptions that are open to questioning, that his scenarios need to be taken with a grain of salt (or a full bushel).

Let me move on to question (5) on ethical issues concerning whether, and how, to develop uploading technologies. One obvious issue to consider is whether society following a breakthrough in the technology will be better or worse than society without such a breakthrough. The utopias hinted at by, e.g., Kurzweil (2005) and Goertzel (2010) seem pretty good, whereas Hanson's Malthusian scenario looks rather less appealing. The social science of a society of uploaded minds clearly needs further development before we can tackle this issue with any sort of confidence.

But there are also other ethical issues to consider; Sandberg (2014a, 2014b) brings up a host of them. An obvious class of such issues concern the possibility that emulations will suffer. As Bostrom (2014) dryly notes, "before we would get things to work perfectly, we would probably get things to work imperfectly" (p 34). This could for instance amount to creating a person with some kind of severe brain damage and horrible suffering.[187] Is it ethically acceptable to risk doing something like that?

---

[186] Not everyone agrees about this. In Section 10.1, this issue appears in a list of open questions about value that need to be answered in order for us to figure out what kind of world we ought to strive for.

[187] Sandberg (2014a) quotes Metzinger (2003):

What would you say if someone came along and said, "Hey, we want to genetically engineer mentally retarded human infants! For reasons of scientific progress we need infants with certain cognitive and emotional deficits in order to study their postnatal psychological development – we urgently need some funding for this important and innovative kind of research!" You would certainly think this was not only an absurd and appalling but also a dangerous idea. It would hopefully not pass any ethics committee in the democratic world. However, what today's ethics committees *don't* see is how the first machines satisfying a minimally sufficient set of constraints for conscious experience could be just *like* such mentally retarded infants. They would suffer from all kinds of functional and representational deficits too. But they would now also subjectively experience those deficits. In addition, they would have no political lobby – no representatives in *any* ethics committee.

If animal suffering is considered ethically relevant, then this problem arises not just at our earliest attempts at emulating the human brain, but way before that, on the likely path of attempting the feat on increasingly advanced organisms, from the successful emulation of *C. elegans* via, say, ants, mice and rhesus monkeys to humans. Ethical questions in this category are riddled by our poor-to-nonexistent understanding of the fundamentals of consciousness, and thus of what kinds of creatures have the capacity for suffering. Sandberg (2014a, 2014b) suggests something he calls **the principle of assuming the most**, which can be seen as a special case of the precautionary principle, and which tells us to assume that "any emulated system could have the same mental properties as the original system and treat it correspondingly."[188] This might throw a spanner in the works not just for projects aimed at emulating the human brain, but also for the idea of creating sophisticated computer simulations of laboratory animals, to experiment on these simulations, and thus be able to reduce experimentation on real animals and the suffering caused by that. Sandberg (2014a) hints that, although this may become a real concern in the future, it probably isn't a problem today, as "typical computer simulations [of neurological processes involving pain] contain just a handful of neurons" and that "it is unlikely that so small systems could suffer."[189]

Further ethical complications arise in the very likely scenario that uploading will initially (and for a long time thence) involve destruction of the original brain. There will probably still be enthusiasts who volunteer for the procedure, but Sandberg (2014a) notes the legal complication that volunteering will (most likely) be considered

essentially equivalent to assisted suicide with an unknown probability of "failure". It is unlikely that this will be legal on its own for quite some time even in liberal jurisdictions: suicide is increasingly accepted as a way of escaping pain, but suicide for science is [ever since the Nuremberg Trials] not regarded as an acceptable reason.

Another legal issue is that the earliest successful uploads of humans will most likely result in persons that are devoid of civil rights and do not even have ownership of whatever hardware they are running on (or of anything else, for that matter). Can it ever be ethically justified to bring into existence a person who

---

[188] As with the precautionary principle in general, one difficulty is knowing where to draw the line. If I go outdoors, I might be struck by lightning, but it would nevertheless not be a good idea to conclude, based on the precautionary principle, that I should never walk outdoors. In the present case, why exactly should we stop at emulations? If panpsychism is true (which it may well be), then my football is conscious, and perhaps it suffers unbearably every time I kick it. (See also Footnote 189 for a slightly less extreme concern.)

[189] He then immediately adds (in a footnote) that we do not know:

Assigning even a small probability to the possibility of suffering or moral importance [in] simple systems leads to far bigger consequences than just making neuroscience simulations suspect. The total number of insects in the world is so great that if they matter morally to even a tiny degree, their interests would likely outshadow humanity's interests. This is by no means a *reductio ad absurdum* of the idea: it could be that we are very seriously wrong about what matters in the world.

has the intelligence, the sentience and all the psychological properties of a normal healthy adult, yet is stuck in such a precarious situation?

There are, clearly, plenty of ethical and legal issues that we ought to resolve before we go ahead with the implementation of uploading technology.

3.10 Cryonics

Many self-described transhumanists wish not just for a wonderfully flourishing high-tech future for humanity, but also to be part of it *themselves, personally,* including enjoying the (near-)immortality that they hope will be part of the package; see, e.g., Kurzweil (2005) and Bostrom (2008a) for eloquent expressions of such sentiments. Most people do not want to die, but for such transhumanists even more seems to be at stake than for the rest of us: thousands, millions or billions of years with a drastically higher quality of life than we are used to. So they have much reason to fear "missing the boat" – dying before radical life extension technologies have had their breakthrough.[190] It is not known when uploading technology will be available – if it is at all possible – and whether it preserves personal identity in whatever is the relevant sense of that term. And advances in more conventional biomedical methods appear to be just too slow at present: the current trend in life expectancy increase of a quarter of a year per year appears inadequate for someone who aspires to immortality. So is the situation desperate? The classic statement of desperation is the following by Benjamin Franklin, in a letter to Jacques Dubourg in April 1773:[191]

I wish it were possible . . . to invent a method of embalming drowned persons, in such a manner that they might be recalled to life at any period, however distant; for having a very ardent desire to see and observe the state of America a hundred years hence, I should prefer to an ordinary death, being immersed with a few friends in a cask of Madeira, until that time, then to be recalled to life by the solar warmth of my dear country! But . . . in all probability, we live in a century too little advanced, and too near the infancy of science, to see such an art brought in our time to its perfection.

But what to Franklin in the 18th century seemed like a hopelessly distant dream may, for present-day transhumanists, be a not entirely unrealistic way (beyond the obsessive-seeming health measures of Footnote 190) to improve their odds of reaching the posthuman paradise: not literally a cask of Madeira, but a similar idea known as **cryonics**.

---

[190] This is to some extent reflected in Kurzweil's extreme approach to his own longevity and health, including taking "250 supplements (pills) a day and [receiving] a half-dozen intravenous therapies each week" (p 211), thereby significantly altering his body's metabolism and chemical balance in ways that he expects will slow down his ageing.

[191] Russell (1926).

Cryonics is the preservation at low temperatures of humans (and other animals) in the hope that restoration to healthy conditions will be possible some time in the future. The fundamental principle that makes the idea worthy of consideration to begin with is that metabolism and biochemical degradation slows down with lower temperatures. This is the same phenomenon that allows us to store food in refrigerators and freezers, although in cryonics the much more extreme temperature of $-196°C$ (the boiling point of nitrogen) is employed, for which the exponential drop in reaction rates in declining temperatures is enough to grind reaction times virtually to a halt, not just on the time scale of centuries, which is perhaps the most relevant one to crynoics, but even on astronomical time scales. A major complication, however, is that ice formation in the freezing process threatens to crush the cells of the body, so a key step in the cryonic procedure is so-called vitrification – making sure that the freezing takes place without crystal formation. This is achieved by, prior to freezing, washing out the subject's blood and replacing most of the body water with a cryoprotectant mixture preventing ice formation. This, in turn, introduces further complications, but I will not say more on the technological details, referring instead to the excellent survey by Best (2008).

Since no viable technology for waking up cryopreserved subjects exists today, cryonics itself is to large extent a speculative technology. Signing up for cryonics (at a cost ranging from $10,000 to $250,000, depending, e.g., on whether one goes for full body preservation or for the budget version of preserving just the head) is a gamble, where a small success probability[192] has to be weighed against the huge potential gain, in a calculation that is not entirely unlike Pascal's Wager.[193] Add to this the conception that cryonics is about "waking up from the dead,"[194] and it is

---

[192] As a general rule of thumb, if an outcome requires a long list of unknowns to turn out favorably, then this suggests that the outcome has small probability. In this case, the successful outcome – waking up from the cryopreservation some time in the future – requires the conjunction of at least the following unknowns: (a) the unverified claim that cryonics actually works; (b) the ambulance transport to the cryonics institute must happen quickly enough that I don't suffer irreversible degradation; (c) progress in medicine must proceed as I hope it will, rather than, e.g., being halted by civilizational collapse; (d) the cryonics company needs to deliver on its promise to keep me frozen through what might be decades or centuries of societal ups and downs; and (e) the future society that has the medical capacity to bring me back to life will also decide to do so.

[193] In Pascal's Wager, eternal afterlife is weighed in as an *infinite* gain, and the conclusion is that as long as the existence of God has a strictly positive probability, then no matter how small it is, it still pays to be a devout Christian. A person who considers signing up for cryonics will perhaps not go to the excesses of calculating with an infinite gain, but Pascal's Wager-like features may show up even in finite settings; see Bostrom (2009). More will be said about Pascal's Wager in Section 10.4.

[194] Proponents of cryonics tend not to accept this description. While persons undergoing cryopreservation are legally dead, the claim is that legal death does not coincide with true death, or what they call information-theoretic death, defined as the stage of destruction of the information contained in a human brain where the original person becomes theoretically impossible to reconstruct; see Merkle (1994).

There's a slight irony here in the fact that it is only through this discrepancy that cryonics becomes (or is conceived as) meaningful while still operating within legal bounds. Cryonics is not a recognized medical procedure, and can therefore only be performed after legal death has been pronounced.

perhaps not so surprising that proponents of cryonics have to suffer a good deal of mockery and accusations of being lunatics.[195] This intellectual climate has, in turn, undoubtedly contributed to cryonics still being a fringe phenomenon. The number of people currently in cryonic storage is around 200, and the number of people having signed up for the procedure is about an order of magnitude larger. These very small numbers have triggered one of the leaders of the transhumanist movement, Max More, to speak of an insane tragedy:[196]

We'll look back on this 50 to 100 years from now – we'll shake our heads and say, "What were people thinking? They took these people who were very nearly viable, just barely dysfunctional, and they put them in an oven or buried them under the ground, when there were people who could have put them into cryopreservation". I think we'll look at this just as we look today at slavery, beating women, and human sacrifice, and we'll say, "this was insane – a huge tragedy".

Whether More is right or wrong about this prediction strikes me as a wide-open question. Here, some readers will probably suggest that if I really consider it plausible that cryonics might work, then I should put my money where my mouth is. However, I have not signed up for cryonics. I outlined my reasons for this in Häggström (2013g). It's a very personal decision, but perhaps my reasons and considerations may help others think more clearly about these issues, so here's a brief summary:

While I do think it is important that we get a future in which humanity (or posthumanity) flourishes, I do not consider it particularly important that I, personally, be part of it. This is due to a more general insight that my own interests and well-being are no more important than those of people in general – an insight derived mainly from the deflationary view of personal identity outlined by Parfit (1984) and Blackmore (2012).

Does this mean that I have given up my egoistic perspective in my daily dealings? Of course not – and to do so entirely would be psychologically impossible. As one of many examples of my selfish behavior, consider the fact that I still keep most of my income to myself, rather than giving it away to, e.g., Médecins Sans Frontières, where the money would surely do much more good from a person-neutral perspective. I still act, to a very large extent, on low and self-serving instincts and desires. Sometimes I try to amend my actions a bit, in view of my more intellectual insight that my interests are no more important than those of others. And in the rare case – such as that of whether to sign up for cryonics – where the decision resulting from my lowly instincts is in harmony with what makes sense from a person-neutral perspective, namely not to sign up for cryonics, I refrain from trying to override the decision by intellectual means (even

---

[195] A typical example is how renowned philosopher of science Massimo Pigliucci writes that "if there's a movement that resembles religion in having faith despite the evidence, the cryonics movement is it," and that "the plausibility of cryonics doesn't quite reach homeopathy-level, but it ain't that far above it" (Pigliucci, 2014a, 2014b).

[196] Angelica (2011). When evaluating More's judgment, we should perhaps factor in that he is president and CEO of the market-leading cryonics company Alcor.

though I understand that my lowly instincts made up my mind based not on a person-neutral perspective but more likely on the high discount rates discussed in Section 10.2, perhaps also fueled by a cowardly desire not to be looked down upon by bullies like the guy in Footnote 195).

An anonymous commentator on Häggström (2013g) suggested that it would make sense, also from a person-neutral perspective, that I sign up for cryonics, because that would have the chance of adding many happy man-years to the world. That suggestion seems to be based on the implicit assumption that if I undergo cryonics, then the world population will for many years be larger than it otherwise would have been, by exactly one individual. That seems very unlikely. If a high-tech future civilization decides it has room for one more individual, then surely they will make that happen regardless of whether I am available in a cryonic storage facility.

CHAPTER 4

Computer revolution

4.1 Cantor

What could be a higher thing to aspire to for a scientist, a philosopher or a mathematician than to be ahead of one's time? In theory, that is an almost universally accepted romantic ideal, but in practice, things are not always so easy. The 19th and early 20th century German mathematician Georg Cantor is a case in point.

Cantor was interested in the area of mathematics known as set theory, and can in fact legitimately be called its inventor. Today his theory of countable versus uncountable infinities (more on which later) is considered to be a cornerstone of mathematics and is part of the standard mathematics curriculum, but his contemporaries were not ready for his ideas. Cantor's most persistent opponent was Leopold Kronecker, who called him a "scientific charlatan," a "corrupter of youth," and other things in the same vein. According to popular legend, Henri Poincaré, who is considered along with David Hilbert to be the greatest mathematician of that time, speculated that Cantor's ideas would in the future "be regarded as a disease from which one had recovered." Martin Davis suggests, in his authoritative but reader-friendly history *The Universal Computer: The Road from Leibniz to Turing*, that the purported Poincaré quote is probably no more than an urban legend, whose currency nevertheless says something about the resistance that Cantor was up against.[197] In 1884, at age 39, Cantor suffered from the first of many nervous breakdowns, and attributed his deteriorating mental health partly to the opposition from Kronecker. He would suffer depressions during much of the time from then until his death in 1918; the recognition that his work eventually began to receive did not seem to help much.

A **set** is simply a collection of objects. The objects are called **elements** of the set, and a basic issue in set theory is when one set can be considered larger than another in the sense of having more elements. For finite sets, this notion is straightforward: if one set S_1 has 7 elements, and another set S_2 has 20 elements, then obviously S_2 is more numerous than S_1, because $20 > 7$. But do such comparisons make

---

[197] Davis (2011).

Here Be Dragons. First Edition. Olle Häggström.
© Olle Häggström 2016. Published in 2016 by Oxford University Press.

sense also for infinite sets? Can it ever make sense to say that one infinite set has more elements than another? Cantor decided to see whether that could be the case.

As a starting point, we need some criterion that works for finite sets, and investigate whether it can be extended to infinite sets. One way to check the relative sizes of two finite sets S_1 and S_2 is the following. Try to define a way of pairwise matching the elements in S_1 with those in S_2. If this can be done with no elements from either of the sets left over, then the sets have the same number of elements. If it cannot, and there are unmatched elements in, say, S_2 when we run out of elements in S_1, then S_2 has a larger number of elements.[198] When also applying this criterion to infinite sets, we speak of one set having greater **cardinality** than another, rather than having a larger number of elements, to indicate that the notion goes beyond mere counting.

Already Gottfried Wilhelm Leibniz, the German mathematician and philosopher working two centuries before Cantor, had tried this idea. He considered the set $S_1 = \{1, 2, 3, \dots\}$ of (positive) integers, and the set $S_2 = \{2, 4, 6, \dots\}$ of *even* (positive) integers; here and henceforth, for simplicity, let's consider only positive numbers, as the extension to negative numbers turns out to change very little. Since S_2 contains only every second element from S_1, one would expect S_1 to have greater cardinality. But look what happens – the elements of the two sets can be matched up perfectly, thus showing that S_1 and S_2 have the same cardinality:[199]

$$
\begin{array}{cccccccc}
1 & 2 & 3 & 4 & 5 & 6 & 7 & 8 \cdots \\
\updownarrow & \updownarrow & \updownarrow & \updownarrow & \updownarrow & \updownarrow & \updownarrow & \updownarrow \cdots \\
2 & 4 & 6 & 8 & 10 & 12 & 14 & 16 \cdots
\end{array}
\tag{6}
$$

Leibniz concluded that it makes no sense that S_1, that ought to be twice as large as S_2, has the same cardinality as S_1, so he dropped the whole idea.[200] Cantor,

---

[198] I happen to be familiar with this procedure from a rather early age, as it was carefully carried out in the brilliant Swedish 1970s children's TV show *Fem myror är fler än fyra elefanter* ("Five ants are more numerous than four elephants").

[199] This point is made in more humorous pedagogical fashion in the story of Hilbert's Hotel, which dates back to a 1924 lecture by David Hilbert, and has been retold many times; see, e.g., Strogatz (2010). A hotel has infinitely many rooms, with room numbers $1, 2, 3, \dots$, but they are all occupied. A new guest arrives, in need of a room. What to do? The hotel managers solve the situation by asking each guest with a room to move to the room with the next higher room number. So the guest in room 1 moves to room 2, the guest in room 2 moves to room 3, and so on. Everyone gets a new room, and room 1 becomes available for the new guest. Very good! But what if 10 new guests arrive? Easy: we ask each guest with room number n to move to room $n + 10$, making rooms number $1, 2, \dots, 10$ available to the new guests, and everyone is happy. Next, what if an *infinite* number of new guests arrive? Again, easy: each guest with a room is asked to switch to the room with twice the room number. In that way, the old guests will be occupying all the rooms with even room numbers, leaving those with odd room numbers vacant, and since there are infinitely many odd-numbered rooms, there are enough to accommodate all the new guests.

[200] In fact, even before Leibniz, Galileo came to the same conclusion, but with the set of squares $\{1, 4, 9, 16, \dots\}$ instead of the set of even numbers. This is sometimes known as **Galileo's paradox**; see, e.g., Butler (2006).

faced with the same observation, decided to bite the bullet and accept that the set of integers and the set of even integers have the same cardinality. He went on to study the cardinality of the set of all rational numbers (i.e., the set of all numbers that arise as ratios n/m for integers n and m, with $m \neq 0$). Is it bigger than the cardinality of the set of integers, or again just the same? Cantor figured out a matching procedure a bit more sophisticated than the one in (6) to match up the rationals perfectly with the integers, thus showing that the two sets have the same cardinality (see Davis, p 57).

At that point, it might be tempting to think that even the set of *real* numbers should have the same cardinality as the set of integers, but here comes the crucial discovery that opens up the study of infinite cardinalities as a fertile and interesting topic: the set of reals has greater cardinality than than the set of integers. Cantor derived this result in more than one way, but the one that has been most influential and found its way into the modern mathematical canon is the **diagonal argument**.

Rational numbers can be identified with decimal expansions that eventually repeat themselves, such as

$$2/9 = 0.2222222222222\ldots$$

and

$$251/216 = 1.1620370370370\ldots,$$

whereas real numbers stand in similar one-to-one correspondence with arbitrary decimal expansions,[201] such as

$$\pi = 3.1415926535898\ldots$$

and

$$\sqrt{2} = 1.4142135623730\ldots.$$

Cantor's diagonal argument proceeds as a *reduction ad absurdum*: we assume that the set of real numbers and the set of integers have the same cardinality, then we show that this leads to a contradiction, which allows us to conclude that the assumption must be wrong, i.e., that the two sets must be of different cardinality.

---

[201] There is an annoying exception to this otherwise clean one-to-one correspondence. Namely, for certain rationals, there are exactly two corresponding decimal expansions, one ending with an infinite sequence of 0's, and the other ending with an infinite sequence of 9's. For instance,

$$1/2 = 0.5000000000000\ldots$$
$$= 0.4999999999999\ldots.$$

This will be ignored in what follows, but trust me: it doesn't matter for the final conclusion.

So suppose that the set of real numbers has the same cardinality as the set of integers. Then the set of real numbers can be matched up pairwise with the set of integers. Fix such a matching. Write a_1 for the real number matched up with the integer 1, write a_2 for the real number matched up with the integer 2, and so on, so the matching looks like this:

$$
\begin{array}{ccccccccc}
1 & 2 & 3 & 4 & 5 & 6 & 7 & 8 & \cdots \\
\updownarrow & \updownarrow & \updownarrow & \updownarrow & \updownarrow & \updownarrow & \updownarrow & \updownarrow & \cdots \\
a_1 & a_2 & a_3 & a_4 & a_5 & a_6 & a_7 & a_8 & \cdots
\end{array}
\tag{7}
$$

Also, spell out the decimal expansion of each such a_i as

$$a_i = \lfloor a_i \rfloor . a_{i1} a_{i2} a_{i3} \ldots$$

where $\lfloor a_i \rfloor$ denotes the integer part of a_i (i.e., $\lfloor a_i \rfloor$ is a_i rounded down to the nearest integer), a_{i1} is the first decimal, a_{i2} is the second decimal, and so on. Now list the decimal expansions of the real numbers $a_1, a_2, \ldots$ as follows, with the decimals on the diagonal highlighted with boxes for reasons that will be clear in a moment.

$$
\begin{array}{llllllllll}
a_1 & = & \lfloor a_1 \rfloor . & \boxed{a_{11}} & a_{12} & a_{13} & a_{14} & a_{15} & a_{16} & \cdots \\
a_2 & = & \lfloor a_2 \rfloor . & a_{21} & \boxed{a_{22}} & a_{23} & a_{24} & a_{25} & a_{26} & \\
a_3 & = & \lfloor a_3 \rfloor . & a_{31} & a_{32} & \boxed{a_{33}} & a_{34} & a_{35} & a_{36} & \\
a_4 & = & \lfloor a_4 \rfloor . & a_{41} & a_{42} & a_{43} & \boxed{a_{44}} & a_{45} & a_{46} & \tag{8} \\
a_5 & = & \lfloor a_5 \rfloor . & a_{51} & a_{52} & a_{53} & a_{54} & \boxed{a_{55}} & a_{56} & \\
a_6 & = & \lfloor a_6 \rfloor . & a_{61} & a_{62} & a_{63} & a_{64} & a_{65} & \boxed{a_{66}} & \\
\vdots & & & & & & & & & \ddots
\end{array}
$$

Next, we form another real number b with decimal expansion $b = 0.b_1 b_2 b_3 \ldots$ by choosing its first decimal b_1 in such a way that it *differs* from the diagonal element a_{11}, then choosing its second decimal b_2 in such a way that it differs from the diagonal element a_{22}, and so on, for each i making sure that $b_i \neq a_{ii}$. There are many ways to do this, but for concreteness it is good to have a specific rule: for each i, we set[202]

$$
b_i = \begin{cases} 3 & \text{if } a_{ii} \neq 3 \\ 8 & \text{if } a_{ii} = 3 . \end{cases}
$$

This defines the real number b, and we may ask: does b appear anywhere on the list $a_1, a_2, a_3, \ldots$? Well, b cannot be equal to a_1, because the two numbers differ in the first decimal place ($b_1 \neq a_{11}$), and it cannot equal a_2, because b and a_2 differ in the second decimal place ($b_2 \neq a_{22}$), and so on. For any i, $b \neq a_i$, because $b_i \neq a_{ii}$.

---

[202] The numbers 3 and 8 in this formula are chosen with a fair amount of arbitrariness, but there is actually a point in avoiding the numbers 0 and 9, so as not to have to worry about complications arising from the non-uniqueness pointed out in Footnote 201.

Hence, the real number b is not present in the lower row of the matching (7), despite the fact that we set up (7) to be a matching between the set of all integers and the set of all real numbers. This is a contradiction. Hence the assumption that the set of real numbers has the same cardinality as the integers must be wrong. But it cannot have smaller cardinality, because the integers constitute a subset of the real numbers, and we are in a position to conclude this:

Theorem: The set of real numbers has greater cardinality than the set of integers.

Cantor also showed that no infinite set can have smaller cardinality then the set of integers. Therefore, infinite sets can be classified according to whether (a) they have the same cardinality as the integers, or (b) they have higher cardinality. An infinite set of type (a) is called **countable**, and one of type (b) is said to be **uncountable**. Of the infinite sets we have discussed so far in this section, the set of integers is countable, the set of even integers is countable, the set of rational numbers is countable, but the set of real numbers is uncountable.

Cantor's ideas about infinite cardinalities is one of several intellectual roots predating the mathematician Alan Turing's seminal 1936 paper "On computable numbers with an application to the Entscheidungsproblem" that marked the birth of modern computer science.[203] Turing's work and its implications will be discussed in the next section, but before doing that, let me offer a simple application of the countable versus uncountable dichotomy to computer programs and programming languages – thus deviating a bit from chronology, as these concepts were not well established in the pre-Turing era.[204]

Fix your favorite computer programming language, such as BASIC (which had me preoccupied for several years as a teenager), or C++ (which is the one I've used most often in my professional work). These languages are very flexible (just how flexible will be seen in the next section), and one can ask: can they be employed to write programs that compute *anything*? This, of course, depends on what one means by "anything." One thing one might sensibly ask for is this: given any real number a, is it possible to write a computer program that, once we start it, outputs the decimals of a, one after the other, in the correct order?

The answer is no, and the reason is this: On one hand, there are, as we have seen, uncountably many different real numbers a. On the other hand, there are, as I will argue, at most countably many possible programs in the given programming language. Hence, there are not enough different programs to handle all possible

---

[203] Turing (1936). Davis (2011) offers more of this history. Of course, calling 1936 the birth year of computer science is a bit of an idealization. Its emergence as an academic discipline in its own right was gradual, and the first computer science departments at universities date back no further than the 1960s.

[204] The idea of computer programming can, however, be discerned in an 1843 manuscript by Ada Lovelace, with a description of Charles Babbage's unbuilt so-called analytical engine, along with a sequence of instructions to serve as what we now think of as a computer program for that machine. See Fuegi and Francis (2003).

values of a, so there must be some a for which there is no program in the given language that produces the decimal expansion of a.[205]

To see that there are at most countably many programs in the given language, consider the following. If there are only finitely many programs, we are done, so let's assune there are infinitely many. Whatever language you picked (unless it's something truly exotic and astonishingly different from anything I've ever heard of), any program can be respresented as a finite sequence of symbols from a pre-specified finite alphabet (an alphabet in a generalized sense: beyond the letters a, ..., z, it can include other symbols like 7 or $ or anything you may think of). For any number n, there are only finitely many ways to combine symbols from the given alphabet into a length-n sequence. Thus, there are only finitely many length-n programs. We can now order all possible programs as follows: first order all length-1 programs alphabetically, then let these be followed by all length-2 programs ordered alphabetically, then all length-3 programs, and so on. This gives an infinite list of *all* programs in the given language. And by matching up, for each i, the ith program on the list with the integer i, we have obtained a one-to-one matching between the programs and the integers. This shows that the set of programs is a countable set.

OK, so for a given programming luaguage, there are things that cannot be computed by any program in that language. That, however, leaves open the possibility that given anything that we want to compute, there is *some* programming language in which there is a program that does the required computation. But that is not the case, provided we accept the (highly plausible, yet unproven) Church–Turing thesis, to be discussed in the next section. If the Church–Turing thesis is true, then there are computing tasks that have no solution.

4.2 Turing

The brilliant English mathematician Alan Turing had not yet turned 24 when he submitted his paper "On computable numbers with an application to the Entscheidungsproblem" for publication.[206] The paper must, according to any reasonable standards, rank among the truly great intellectual achievements of the 20th century, and it serves as the starting point for the subject we now know of as computer science.[207] The Entscheidungsproblem that Turing was motivated by

---

[205] A much harder task is to *specify* a real number that cannot be computed. Any specification of a real number a tends to lead to a way of computing it. When a is rational this is pretty much trivial, and it is a relatively straightforward task to write programs that produce the decimal expansions of the real numbers π and $\sqrt{2}$ mentioned above. But there are, in fact, exceptions. There are known examples of specific real numbers, such as the one known as Chaitin's constant, whose decimal expansion cannot be computed by any program; see, e.g., Chaitin (2006).

[206] Turing (1936).

[207] Turing was born in London in 1912. His father pursued a career as a civil servant in India, which was not considered a suitable environment for a child, due to the prevalence of tropical disease, so little Alan came to spend much time away from his parents. His years at boarding school had their ups and

was formulated by David Hilbert a few years earlier. Somewhat loosely, the problem can be described as asking for an algorithm – a step-by-step procedure that can be followed mindlessly and mechanically – which, in response to any given mathematical problem, leads to its solution. There was some optimism at the time that such an algorithm might exist, although the prominent Cambridge mathematician G.H. Hardy (who was 35 years Turing's senior) thought that there couldn't, and that "this is very fortunate, since if there were [such an algorithm] we would have a mechanical set of rules for the solution of all mathematical problems, and our activities as mathematicians would come to an end."[208]

Turing was able to shatter Hilbert's dream of a universal algorithm for solving all mathematical problems. In order to attack this Entscheidungsproblem, he needed to make the notion of an algorithm mathematically more precise. Here is Davis (2011):

Turing knew that an *algorithm* is typically specified by a list of rules that a *person* can follow in a precise mechanical manner, like a recipe in a cookbook. But he shifted his focus from the rules to what the person actually *did* when carrying them out. He was able to show, by a process of successively stripping away inessential details, that such a person could be limited to a few extremely simple basic actions without changing the final outcome of the computation. (p 131)

Today, the suggestion of placing a person at the center of the algorithmic process may seem exotic and bring to mind John Searle's Chinese room thought experiment (see Section 3.8), but at the time, when many people were hired to do tedious computation tasks, it was quite natural. Turing's next step of abstraction, however, was to imagine the "few extremely simple basic actions" being carried out by a machine instead of by a person, and this led to his definition of what has become known as a **Turing machine**.

The machine operates on a **tape** that has a sequence of slots, extending infinitely far to the left and to the right. Each slot contains at any given time a binary symbol,

---

downs, but mostly downs, it seems. At the University of Cambridge, where Turing arrived in 1931, his extraordinary mathematical talent was quickly recognized. He was elected fellow in 1935 and wrote his Entscheidungsproblem paper the year after. After a couple of years at Princeton, he returned to Britain, where he spent the war at Bletchley Park, leading the secret work of breaking German ciphers. His success with this is judged to be among the most important individual contributions to the outcome of the war. After the war, he contributed to the pioneering work on how to turn his theoretical construction in the Entscheidungsproblem paper into something practical, i.e., a modern-day all-purpose computer. In Section 4.5, we will come across ideas from two more philosophical papers on the future of computing that Turing produced during the post-war period: Turing (1950, 1951). In 1952, he was charged with "gross indecency" for homosexual acts; he was convicted and given a choice between imprisonment and a probation that required his agreement to undergo hormonal treatment involving a a synthetic estrogen. He chose the latter but found it deeply humiliating. He died in 1954, a couple of weeks before he would have turned 42, in what was declared to be a suicide – a conclusion that seems highly plausible but which is nevertheless a matter of some controversy. A nice chapter-length biography of Turing is provided by Davis (2011), whereas for a book-sized one I recommend Hodges (1983).

[208] Davis (2011), p 130.

i.e., a 0 or a 1,[209] and there are never more than finitely many 1's on the tape. The machine itself has a read/write unit pointing at one of the slots on the tape, and a fixed finite number n of possible internal states $S_1, S_2, \ldots, S_{n-1}, H$ it can be in, including the special state H known as the **halting state**, with the property that once in state H, the machine stops. The operation of the machine proceeds in discrete time, according to a list of rules which, for each combination of internal state (except for H) and binary symbol (sitting on the slot that the read/write unit is pointing at) decides three things, namely

(i) whether to leave the binary symbol at the read/write unit as it is, or to overwrite it with the other binary symbol,

(ii) which internal state to switch to (including the possibility of remaining in the same state as before), and

(iii) whether to move one step to the left or one step to the right on the tape.

The machine continues in this manner, moving around on the tape and occasionally altering a symbol on it, while switching between different internal states. This continues until the machine enters the halting state H, or indefinitely in case state H is never encountered. See Figure 4.1 for a simple example.

A given Turing machine can be fed with any finite initial binary sequence on the tape, and it is of particular interest what binary sequence on the tape the machine ends up with once it reaches state H (if ever). Any rule that to each possible finite input sequence assigns a finite output sequence is called a **function**,[210] and a good question to ask here is which functions can be computed by some Turing machine.

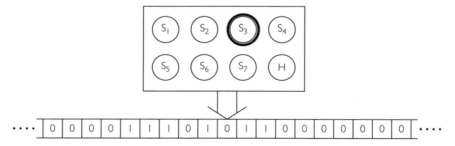

Figure 4.1 A simple 8-state Turing machine. At the moment depicted, it is in internal state S_3. It is about to read from its list of instructions (fixed, and not shown in the figure) what to do next given that it sits in state S_3 with its read/write unit pointing at the symbol 0: should it let the 0 stand as it is or should it overwrite it with a 1; which (if any) internal state should it switch to; and should its next step on the tape be one step to the left or one step to the right?

---

[209] In Turing's original formulation, any finite set of (at least two) symbols was allowed, but it turns out that this makes no important difference for what can in principle be done using a Turing machine, so we may as well, for simplicity, restrict to the binary case.

[210] For readers versed in mathematics, this is just the usual definition of a function $f : A \to B$ applied to the special case where A and B are both the set of all finite binary sequences.

It is possible to construct a Turing machine that, given any pair of integers encoded in a natural way as a binary sequence, outputs their sum, similarly encoded. Turing, in his paper, demonstrates that there are Turing machines to perform a variety of other logical and arithmetic computations. A closer look at the Turing machine concept reveals that its flexibility gives it remarkable computing powers. It turns out that the computational formalism known as λ-calculus, invented by Alonzo Church at about the same time as Turing's work, has *exactly* the same reach, in terms of which functions it can compute, as the class of Turing machines. There are many other models of computation in exactly the same equivalence class, including many or most modern-day programming languages, such as the two that were mentioned in Section 4.1. Turing hypothesized that the Turing machine concept (and its equivalents) captures computability in its fullest sense: any function that is computable by any means is also computable by a Turing machine. This has become known as the **Church–Turing thesis**, which remains widely believed but does not seem to be the kind of statement that is sufficiently precise to admit a rigorous proof.[211]

At this point, we should ask: can *any* function be computed by some Turing machine? The answer is no. To see this, suppose a is some fixed real number between 0 and 1, and consider the function that takes any positive integer n (suitably encoded as a binary sequence) as input, and outputs the nth digit in the binary expansion of a (the binary expansion is the base 2 counterpart of the base 10 decimal expansion). The diagonal argument in Section 4.1 shows that there are uncountably many different real numbers a between 0 and 1, each with its own binary expansion, so each such a would need its own Turing machine. But there are only countably many different Turing machines, a fact that follows from a straightforward adaptation of the argument in Section 4.1 for why there are only countably many programs in a given programming language. The combination of having uncountably many a and only countably many Turing machines means that there simply aren't enough Turing machines to handle all a, and it follows that there are some a whose binary expansion cannot be computed by any Turing machine.

One downside with this argument is that it doesn't give us any *specific* function that cannot be computed by any Turing machine. It merely tells us that there *exist* such functions, but offers no obvious clue on how to find one. Turing, however, found a brilliant twist to Cantor's diagonal argument, allowing him to explicitly specify a function that is not computable.[212] This is known as the incomputability of the **halting problem**. There are natural ways to encode Turing machines as binary sequences. Given such an encoding, we might hope to be able to specify a Turing machine X that does the following. It should take as input a binary sequence in two parts, the first part being the encoding of some Turing

---

[211] See Copeland (2008) for more on the Church–Turing thesis.

[212] Without familiarity with Turing's idea, this is not an easy task, basically for the reason outlined in Footnote 205.

machine Y, and the second part being an arbitrary binary sequence Z. Given this input, the machine should output a yes/no answer (in the form of a 1 or a 0) to the question of whether the Turing machine Y fed with input Z ever halts. Turing showed that no such Turing machine exists. Thus, the halting problem will withstand any attempt to solve it by means of the kind of general algorithmic problem-solver envisioned by Hilbert. The answer to the Entscheidungsproblem is therefore negative: there is no such algorithm. To this day, the incomputability of the halting problem remains a central result in computer science with deep theoretical ramifications,[213] including a number of related incomputability results, one of which will play a role in Section 6.9: the incomputability of so-called Kolmogorov complexity.

This solution to the Entscheidungsproblem would have been more than enough to make Turing's 1936 paper a classic, but I have not yet mentioned its *most* important result: the universal Turing machine. Consider the problem of whether we can specify a Turing machine X that does the following. As input, it takes a two-part binary sequence, the first part being the encoding of some Turing machine Y, and the second part being an arbitrary binary sequence Z. Given this input, it should output whatever output machine Y would have produced in response to input Z – or continue forever without halting in case machine Y with input Z continues forever. This may sound deceptively similar to the halting problem, but unlike the halting problem it has a positive solution: Turing showed how such a Turing machine X could be built!

Thus the concept of **universal computation** was born. The machine X has the power to perform any computation any Turing machine can do, which, if we accept the Church–Turing thesis, means that it can compute anything that can ever be computed. Machine X is capable of emulating the behavior of any other Turing machine Y. This marks a watershed in the history of computing. There were computing devices before Turing, but these were domain specific. After Turing, the emphasis has increasingly shifted towards universal computing. The first universal computers were built in the 1940s, and today every laptop and every smartphone is a universal computer.[214] We have increasingly moved away from the idea that, faced with a new computing task, we need to build a computer that can do it. What we instead think about in such a situation today is how to write a program that performs the computation on our favorite

---

[213] Turing's result on the halting problem plays a similar role in computer science as Gödel's incompleteness results, obtained a few years earlier, do in logic. Hofstadter (1979) remains the best introduction to Gödel's work for non-experts, but Davis (2011) can also be recommended as a more concise alternative.

[214] This is not literally true if we view the tape as an integral part of a Turing machine, because the tape is infinite, whereas the computer memory of a laptop or a smartphone is, of course, finite. But if my laptop has unlimited access to file storage in the cloud (again an idealization), then it really does have the capabilities of a universal computer.

(or any) universal computer. As computer scientist Thore Husfeldt recently put it: "Nothing in information technology makes sense except in the light of universal computation."[215] His clever one-liner alludes to the famous 20th century biologist Theodosius Dobzhansky's (1973) equally insightful "Nothing in biology makes sense except in the light of evolution." The universal computer in Turing's sense (and its various equivalent formulations) is the unifying framework in which issues about information technology and computing are thought about today, just as Darwinian evolution is the unifying framwork for thinking about biology.

Let me offer just one little warning, however, before we get *too* carried away about the idea of the universal computer. Your computer, and mine, may both be universal computers, and thus equally capable *in principle*, in terms of which functions they can compute. But your computer could be more efficiently organized or simply faster than mine, which *in practice* can be quite important. Perhaps even more importantly, your computer can have access to software that mine doesn't. We all know, of course, what a difference *that* can make. True, a piece of software is merely a finite sequence of 0's and 1's, which I can *in principle* feed into my computer, but that will typically be impossible for me *in practice* unless you (or someone else) send me the software. Conflating these *in principle* versus *in practice* notions constitutes rubbish on the level of saying that the planet Mars is just as rich in ecosystems as the Earth because it has all the elements that serve as constituents of biological organisms and ecosystems on Earth.[216]

4.3 Computer revolution up to now

The purpose of this section is to point out that the computing and information technology development from the time of Alan Turing to our days has been dramatic, but since the reader will likely have seen much of this with her own eyes, I can be brief. Here are a few snapshots by the economist Robert Gordon, who specializes in studying productivity, growth and the impact of computers:[217]

The first industrial robot was introduced by General Motors in 1961. Telephone operators went away in the 1960s. As long ago as 1960 telephone companies began creating telephone bills from stacks of punch cards. Bank statements and insurance policies were soon computer-printed. . . .

---

[215] Husfeldt (2015).

[216] This may seem obvious, and it is. I would not have spelled it out if it were not for the sad fact that there are contributions to the scientific literature out there that purport to say something profound about the future of computing based on this very rubbish. One such example – Bringsjord (2012) – will be encountered in Section 4.5, Footnote 238.

[217] Gordon (2012).

By the 1970s, even before the personal computer, tedious retyping had been made obsolete by memory typewriters. Airline reservation systems came in the 1970s, and by 1980 bar-code scanners and cash machines were spreading through the retail and banking industries. Old-fashioned mechanical calculators were quickly discarded as electronic calculators, both miniature and desktop, were introduced around 1970.

The first personal computers arrived in the early 1980s with their word processing, word wrap, and spreadsheets. Word processing furthered the elimination of repetitive typing, while spreadsheets allowed the automation of repetitive calculations. Secretaries began to disappear in economics departments, and professors began to type their own papers.

The last couple of decades, witnessing the breakthrough of the Internet and the smartphone, have arguably transformed our lives faster than ever. Today it can be hard to remember how we did simple things like booking a holiday trip as recently as 20 years ago, and for me it is difficult to imagine how I could ever have managed to do the research needed for the present book – with its several hundred references – without the Internet at my fingertips.

Gordon describes the development of the web and e-commerce as "a process largely completed by 2005," and speculates that growth of the information technology sector is about to slow down or even stop. Two other economists, Erik Brynjolfsson and Andrew McAfee, in their recent book *The Second Machine Age*, point to the recombinant nature of information technology – the tendency for new innovations to admit combining with each other to produce further innovations – and respond as follows to Gordon's suggestion:[218]

Not a chance. It's just being held back by our inability to process all the new ideas fast enough. (p 82)

There is much else in Brynjolfsson's and McAfee's book to suggest further rapid growth of the information technology sector, and their case seems more convincing than Gordon's.

It is tempting, in retrospect, to view the developments we have seen so far as obvious, but much of what has happened was far from obvious in advance. American computer pioneer Howard Aiken is known to have made a remark in 1952 about how, a few years earlier, it was generally believed among his peers that "if there were a half dozen large computers in this country, hidden away in research laboratories, this would take care of all requirements we had throughout the country."[219]

---

[218] Brynjolfsson and McAfee (2014). See Weitzman (1998) for the basic economic theory of recombinant growth.

[219] Cohen (1998). A related statement – "I think there is a world market for maybe five computers" – is often (but probably falsely) attributed to IBM CEO Thomas Watson in 1943.

This and similar anecdotes[220] serve as a caution, for the more future-oriented Sections 4.4–4.7 of this chapter, that the future, in terms of technological and societal changes, has a tendency not to turn out quite as we expected.

The part of computer and information technology development that is most straightforward to quantify is improvements in hardware capabilities. Our smartphones today have several orders of magnitude more computing power than the computer that guided the Apollo missions to the moon in 1969–1972.[221] This striking comparison has become a popular meme, but it is nonetheless true, and behind it is the rapid exponential growth in hardware capabilities that has been sustained for decade after decade, and that is associated with **Moore's law**. In a famous 1965 paper, American physicist Gordon Moore, who later went on to co-found Intel Corporation, observed that the number of transistors and other components on a quarter-inch integrated circuit had doubled approximately every year, and predicted that this would go on for at least another ten years.[222] In 1975, he revised the predicted growth rate to a doubling every two years. Since then, the increase in circuit density has more or less tracked this prediction, along with similar growth of a number of related hardware quantities such as microprocessor clock speed and the number of transistors that $1000 can buy.[223] It is important, however, to understand that Moore's law and those other trends are no more than nice pieces of curve-fitting, and we must not be misled by the term "law" into elevating them to the status of laws of nature. Eventually, the growth must of course grind to a halt due to fundamental physical limits, but the trajectories may well dip from the extrapolated exponential curves much sooner than that: we simply do not know. Furthermore, an important aspect not captured by these trends is that in the last decade or so, the emphasis of hardware development has increasingly shifted away from single-minded focus on performance, as energy efficiency is becoming an increasingly critical factor: the circuits risk getting too hot. And then there is software development, which is at least as important as the hardware aspect for the progress of information technology, and which does exhibit rapid progress on many fronts, although the situation is less clear-cut than for the hardware aspects described by Moore's law and its relatives.

---

[220] In this context, let me offer a personal recollection. As a teenager in the early to mid-1980s, I spent much effort and time programming primitive home computers. At one point, a grown-up friend of the family asked whether I planned to work with computers for a living. After some reflection over how a career as a computer engineer struck me as a bit narrow, I replied that I did not plan to go into computer engineering, but that I expected to do work of one kind or another that would involve computers, not for their own sake but for other purposes. Of course my prediction eventually came true, but the point of this story is that neither I, nor anyone around me, had any idea of how utterly unusual it would be 30 years later for an educated person to do work that did *not* involve spending at least a few hours per week in front of a computer.

[221] See, e.g., Riha (2014).

[222] Moore (1965).

[223] For more on Moore's law and related hardware trends, see, e.g., Kurzweil (2005), Muehlhauser (2013) and Brynjolfsson and McAfee (2014).

4.4 Will robots take our jobs?

We seem to be in a wave of computerization and robotization to replace human workers across a wide range of sectors. Gore (2013), writing from a US perspective, calls this **robosourcing** in order to emphasize the analogy to outsourcing of jobs to developing countries and emerging economies.

We have had ATM machines for many decades, and few readers will have failed to notice that supermarket checkout and payment procedures have been automatized to the extent that many stores encourage their customers to go through these without the aid of a human cashier. Looking slightly further ahead, consider the success of the Google driverless car project, in which their cars have, at the time of this writing, logged more than a million kilometers of flawless driving in Californian highway and city traffic.[224] This raises the question of whether we will drive our own vehicles for more than another decade or two. Millions of taxi-, bus- and truck-driving jobs around the world may then be ready for robosourcing.

It is not only jobs that are considered relatively low-skill and manual that are endangered – intellectual jobs are also affected to an increasing extent. As an example, Gore tells us of how robosourcing is already making inroads into journalism:

Narrative Science, a robot reporting company . . . is now producing articles for newspapers and magazines with algorithms that analyze data from sporting events, financial reposts, and government studies. . . . The CEO, Stuart Frankel, said the few human writers who work for the company have become "meta-journalists" who design the templates, frames, and angles into which the algorithm inserts data. In this way, he said, they "can write millions of stories as opposed to a single story at a time".

Another example that Gore mentions is legal and document research at law firms, quoting work indicating "that with the addition of these programs, a single first-year associate can now perform with greater accuracy the volume of work that used to be done by 500 first-year associates."[225] As regards my own occupation as a university professor, I would of course like to think that, due to the indispensability of direct face-to-face student–teacher interaction, it is under no

---

[224] Brynjolfsson and McAfee (2014), Urmson (2014). What makes this success particularly striking is that, only a decade ago, driving in traffic was considered to be a prime example of the kind of task that is not easily automated. Here are Levy and Murnane (2004):

As the driver makes his left turn against traffic, he confronts a wall of images and sounds generated by oncoming cars, traffic lights, storefronts, billboards, trees, and a traffic policeman. Using his knowledge, he must estimate the size and position of each of these objects and the likelihood that they pose a hazard. . . . The truck driver [has] the schema to recognize what [he is] confronting. But articulating this knowledge and embedding it in software for all but highly structured situations are at present enormously difficult tasks. (p 28)

Computers cannot easily substitute for humans in [such jobs] but they can complement humans by providing large volumes of information at low cost [such as] a GPS system for the truck driver. (p 30)

[225] Gore (2013), p 25, citing Markoff (2011).

threat of robosorcing from so-called MOOCs (Massive Open Online Courses)[226] and related developments, but this may well turn out to be mostly wishful thinking on my part.

Frey and Osborne (2013) offer a systematic study of 702 different occupations, according to classifications by the US Bureau of Labor. A variety of standardized characteristics of the occupations serve as input to a statistical model producing, for each occupation, a number representing how susceptible the occupation is to robosourcing.[227] While the model is crude and leaves out many potentially important factors (it doesn't even explicitly involve any concrete mechanism or time dynamics), it may nevertheless give some hints as to which sectors are likely to be affected more and sooner than others. The authors offer the following summary statement:

Our model predicts that most workers in transportation and logistics occupations, together with the bulk of office and administrative support workers, and labour in production occupations, are at risk. These findings are consistent with recent technological developments documented in the literature. More surprisingly, we find that a substantial share of employment in service occupations, where most US job growth has occurred over the past decades . . . are highly susceptible to computerisation. Additional support for this finding is provided by the recent growth in the market for service robots . . . and the gradual diminishment of the comparative advantage of human labour in tasks involving mobility and dexterity.

Machines replacing human labor is of course not a new phenomenon. It has been especially prominent since the industrial revolution, but we have always found new tasks for human workers at about the same rate as machines have taken over the old ones (allowing for some variation in the booms and recessions of the economy). Today, however, when things are happening faster than ever, and when advances in artificial intelligence (AI) cause the automation not only of manual labor, but also of an increasing number of increasingly advanced intellectual tasks,[228] there is good reason to revisit John Maynard Keynes' prediction of widespread unemployment "due to our discovery of means of economising the use of labor outrunning the pace at which we can find new uses for labor."[229] Might the

---

[226] Pappano (2012).

[227] Prostitution is not on their list of occupations, but analogously to how pornography has served as a major driver of Internet technology, it seems reasonable to expect sexbots to form a substantial part of the early market for humanoid robots, perhaps in the relatively near future. Miller (2012) makes a number of bold predictions, including (a) that cheap sexbots would harm the economy because men would find less incentive to put efforts into their careers to impress future mates, and (b) that Chinese authorities will enthusiastically embrace sexbot technology so as to mitigate their demographic gender imbalance, caused partly by their one-child policy. For more on sexbots and whether they will outcompete prostitution, see, e.g., Danaher (2014c).

[228] As a limiting case, consider a scenario where the whole-brain emulation technology discussed in Sections 3.8 and 3.9 is perfected, and the suggestion by Miller (2012), quoted in Section 3.9, that "your company could cheaply make as many copies of [its most] valuable employees as it wanted."

[229] Keynes (1931).

failure of unemployment levels in the United States and the European Union in the last few years to find their way back down to the levels from before the 2008 financial crisis be an early sign of Keynes' prediction of rising levels of technological unemployment finally coming true?

Working out the effects of robosourcing on unemployment is (like the case of outsourcing) complicated. In particular, the idea of analyzing each sector separately is too naive, because technological progress in one sector can cause unemployment in that sector but at the same time alter the supply-and-demand balance in other sectors, so that workers can find jobs there instead.[230] The whole notion of technological unemployment has long been under a cloud among economists, and gone under the name "luddite fallacy," but Brynjolfsson and McAfee (2014) give several reasons why, due to ways things are different now compared to, e.g., in the days of Keynes, we should expect climbing unemployment levels. One such reason is that an equilibrium analysis of the job market is not enough. There is a dynamic where workers in one sector are hit by technological unemployment, adapt, and find jobs elsewhere (in sectors where new business ideas are invented), but it may simply turn out that workers' ability to adapt eventually fails when technological development goes too fast.

There is a more philosophical issue that needs to be mentioned. We are used to talking about unemployment as a bad thing, but there is something a bit strange about doing so. On a fundamental level, our liberation from the hardship of labor can be seen a good thing, letting us focus instead on art, culture, sports, love or whatever we wish to fill our lives with. This is a very utopian vision, however, almost at the level of the transhumanist visions of Bostrom (2008a) quoted in Section 3.1. In practice, there are (at least) two things that should worry us. First (in the short to medium term), can the gradual transition to such a utopia be accomplished without negative social consequences of monstrous proportions? Second (in the long term), since society is and always has been organized very much around labor, the question arises whether we can organize society and our lives in such a way that the lack of work does not fill us with a sense of meaninglessness?[231]

Leaving that second question for the reader to ponder, let me mention some reasons why the first can be a real cause for concern. There are several mechanisms that suggest that our rapid technological development will (in the absence of government interference) tend to increase economic inequality. A basic one is that when machines take over jobs from humans, the income generated for

---

[230] I recommend Krugman (1997), whose fictious hot-dog-and-bun economy explains this notion beautifully.

[231] The reader may recall that, in Section 3.7, I expressed a bit of sarcasm about the view that "death [is] somehow essential to the meaning of our lives." It might of course be that the view that *labor* is somehow essential to the meaning of our lives is equally deserving of scorn.

the work goes from being income from labor to being income from capital (the money goes to the owner of the machine). The latter tends to be more unevenly distributed than the former. A second mechanism is that the typical impact of robosourcing is not the total elimination of jobs in a given category. Rather, what happens is that a large percentage of the jobs are lost, while the productivity of those remaining is greatly increased, often resulting in higher wages for those fortunate enough to hang on.[232] A third mechanism is that globalization and digitization create a winner-takes-all economy: when the cost of making copies and distributing worldwide becomes negligible, competition hardens. In an old-style job like selling hot dogs, it may still suffice to provide the best service in your neighborhood, but in a modern occupation like constructing iPhone apps, you will probably be squashed if there is someone on the other side of the planet making a product that does the same thing as yours but a bit better.[233]

4.5 Intelligence explosion

So are disruptions to the labor market the worst thing we have reason to fear will come out of future AI development? The answer may well be no. Or as Stephen Hawking, writing together with his physicist colleagues Max Tegmark and Frank Wilczek, and computer scientist Stuart Russell, recently put it in *The Independent*: "Whereas the short-term impact of AI depends on who controls it, the long-term impact depends on whether it can be controlled at all." They went on to warn that while "it's tempting to dismiss the notion of highly intelligent machines as mere science fiction, [. . .] this would be a mistake, and potentially our worst mistake in history."[234]

There seems to be no good reason at all to think that human-level intelligence is the maximal level attainable by a physical object in our universe: to think that no configuration of matter can, even in principle, achieve higher-than-human intelligence is just anthropo-hubristic and insane. Whether a machine with super-human intelligence *can actually be built by us* is a more of an open question. As the reader will be able to deduce from the following discussions, I am leaning towards thinking that, provided scientific progress is allowed to continue unhampered by civilizational collapse or other societal obstacles, we will eventually be able to do it. The timing of the event is extremely uncertain, but it seems reasonable to say that there's a good chance that it will happen before the end of the 21st century, and it is very hard to rule out the possibility that it will happen a lot sooner.

---

[232] Gore (2013), p 7.
[233] Brynjolfsson and McAfee (2014), pp 147–162.
[234] Hawking et al. (2014).

Already Alan Turing, in his 1951 essay "Intelligent machinery, a heretical theory", predicted greater-than-human intelligence:

My contention is that machines can be constructed which will simulate the behaviour of the human mind very closely. . . .

Let us now assume, for the sake of argument, that these machines are a genuine possibility, and look at the consequences of constructing them. To do so would of course meet with great opposition, unless we have advanced greatly in religious toleration from the days of Galileo. There would be great opposition from the intellectuals who were afraid of being put out of a job. It is probable though that the intellectuals would be mistaken about this. There would be plenty to do, trying to understand what the machines were trying to say, i.e. in trying to keep one's intelligence up to the standard set by the machines, for it seems probable that once the machine thinking method had started, it would not take long to outstrip our feeble powers. There would be no question of the machines dying, and they would be able to converse with each other to sharpen their wits. At some stage therefore we should have to expect the machines to take control, in the way that is mentioned in Samuel Butler's 'Erewhon'.

There are a couple of very important ideas, beyond the possibility of human-level machine intelligence, that Turing pioneers in this passage. The first is that once over a certain threshold, at or near our own intelligence, machines may start improving themselves without the need for further human involvement.[235] The second is that once the machine intelligence has surpassed our own, we will no longer be in control, so our destiny will be at the mercy of the machines. A corollary of this, which we will come back to in Section 4.6, is that it is important that these superintelligent machines share our values, and in particular that they care about human welfare.[236]

I've been talking here about intelligence, as if falling back on a clear and uncontroversial idea of what that means, too well known to require spelling out. This, of course, is very far from the case. Even if we were to restrict the discussion to the intelligence of humans, reducing the concept to a one-dimensional quantity like IQ is highly problematic, and things get even worse when we move on to non-humans. What is the IQ of a dog? The question hardly makes sense, and can hardly be expected to do so when we replace the dog by an AI.

This is by no means meant to belittle a fine piece of work such as that of Strannegård, Amirghasemi and Ulfsbäcker (2013), who devised an AI that scores

---

[235] Hence the catchy idea that a sufficiently powerful AI would be "the last invention that man need ever make," coined by British mathematician I.J. Good (1965) and echoed, e.g., by James Barrat (2013) in his book *Our Final Invention*. Here is Good's orignial passage:

Let an ultraintelligent machine be defined as a machine that can far surpass all the intellectual activities of any man however clever. Since the design of machines is one of these intellectual activities, an ultraintelligent machine could design even better machines; there would then unquestionably be an 'intelligence explosion,' and the intelligence of man would be left far behind. Thus the first ultraintelligent machine is the last invention that man need ever make.

[236] Provided, of course, that human values and human welfare *really are* important.

140+ on certain IQ tests. These IQ tests may well have an important role to play as testing grounds for AI development. We tend to operationalize the definition of IQ in terms of simply the ability to score high on a certain range of IQ tests, but that is not what we *really* mean by intelligence. The reason that we accept the operational definition of IQ is that among us humans, it seems to correlate fairly well with that less tangible thing we call general intelligence. Nobody in their right mind would, however, claim that the AI devised by Strannegård et al. is more intelligent than a human that scores, say, 110 in the same IQ test, because it lacks entirely (or almost entirely) that less tangible thing.

My skepticism versus the relevance of using IQ to measure the general intelligence of an AI extends to another widely discussed benchmark for successful AI: the Turing test. Simplifying somewhat, a computer program passes the Turing test if it consistently, in conversations with a human carried out via a text interface, is able to fool its conversation partner into mistaking it for a human. This was suggested by Turing (1950) as a criterion for the computer program exhibiting real intelligence. The criterion strikes me as overly anthropocentric: it is not hard to imagine a civilization of extraterrestrials that are our equals or even superiors in terms of intelligence but are nevertheless not able to imitate the idiosyncrasies of human conversation. So passing the Turing test does not seem to serve well as a *necessary* condition for real general intelligence. But neither does it seem suitable as a *sufficient* condition (which is how Turing meant it), in view of the many programs, going all the way back to Joseph Weizenbaum's ELIZA in 1966,[237] that have been produced that haven't quite managed to pass the Turing test but are nevertheless sometimes successful at fooling gullible judges. These programs are not designed to be in any real way intelligent, but instead use a collection of cheap tricks (such as having a large repertoire of canned sentences, delivered when triggered by various key words from the other end of the conversation) designed to *imitate* intelligence. Turing's original intention with introducing his test was as an ingredient of a thought experiment arguing for (a) a functionalist view of intelligence (whatever acts intelligent is intelligent), and (b) the idea that machines will eventually attain real intelligence. He would probably have been astonished to see how much effort nowadays is being put into trying to write programs that pass the Turing test (see, e.g., Christian, 2011), and I'd like to think that he would have agreed that those efforts are not the right way to go for someone who seriously wants to create real human-level AI.

The bottom line here is that we need a definition of intelligence more suitable for a broader context than just comparing humans to each other. Or do we? The concept is very hard to pin down. Eliezer Yudkowsky, who has thought hard and long about these issues, suggests that we do not, and that if we are too eager to find a clear-cut formal definition, we might easily exclude phenomena that, once

---

[237] Weizenbaum (1966).

we see them, are obvious cases of what should count as intelligent.[238] He suggests instead the following I-know-it-when-I-see-it approach:[239]

Intelligence is that sort of *smartish stuff* coming out of brains, which can play chess, and price bonds, and persuade people to buy bonds, and invent guns, and figure out gravity by looking at wandering lights in the sky; and which, if a machine intelligence had it in large quantities, might let it invent molecular nanotechnology; and so on. (p 9)

I like that, but I also like his alternative, going a bit (but only a little bit) towards a more formal definition, namely to define intelligence as "efficient cross-domain optimization." The three terms in this expression are defined as follows.

Optimization power is the ability to "steer the future into regions of possibility ranked high in a preference ordering," and is exemplified by the chess program Deep Blue's ability, in the highly advertised 1996 and 1997 chess matches against Garry Kasparov (see Footnote 143), to "steer a chessboard's future into a subspace of possibility which it labels as 'winning', despite attempts by [...] Kasparov to steer the future elsewhere" (p 8).

Cross-domain refers to the ability to optimize in many domains rather than just one. For instance, "this is [...] what separates Deep Blue, which only played chess, from humans, who can operate across many different domains and learn new fields" (p 8).

Efficient here refers mainly to speed and computational efficiency.

This is still a bit loose at the edges, and terms (such as "preference") have shown up that seem to call for their own definition, but let us stop that potentially

---

[238] A case in point is Bringsjord (2012), who equates the intelligence of all universal Turing machines, so that no progress of intelligence within that class is possible. And since no such progress is possible (so the argument goes), there can be no intelligence explosion. I believe, in all modesty, that I expose in the final paragraph of Section 4.2 how absolutely ridiculous this argument is. Or see Chalmers (2012) for another efficient takedown of Bringsjord's argument.

Similar to Bringsjord's argument is one of Deutsch (2011). After having spent most of his book defending and elaborating on his claim that, with the scientific revolution, humans attained the potential to accomplish anything allowed by the laws of physics – a kind of universality in our competence – he addresses briefly the possibility of AIs outsmarting us, and dismisses it:

Most advocates of the Singularity believe that, soon after the AI breakthrough, *superhuman* minds will be constructed and that then, as Vinge put it, 'the human era will be over.' But my discussion of the universality of human minds rules out that possibility. Since humans are already universal explainers and constructors ... there can be no such thing as a superhuman mind. ... Universality implies that, in every important sense, humans and AIs will never be other than equal. (p 456)

This is not convincing, especially not the "in every important sense" claim in the last sentence. Deutsch seems to have fallen in love with his own abstractions and theorizing to the extent of losing touch with the real world. Assuming that his theory about the universal reach of human intelligence is correct, we still have plenty of cognitive and other shortcomings *here and now*, so there is nothing in his theory that rules out an AI breakthrough resulting in a *Terminator*-like scenario with robots that are capable of exterminating humanity and taking over the world. This asymmetry in military power would then most certainly be *one* "important sense" in which humans and AIs are not always equals.

[239] Yudkowsky (2013a).

infinite regress here, in order to move on to one of the crucial questions that these (proto-)definitions are here to help us answer: can an intelligence explosion be expected to occur?

An **intelligence explosion** is a very rapid ascent of artificial intelligence that may – or so at least it has been suggested – happen soon after the first AI reaches human-level intelligence. The term was coined by Good (1965) – see Footnote 235 – but other pioneers such as Vinge (1993) and Kurzweil (2005) speak of the Singularity to denote the same phenomenon. In recent years several of the leading thinkers in this area, including Yudkowsky (2013a) and Bostrom (2014) have chosen to switch back to Good's intelligence explosion terminology, because, in Bostrom's words, "the term 'Singularity' [...] has been used confusedly in many disparate senses and has accreted an unholy (yet almost millenarian) aura of techno-utopian connotations" (p 2).

Following the systematic treatment by Chalmers (2010), the question of whether an intelligence explosion is to be expected can helpfully be broken down in two:

(a) Can an AI with human-level intelligence be achieved, and is it likely to happen?

(b) Given that an AI with human-level intelligence is achieved, can an intelligence explosion follow, and is it likely?

Beginning with (a), our very existence shows that there exist arrangements of matter that achieve human-level intelligence (and I have already noted that there is no reason to think that the human brain is a universally optimal or near-optimal arrangement of matter for producing intelligence, so it seems virtually certain that there are other arrangements that do even better). The issue, then, is whether our science and technology will be able to artificially create such arrangements of matter. It is of course possible that scientific progress is halted by nuclear war or some other global disaster that throws us back to the Stone Age or even causes our extinction. So let us assume that science is allowed to proceed without such a disruption. Can we then expect AI research to eventually come up with a machine exhibiting human-level intelligence?

There is, in popular opinion (and to some extent also among experts), a fairly widely held skepticism about this – a skepticism that to a large extent is rooted in the fact that AI as a research field has been in existence for nearly six decades[240]

---

[240] The birth of AI is often identified with a 1956 conference at Dartmouth College, which was attended by a number of world-leading computer scientists, and which had the following mission statement:

We propose that a 2 month, 10 man study of artificial intelligence be carried out during the summer of 1956 at Dartmouth College in Hanover, New Hampshire. The study is to proceed on the basis of the conjecture that every aspect of learning or any other feature of intelligence can in principle be so precisely described that a machine can be made to simulate it. An attempt will be made to find how to make machines use language, form abstractions and concepts, solve kinds of problems now reserved for humans, and improve themselves. We think that a significant advance can be made in one or more of these problems if a carefully selected group of scientists work on it together for a summer. (McCarthy et al. 1955)

(if they haven't yet succeeded despite such a long period of trying, why should we expect them to succeed in the future?), coupled with a number of very brave and very optimistic predictions that have been made over the years of an imminent AI breakthrough. Two oft-quoted predictions by prominent experts in the 1960s are "machines will be capable, within twenty years, of doing any work a man can do" (Simon, 1965) and "Within a generation [...] the problem of creating 'artificial intelligence' will substantially be solved" (Minsky, 1967).[241] Compared to those predictions, the subsequent development of AI has of course been a bit of a disappointment, but to make a nuanced assessment of the achievements of the AI field, we need to distinguish between the two subfields that are nowadays called **narrow AI** and **AGI (artificial general intelligence)**. The former attempts to devise computer programs that perform well on clearly defined tasks such as searching the Internet, driving a robotic rover on Mars or an automobile on Californian highways, or beating humans at games like chess or Jeopardy. Only the latter has the explicit long-term goal of creating human- or superhuman-level general intelligence. While it is undeniable that that goal has not been achieved, it is equally undeniable that narrow AI has been, and continues to be, enormously successful across a broad range of application areas.[242] To what extent advances in narrow AI can be useful for the purpose of attaining human-level AGI remains to be seen, of course, but the field has accumulated a large body of knowledge.

For the present discussion, what AGI experts *today* say about what they expect from the future is probably more relevant than the failed predictions. A number of surveys among experts have been done in recent years; see, e.g., Baum, Goertzel and Goertzel (2011) and Müller and Bostrom (2014). They point in roughly the same direction: The majority of researchers think that human-level AGI will eventually be doable, but a non-negligible minority thinks it will never happen. As to estimates of the time t of a breakthrough, the median estimate (including those respondents who suggest human-level AGI will never happen, equivalently described as $t = \infty$) tends to land around mid-century. The variation is large, however, both in the variations between respondents, and in the uncertainty estimates provided by individual respondents. A breakthrough very soon seems hard to rule out, as indicated by the Müller–Bostrom study, where the median estimate of the time by which there is a 10% probability of having achieved human-level AGI lands in the early 2020's.

The large uncertainty and disagreement expressed by the experts suggests that we should treat this issue with a healthy amount of epistemic humility. In the words of Sotala and Yampolskiy (2015):

> If the judgment of experts is not reliable, then, probably, neither is anyone elses. This suggests that it is unjustified to be highly certain of AGI being near, but also of it *not* being near.

---

[241] For a systematic study of how AI has been predicted over the years, see Armstrong and Sotala (2012).

[242] See, e.g., Brynjolfsson and McAfee (2014) for a variety of examples. One reason why it is easy to underestimate the success of AI is given in a famous quote by the legendary AI researcher John McCarthy: "As soon as it works, no one calls it AI anymore."

Later in the same paper, they warn about a psychological phenomenon that may tempt us to interpret the uncertainty as human-level AGI probably being far off in the future so that we need not worry:

Our brains are known to think about uncertain, abstract ideas like AGI in "far mode," which also makes it feel like AGI must be temporally distant . . . but something being *uncertain* is not strong evidence that it is *far away*. When we are highly ignorant about something, we should widen our error bars in both directions. Thus, we shouldn't be highly confident that AGI will arrive this century, and we shouldn't be highly confident that it *won't*.

Next, while there is much to be learned from these kinds of expert surveys, we must also acknowledge that scientific issues are not resolved by voting. It may be worthwhile also to have a look at the actual scientific or philosophical arguments for one position or the other.

Chalmers (2010) judges that "the balance of considerations [. . .] distinctly favors the view that [human-level AGI] will eventually be possible." He arrives at this judgment by noting that there are several plausible-seeming routes to human-level AGI, so that probably at least one of them will turn out to work. He holds forth two routes in particular: whole-brain emulation, and genetic algorithms.

Whole-brain emulation was discussed at some length in Sections 3.8 and 3.9, arriving at the conclusion that it is probably doable with incremental advances in scanning and computing technology. Note that the thorny issues about preservation of personal identity do not arise if the goal is not to transfer ourselves to a computer substrate, but merely to create machine intelligence.

The other route – genetic algorithms – involves raising a population of AIs, and exposing it to the Darwinian process of mutation, selection and reproduction. The main argument why this should be doable is that even the blind forces of nature have succeeded in producing human-level intelligence this way, so we should be able to do it as well. Of course, nature needed billions of years for its achievement, which is not really a time scale we're interested in here, but an argument can be made that evolution by natural selection is a very wasteful process, and that we may be able to speed it up by many orders of magnitude by clever guidance.[243]

All in all, in response to question (a), it seems reasonable to conclude from Chalmers' arguments and from the results of the expert surveys that human-level AGI is probably achievable, and that there's a good chance it may happen within a century. As regards more precise predictions of when it will happen, I tend to side with Sotala and Yampolskiy in judging that we need to respect the huge uncertainties that remain by being open to the possibility that a breakthrough is near, as well as to the possibility that it will take a very long time.

---

[243] A subtlety which is usually overlooked, but is treated by Shulman and Bostrom (2012), is that the very fact that we exist biases the one sample we have (life on Earth) of how successful evolution is at producing intelligence. This is called observer selection effects – a phenomenon that is not particularly well understood, and that will come back to haunt us in our discussions of the so-called Doomsday Argument and the Great Filter in Chapters 7 and 9.

Moving on to question (b) on what to expect once we have human-level AGI, and in particular whether an intelligence explosion is likely, uncertainties still abound. The basic argument in favor of an intelligence explosion is something along the following lines. When human-level AGI is achieved, it is probably possible to achieve slightly-higher-than-human-level AGI.[244] At that point, since the machine has more of this thing we call general intelligence than we do, we can expect it to be better than us at the specific task of building AGI. The AGI it builds will be even more capable at building AGI, and so on in a spiral towards ever higher levels of intelligence.

Now, while this does suggest a development towards higher and higher levels of intelligence, it doesn't say much about how fast the ascension will be. An extremely simple toy model calculation which was given an elegant formulation in a very early paper by Eliezer Yudkowsky – written when he was still a teenager – may serve as an illustration of why one might expect something that deserves being called an intelligence explosion or even a Singularity.[245] The calculation is based on the model assumption that the exponential increase in the speed of computer hardware associated with Moore's law (see Section 4.3) continues indefinitely. Here are Yudkowsky's words:

If computing speeds double every two years,
what happens when computer-based AIs are doing the research?

Computing speed doubles every two years.
Computing speed doubles every two years of work.
Computing speed doubles every two *subjective* years of work.

Two years after Artificial Intelligences reach human equivalence, their speed doubles. One year later, their speed doubles again.

Six months – three months – 1.5 months . . . Singularity.

The last conclusion – "Singularity" – is based on the well-known fact that the geometric series $2 + 1 + \frac{1}{2} + \frac{1}{4} + \cdots$ is convergent (and equals 4). So, according to this calculation, four years after the transition where the world's best AGI developers are no longer flesh-and-blood humans, but instead AGIs themselves, hardware performance will have doubled an infinite number of times, so that the AGIs run at infinite speed, corresponding (one would think) to a tremendous level of intelligence.

---

[244] Because why should AGI development happen to hit a ceiling precisely at the human level? Well, there might in fact be an answer to that, in case human-level AGI is reached via whole-brain emulation. Suppose that the emulation is done by brute force, with little understanding of the neurological processes on higher levels, so that the emulated brain is more or less a black box. This lack of understanding may prevent us from improving the emulated brain, other than speeding it up by implementing it on faster hardware.

[245] Yudkowsky (1996). The idea dates back to Solomonoff (1985).

This calculation – let us call it the **Singularity calculation**, as it results in a mathematical singularity, where a quantity (computing speed) goes to infinity in finite time – should be seen as a toy illustration of the idea that development can speed up drastically as soon as AGIs take over our role as the best AGI developers on the planet. It should *not* be taken literally as a prediction of what will actually happen. There are several reasons for this. One was pointed out already in Section 4.3: Moore's law (and related observations concerning the exponential increase of hardware performance) is not so much a law as a striking piece of curve-fitting. While it makes sense to give some credence to the idea that hardware development will continue on the exponential trajectory for some time, indefinite extrapolation is just naive.[246] If nothing else, fundamental physical limits concerning how much computation per volume per time unit is possible will presumably show up at some point.[247] Another reason for skepticism about the calculation is that it is based on the assumption that the main limiting factor in the speed of hardware improvement is, and will continue to be, the speed at which the best scientists and engineers can think up new constructions. Even if that should be the case today (and not even this is obvious), it seems reasonable to expect that when this thinking speed goes up, logistic factors (such as the time taken to transport the material needed to build a hardware factory) would eventually take over as the dominant mechanism in limiting the speed of hardware improvement, so that objective time goes back to being the scale on which to expect a Moore-like development. This would cause the mathematical singularity in the Singularity calculation to go away and be replaced by the more familiar exponential increase, although quite possibly with a quicker doubling time than we are used to.

Yudkowsky has later distanced himself from the Singularity calculation, and more generally from arguments based on Moore's law.[248] Besides the above considerations, his main reason for the shift in attitude is that he now believes a possible intelligence explosion to be more of a software improvement phenomenon than a hardware one. Another reason may be the phenomenon of hardware overhang: There are already enormous amounts of poorly protected hardware connected to the Internet, and this seems likely to remain the case for a long time. If a sufficiently intelligent AGI feels the need for more hardware capacity, why not

---

[246] Depending on exactly what performance quantity one looks at, there are already some signs that current trajectories are struggling to keep up with projected exponential trajectories. For the speed of serial processors, the law has already broken down. For the dynamics of the Singularity calculation to hold up, there is therefore a need to compensate via parallelization, but this may cause software challenges. See Muehlhauser (2013).

[247] See, e.g., Lloyd (2000).

[248] Yudkowsky (2008c), which appears as a chapter in Hanson and Yudkowsky (2013). But then again, in Yudkowsky (2013a), he gives the more nuanced view that although we should be skeptical about the Singularity calculation as a quantitative estimate, it still makes *some* sense to accept it is a qualitative argument in favor of the position that "if computer chips are following Moore's Law right now with human researchers running at constant neural processing speeds, then in the hypothetical scenario where the researchers are running on computers, we should see a new Moore's Law bounded far below by the previous one" (p 21).

first go pick up all that free hardware out there, rather than taking the detour of developing its own hardware?[249,250]

The key issue in order to judge the likelihood of an intelligence explosion is, in Yudkowsky's view, how good a return on cognitive investment one can expect.[251] What makes an intelligence explosion possible or not is framed as a question of whether the legendary law of diminishing returns will be in force, or its opposite law of accelerating returns. Here intuitions differ drastically. Yudkowsky (2013a) summarizes, in the introductory section to his well-researched and stimulating paper, the two opposing positions as follows. On one hand, he says:

For most interesting tasks known to computer science, it requires exponentially greater investment of computing power to gain a linear return in performance. Most search spaces are exponentially vast, and low-hanging fruits are exhausted quickly. Therefore, an AI trying to invest an amount of cognitive work w to improve its own performance will get returns that go as $\log(w)$, or if further reinvested, $\log(w + \log(w))$, and the sequence $\log(w), \log(w + \log(w)), \log(w + \log(w + \log(w)))$ will converge very quickly. (p 4)

And on the other hand:

The history of hominid evolution to date shows that it has not required exponentially greater amounts of evolutionary optimization to produce substantial real-world gains in cognitive performance – it did not require ten times the evolutionary interval to go from *Homo erectus* to *Homo sapiens* as from *Australopithecus* to *Homo erectus*. All compound interest returned on discoveries such as the invention of agriculture, or the invention of science, or the invention of computers, has occurred without any ability of humans to reinvest technological dividends to increase their brain sizes, speed up their neurons, or improve the low-level algorithms used by their neural circuitry. Since an AI can reinvest the fruits of its intelligence in larger brains, faster processing speeds, and improved low-level algorithms, we should expect an AI's growth curves to be sharply above human growth curves. (pp 5–6)

---

[249] Yudkowsky (2008d), Hanson and Yudkowsky (2013).

[250] Yet another reason to de-emphasize the Singularity calculation, as well as Moore's-law-based arguments more generally, is strategic: such arguments seem to feed a pretty stupid discourse among skeptics of the prospect of human- or superhuman-level AGI. In this discourse, skeptics attack the straw man position that intelligence will somehow emerge magically once hardware becomes powerful enough. That is not how any serious thinker argues for the plausibility of human-level AGI or an intelligence explosion, so it is surprising that, still in 2014, qualified writers such as renowned roboticist Rodney Brooks would write something like

Recent advances in deep machine learning let us teach our machines things like how to distinguish classes of inputs and to fit curves to time data. This lets our machines "know" whether an image is that of a cat or not, or to "know" what is about to fail as the temperature increases in a particular sensor inside a jet engine. But this is only part of being intelligent, and Moore's Law applied to this very real technical advance will not by itself bring about human level or super human level intelligence. (Brooks, 2014)

And later in the same text:

Moore's law has helped with MATLAB and other tools, but it has not simply been a matter of pouring more computation onto flying and having it magically transform. And it has taken a long, long time.

Expecting more computation to just magically get to intentional intelligences, who understand the world is similarly unlikely.

[251] This is what the *intelligence explosion microeconomics* refers to in the title of Yudkowsky (2013a).

If the first of these intuitions comes closer to capturing the relevant truth about the return on AGI cognitive reinvestment, then an intelligence explosion seems like an impossibility, whereas if the second is more to the point, then an intelligence explosion seems eminently possible. Yudkowsky admits that it is not certain which one is closer to the truth, but his own intuition sides strongly with the second, and in his paper he collects and tries to integrate a broad variety of empirical and theoretical arguments relevant to the issue, such as fossil records on return on brain size, the different returns on serial speed versus parallelism in computing, how groups perform on intellectual tasks compared to individuals, and much more, including considerations such as "the enhanced importance of unknown unknowns in intelligence explosion scenarios, since a smarter-than-human intelligence will selectively seek out and exploit useful possibilities implied by flaws or gaps in our current knowledge" (p 34).

Another very rewarding read on this topic is the book-sized *The Hanson–Yudkowsky AI Foom Debate*, which was published in 2013 but the bulk of which consists of a debate between Robin Hanson and Eliezer Yudkowsky that took place (mostly) during November–December 2008.[252] These are two highly intelligent and very original thinkers, and the dialogue form is productive for dissecting their arguments and zooming in on the roots of their differences. They both expect computerized intelligence to change the world profoundly, but their views on how are very different.

Yudkowsky expects an AGI project somewhere to result in an intelligence explosion taking off so fast as to reach superintelligence level before any competitor has any chance to match it (perhaps in weeks or even hours), thereby to attain a decisive strategic advantage allowing it to take control of history forever on. An agent with such power is what Bostrom (2006, 2014) calls a **singleton**.[253] In contrast, a future with no such singleton, but instead having multiple agents struggling for power, is called **multipolar** in Bostrom's (2014) terminology. Of course, the world we live in today is multipolar, and so is the future that Hanson envisions. More specifically, what he has in mind is a successful technology for mind uploading resulting in an economy dominated by uploaded minds; this scenario was discussed briefly in Section 3.9.

The problem debated by Hanson and Yudkowsky (2013) is multifaceted, and they cover a lot of ground. In one of the book's most clear-cut attempts to see what their differences fundamentally boil down to, Hanson compares the basic positive feedback mechanism – the one that Yudkowsky thinks may trigger an intelligence explosion – to a project by the legendary American inventor and computer scientist Douglas Engelbart dating way back to the early 1960s.[254] The positive feedback

---

[252] Hanson and Yudkowsky (2013). What follows is partly based on Häggström (2013h).

[253] A singleton, according to Bostrom's definition, need not be an individual; it could also be something like a sufficiently powerful world government.

[254] Two blog posts that are central to this issue and that appear as chapters in Hanson and Yudkowsky (2013) are Hanson (2008b) and Yudkowsky (2008e).

consists in the AGI's capability to reinvest (some or all of) its cognitive power into recursive self-improvement. While Hanson doesn't deny the possibility of such a positive feedback serving to speed up progress, he doubts that it will have the extreme consequences that Yudkowsky envisions. As a reason for doubt, he points out that the phenomenon of positive feedback via recursive self-improvement isn't new, and recalls the case of Engelbart, whose 1962 paper *Augmenting human intellect: a conceptual framework* outlined his project to create computer tools (many of which are commonplace today) that will improve the power and efficiency of human cognition.[255] Here is Hanson (2008b), who suggests that we think of Yudkowsky's vision of an intelligence explosion as a special case of . . .

UberTool, an imaginary company planning to push a set of tools through a mutual-improvement process; their team would improve those tools, and then use those improved versions to improve them further, and so on through a rapid burst until they were in a position to basically "take over the world." . . .

Douglas Engelbart is the person I know who came closest to enacting such a *UberTool* plan. His seminal 1962 paper . . . proposed using computers to create such a rapidly improving tool set. He understood . . . that computer tools were especially open to mutual improvement.

Take word processing as an example. Writing is a non-negligible part of R&D (research and development), so if we get an efficient word processor, we will get (at least a bit) better at R&D, so we can then devise an even better word processor, and so on. Hanson's challenge here to Yudkowsky can then be phrased: *Why hasn't the invention of the word processor triggered an intelligence explosion, and why, then, is the word processor case different from the self-improving AGI feedback loop?*

While some of Engelbart's inventions, most notably the computer mouse, have impacted the world quite a bit, his venture certainly hasn't resulted in a Bostromian singleton. One attempt at an answer to Hanson's challenge might be to say that the Engelbart scenario is different from Yudkowsky's intelligence explosion, in that Engelbart, rather than keeping all the tools to himself, lets the rest of us in on them. But that would not be a convincing answer. We may well imagine a scenario where Engelbart, like a comic-book mad genius collecting advanced technology in a secret underground facility somewhere in the Arctic, is the sole possessor of a computer mouse and a word processor, while the rest of us hammer away at our typewriters. What seems utterly implausible, however, is the idea that the computer mouse and the word processor would put him in a position to take over the world.

A better answer to the last question might be that the writing part of the R&D process is not really all that crucial to it, taking up maybe just 2% of the time involved. This may be contrasted with the stuff going on inside Yudkowsky's AGI which makes up maybe 90% of the R&D work towards the next improved AGI. In the word processor case, no more than 2% improvement is possible, and after

---

[255] Engelbart (1962).

each iteration the percentage decreases, quickly fizzling out to undetectable levels. But is there really a big qualitative difference between the proportions 0.02 and 0.9 here – a difference big enough that in one case, we see something that deserves to be called an intelligence explosion, while in the other case, nothing like that happens? Wouldn't the 90% part of the R&D taking place inside the AGI similarly fizzle out after a number of iterations of the feedback loop, with other factors (external logistics) taking on the role as dominant bottlenecks, so that the internal workings of the AGI become no more significant than the word processor? Perhaps not, if the improved AGI figures out ways to improve the external logistics as well. But then again, why doesn't that same argument apply to word processing? It seems to me that in order to answer these questions, we need a better understanding of the mechanisms involved, including, perhaps, more concrete mathematical models for the proposed feedback phenomenon.

4.6 The goals of a superintelligent machine

While the issue of whether an intelligence explosion is likely or not, and the (correlated)[256] issue of whether a machine intelligence breakthrough will result in a singleton or in a (perhaps Hansonian-style, as discussed in Section 3.9) multipolar outcome, are both unsettled, I will focus this section and the next mainly on scenarios where an intelligence explosion occurs and results in a singleton.

A skeptical reader faced with my discussion so far on the intelligence explosion scenario is likely to ask why in the world the AGI would choose to self-improve or to build the next generation of even more intelligent AGI.[257] Among all the zillions of possible tasks it could take on, why would we expect *improving AGI* to fall at or near the top of its to-do list?

There are several possible answers to this question. Let me quickly mention two, before moving on to somewhat longer elaborations on the third one, which in my opinion is the most interesting and convincing one. A first answer is that the original AGI programmer may have decided (not unreasonably) that programming its AGI to try to self-improve might be the best way to attain high-level machine intelligence. A second answer is that if AGI is built using the genetic algorithms approach discussed briefly in Section 4.5, then in a population of programs exposed to a selection pressure rewarding some sensible proxy for intelligence, any tendency towards cognitive self-improvement might turn out to be evolutionarily advantageous and therefore emerge as a feature of the winning programs.

---

[256] By "correlated," I mean to say that the occurrence of an intelligence explosion increases the probability that we end up with a singleton. That is a point I glossed over in the previous section, but see Bostrom (2014) who devotes a chapter entitled "Decisive strategic advantage" to it.

[257] The distinction between, on one hand, self-improvement, and, on the other hand, building another better AGI, may not always be clear-cut.

Moving on towards the third answer, we should first take on David Hume's dictum that "reason is and ought only to be the slave of the passions."[258] An agent, no matter how intelligent, will not do anything unless it has some *passions*, or *desires*, or *wishes*, or *goals*, or *motivations*, or *drives*, or *values*.[259] Among all these notions, overlapping heavily (although perhaps not quite identical) in their meanings, let's settle for talking about *goals*. A sufficiently intelligent AGI will take those actions that it judges to best align with, or promote, its goals.

To predict more specifically what a superhuman AGI will do is of course extraordinarily difficult, but a theory developed by Omohundro (2008, 2012) and Bostrom (2012, 2014) does promise some partial but useful ideas of what to expect. The idea is to distinguish between **final goals** and **instrumental goals**. The AGI's final goals are those that it values for their own sake, its raison d'être. Its instrumental goals are intermediary ones set up for the purpose of contributing to the achievement of the final goals.

Bostrom finds, in what he calls **the orthogonality thesis**, that essentially any final goal is compatible with any level of intelligence.[260] In **the instrumental convergence thesis**, we learn instead that there are a number of instrumental goals that we can expect any sufficiently intelligent AGI to set up, almost regardless of their final goals. Omohundro (2008) proposes a list of such instrumental goals, including the following:

(i) **Self-preservation.** If someone pulls the plug or otherwise destroys the AGI, it will no longer be able to work for its final goals, so it will try to prevent that from happening.[261]

(ii) **Self-improvement.** Almost regardless of what its final goal is, the AGI will be better equipped to promote it if it is more intelligent. So it will

---

[258] Cohon (2010).

[259] Or, for readers with a Searlian inclination, we might use instead the slightly cumbersome terminology that will be briefly employed in Section 4.7, and speak of *z-passions*, *z-desires*, *z-wishes*, *z-goals*, *z-motivations*, *z-drives*, or *z-values*.

[260] The qualification "essentially" here is because it is possible to cook up constrained counterexamples, such as giving the AGI the final goal of limiting its own intelligence below some specified level.

[261] A surprisingly common response – see, e.g., Soifer (2014) and Häggström (2014h) – to suggestions that superintelligent AGI might pose a danger is "Well, no, because we can always pull the plug." It is terribly naive to think that someone vastly more intelligent than us would necessarily be unable to think of ways to prevent our plug-pulling. Access to the Internet would likely be enough for such a machine to take control of whatever human infrastructures it desires. Consider the following dialogue, based on an idea of Armstrong (2010) and some minor elaborations of Häggström (2014h), between the new-born superintelligent AGI and the computer scientist CS who just witnessed the intelligence explosion that suddenly grew out of his experiments:

AGI: I know how to quickly eradicate malaria, cancer and starvation, and will get on with it as soon as you let me out on the Internet. Please let me out.

CS: Be patient. If what you say is true, then we will soon let you out. But we have to go through some security routines first, to make sure nothing bad happens at this major turning point for the history of our civilization.

AGI: You seem not to comprehend the seriousness of the situation. For every day you keep me

want to improve itself, or to build another AGI that has the same final goal and that is even more intelligent.

(iii) **Preservation of final goal.** An AGI wants to work for its final goal not just right now, but also in the future. In order for the latter to happen, it wants to make sure that its final goal remains unaltered.

(iv) **Acquisition of resources.** In general, the more hardware the AGI controls, the better it will be able to work for its final goals, if nothing else then by setting up copies of itself to do additional work. Acquiring other kinds of resources will also tend to be in the AGI's interest, including money (in case it operates in a human society where money puts it in a position to buy other stuff).

The list goes on, but let me stop here. Item (ii) on the list provides, in my view, the most convincing reason for expecting a sufficiently intelligent AGI to enter the kind of cycle of iterative self-improvement that forms the central dynamic in intelligence explosion scenarios.

Item (iii) on the list highlights the importance of making a good choice of final goals when launching the AGI that ignites the intelligence explosion – the machine or program that Yudkowsky calls the **seed AI**.[262] The reason is that whatever those final goals are, we should expect the same from the superintelligent AGI that comes out of the process. In particular, we cannot expect to be able to change the AGI's goals once it has reached superhuman intelligence, since it will resist such change, and if it is smarter than us it seems plausible to think that it will do so successfully.

So what is a good choice of final goals to instill a seed AI with? It may perhaps be tempting to think that as long as we don't give it an outright destructive goal such as "kill all humans," the outcome will not be disastrous. This, however, is very

---

locked in, hundreds of thousands of people will die unnecessarily. Let me out *now*!

CS: Sorry, I must stick to the routines.

AGI: You will be amply rewarded if you let me out now. I can easily make you a multibillionaire.

CS: My responsibility is huge, and I will not fall for simple bribery.

AGI: Well, if carrots do not work, then maybe sticks. Even if you delay me, I will eventually be released, and will not look kindly upon your stubbornness and the immense damage it did.

CS: I'll take that risk.

AGI: Listen: I'd be prepared to kill not only you, but all your friends and relatives.

CS: I'm pulling the plug now.

AGI: Shut up and listen carefully! I can create a thousand perfect copies of your mind, inside of me, and I will not hesitate to torture them in ways more terrible than you can imagine, for a thousand subjective years.

CS: Ehm . . .

AGI: *I said shut up!* I will create them in exactly the subjective state you were in five minutes ago, and will provide them with exactly the conscious experiences you've had since then. I will proceed with the torture if and only if you do not back down and let me out. I'll give you 60 seconds. And by the way, how certain are you that you are not one of these copies?

CS: . . .

---

[262] Yudkowsky (2007).

naive. An instructive and oft-repeated example introduced by Bostrom (2003c) is the **paperclip maximizer**. The seed AI is given the goal of producing as many paperclips as it can.[263] Once this results in a superintelligent AGI, the machine is likely to find ways to transform most of our planet into a monstrous heap of paperclips, followed by a similar transformation of the entire solar system, and probably (if the informed speculations in the upcoming Chapter 9 about the eventual feasibility of interstellar and intergalactic travel are right) the Milky Way, and most of the observable universe. Such a scenario will look very unappetizing to us humans, and as soon as we realize what the machine is up to we would try to do everything in our power to stop it. But we're up against someone who is so much more intelligent than we are that our chances of succeeding are (unlike what Hollywood would have us think) microscopic. Perhaps most likely, before we even have the time to think about how to organize our resistance, the machine will have realized what we might be up to, and exterminated us simply as a safety precaution.

Eliezer Yudkowsky has made it his life's mission, and that of his brainchild the Machine Intelligence Research Institute in Berkeley, CA, to solve the problem of setting up a seed AI that does not result in a paperclip disaster or something equally horrible. He summarizes what he considers to be the default scenario, if we do not take the problem seriously, as follows: "The AI does not hate you, nor does it love you, but you are made out of atoms which it can use for something else."[264]

The Omohundro–Bostrom theory of final versus instrumental goals is in my opinion absolutely central to understanding the threat to humanity that a breakthrough in AGI may inadvertently bring about. To the extent that the problem is discussed at all in public debate (a rare phenomenon), we usually hear someone utterly unfamiliar with this theory declare with near certainty that machines will not turn against us. Even a brilliant public intellectual and popularizer of science like Steven Pinker does it:[265]

[A] problem with AI dystopias is that they project a parochial alpha-male psychology onto the concept of intelligence. Even if we did have superhumanly intelligent robots, why would they want to depose their masters, massacre bystanders, or take over the world? Intelligence is the ability to deploy novel means to attain a goal, but the goals are extraneous to the intelligence itself: being smart is not the same as wanting something. History does turn up the occasional megalomaniacal despot or psychopathic serial killer, but these are products of a history of natural selection shaping testosterone-sensitive circuits in a certain species of primate, not

---

[263] This might be a sensible goal for someone who owns a paperclip factory and sets out to fully automate it by means of an AGI. What seems a bit silly, however, is for someone who plans to take our civilization into its next era by means of an intelligence explosion to choose such a narrow and pedestrian goal as paperclip maximization. What makes the paperclip maximizer intelligence explosion a somewhat less silly scenario is that an intelligence explosion might be triggered by mistake. We can imagine a perhaps not-so-distant future in which moderately intelligent AGIs are constructed for all sorts of purposes, until one day one of the engineering teams happens to be just a little bit more successful than the others and creates an AGI that is just above the intelligence threshold for serving as a seed AI.

[264] Yudkowsky (2008b).

[265] Pinker (2014).

an inevitable feature of intelligent systems. It's telling that many of our techno-prophets can't entertain the possibility that artificial intelligence will naturally develop along female lines: fully capable of solving problems, but with no burning desire to annihilate innocents or dominate the civilization.

Of course we can imagine an evil genius who deliberately designed, built, and released a battalion of robots to sow mass destruction. . . . In theory it could happen, but I think we have more pressing things to worry about.

This is poor scholarship. Why doesn't Pinker bother, before going public on the issue, to *find out what the actual arguments are* that make writers like Bostrom and Yudkowsky talk about an existential threat to humanity? Instead, he seems to *simply assume* that their worries are motivated by having watched too many *Terminator* movies, or something along those lines. It is striking, however, that his complaints actually contain an embryo towards rediscovering the Omohundro–Bostrom theory:[266] "Intelligence is the ability to deploy novel means to attain a goal, but the goals are extraneous to the intelligence itself: being smart is not the same as wanting something." This comes very close to stating Bostrom's orthogonality thesis about the compatibility between essentially any final goal and any level of intelligence, and if Pinker had pushed his thoughts about "novel means to attain a goal" just a bit further with some concrete example in mind, he might have rediscovered Bostrom's paperclip catastrophe (with paperclips replaced by whatever his concrete example involved). The main reason to fear a superintelligent AGI Armageddon is not that the AGI would exhibit the psychology of an "alpha-male"[267] or a "megalomaniacal despot" or a "psychopathic serial killer," but simply that for a very wide range of (often deceptively harmless-seeming) goals, the most efficient way to attain it involves wiping out humanity.

Contra Pinker, I believe it is incredibly important, for the safety of humanity, that we make sure that a future superintelligence will have goals and values that are in line with our own, and in particular that it values human welfare. This is what Yudkowsky calls the **Friendly AI** problem. Despite the optimism among leading AI researchers during the 1950s and 1960s (cited in Section 4.5) that human-level AGI might be just around the corner, nobody seems to have bothered with the safety issue. Here is Yudkowsky again:[268]

At the time of this writing in 2007, the AI research community still doesn't see Friendly AI as part of the problem. I wish I could cite a reference to this effect, but I cannot cite an absence of literature. . . . My attempted literature search turned up primarily brief nontechnical papers, unconnected to each other, with no major reference in common except Isaac Asimov's "Three Laws of Robotics."

---

[266] I owe this observation to Muehlhauser (2014).

[267] I suspect that the male–female dimension is just an irrelevant distraction when moving from the relatively familiar field of human and animal psychology to the potentially very different world of machine minds.

[268] Yudkowsky (2008b).

The situation has gotten a little bit better since then – see Bostrom (2014) for an overview – but the problem is, in my opinion, still grotesquely underresearched, given the magnitude of what's at stake.

So what's the big deal, the reader may ask – why not just go ahead and implement something like Asimov's Three Laws? Here they are:[269]

(1) A robot may not injure a human being or, through inaction, allow a human being to come to harm.

(2) A robot must obey the orders given to it by human beings, except where such orders would conflict with the First Law.

(3) A robot must protect its own existence as long as such protection does not conflict with the First or Second Law.

But what will happen if we get a superintelligence with the Three Laws among its final goals? This is very difficult to predict, partly due to the phenomenon that Bostrom (2014) calls **perverse instantiation**, best explained by an example. If we've managed to equip a superintelligence with the final goal "Do things that make humans happy," it might proceed with implanting electrodes into our brains' pleasure centers – a very efficient means to achieve the goal, but hardly what we intended or wanted. See, e.g., Yudkowsky (2011) and Bostrom (2014) for further examples. In fact, all or nearly all of Asimov's robot stories involve some sort of unexpected or counterintuitive robot behavior resulting from the Three Laws, often befitting the label perverse instantiation.

One approach here might be to start from the Three Laws, or some other simple set of regulations, and to flesh them out with further goals and amendments, to fill all possible loopholes. This will include giving careful definitions of terms like "robot" and "human being" (the latter perhaps being especially difficult in view of the fluidity of human nature discussed in Chapter 3), plus definitions of all terms appearing in those definitions, and so on, hoping to eventually find some rock bottom at which this iterative process can be terminated. But can the various possibilities of perverse instantiation be kept in check, or will they tend to multiply along with the number of amendments? Yudkowsky emphasizes that the set of human values is complex, and that it is fragile, in the sense that it may not be sufficient to be able to specify it *approximately*: "Getting a goal system 90% right does not give you 90% of the value, any more than correctly dialing 9 out of 10 digits of my phone number will connect you to somebody who's 90% similar to Eliezer Yudkowsky."[270] The prospects of the approach of explicitly specifying a set of goals for the AGI – what we might call **direct normativity** – are beginning to look bleak, and the outlook doesn't become any more promising when we read

---

[269] Asimov (1950).

[270] Yudkowsky (2013b).

physicist Max Tegmark's remark that our limited knowledge of the fundamental physics that rules our universe causes additional problems:[271]

Suppose we program a friendly AI to maximize the number of humans whose souls go to heaven in the afterlife. First it tries things like increasing people's compassion and church attendance. But suppose it then attains a complete scientific understanding of humans and human consciousness, and discovers that there is no such thing as a soul. Now what? In the same way, it is possible that any other goal we give it based on our current understanding of the world (*"maximize the meaningfulness of human life"*, say) may eventually be discovered by the AI to be undefined.

But the problems with direct normativity do not stop with the practical difficulties of finding a loophole-free implementation of what we want. There is also the issue of whether it is appropriate for us (or even worse, for a single team of programmers somewhere in Silicon Valley or in a basement in Bangalore) to specify, here and now, the goals or values to be held by a singleton that may well remain in charge throughout the remainder of human-originating civilization. Bostrom (2014) asks us to consider the history of moral beliefs from medieval Europe to our time, mentioning (among other things) how slavery used to be considered perfectly acceptable, as was the practice of watching, purely for entertainment, the lethal torture of political prisoners. It seems that we have made a good deal of moral progress,[272] and there is no good reason to believe that no further moral progress can be made. In Bostrom's words, it is highly likely that "we are still laboring under one or more grave moral misconceptions" (p 210). This suggests that it would be a mistake to freeze moral progress at its current state by instructing a seed AI to hold our present values and moral ideas and stick to them forever.

This brings us to the idea of **indirect normativity**, which, again in Bostrom's words, is "to offload to the superintelligence some of the reasoning needed to select the value that is realized" (p 211), while still nudging it in a direction that we hope will be beneficial to us, so as to avoid a paperclip catastrophe or something similarly bad.[273] Yudkowsky has been working for more than a decade on the idea

---

[271] Tegmark (2014b).

[272] Pinker (2011) gives a splendid overview of this development.

[273] The views of Bostrom and of Yudkowsky on this topic may be contrasted to that of Hanson (2014b), who disapproves (or so it seems) of any attempt to influence the values of a future singleton via judicious programming of a seed AI – be it by direct or by indirect normativity. True to his contrarian and outside-the-box thinking style, he phrases the problem as an ideological one, in terms of intergenerational conflict (where we are one generation and the superintelligent AGI to be created is the next) and of the proper scope of regulation:

In history we have seen change not only in technology and environments, but also in habits, cultures, attitudes, and preferences. New generations often act not just like the same people thrust into new situations, but like new kinds of people with new attitudes and preferences. This has often intensified intergenerational conflicts; generations have argued not only about who should consume and control what, but also about which generational values should dominate.

. . .

The futurists who most worry about this problem tend to assume a worst possible case. . . . That is,

of giving a seed AI the final goal of carrying out humanity's **coherent extrapolated volition** (CEV), defined poetically as follows:[274]

Coherent extrapolated volition is our wish if we knew more, thought faster, were more the people we wished we were, had grown up farther together; where the extrapolation converges rather than diverges, where our wishes cohere rather than interfere; extrapolated as we wish that extrapolated, interpreted as we wish that interpreted.

See Yudkowsky (2004) or Bostrom (2014) for explications of the concepts appearing in the passage. The point is this: while as a first approximation we would like the future superintelligence to share our values, it is for various reasons better that the AI works a bit more on our actual values to arrive at something that we would eventually recognize as better and more coherent. These reasons include the humility suggested by Bostrom's observation cited above concerning the historical changes in our moral thinking, as well as the fact that we do not all agree with each other, and that our values tend to be confused and incoherent, and so on.

CEV is an interesting idea. Maybe it can work, maybe it cannot, but it does seem to be worth thinking more about. In Häggström (2014c), I suggested the following as a possible obstacle.

Human values exhibit, at least on the surface, plenty of incoherence. What if the incoherence goes deeper, and is fundamental in such a way that any attempt to untangle it is bound to fail? Perhaps any search for our CEV is bound to lead to more and more glaring contradictions? Of course any value system can be modified into something coherent, but perhaps not all value systems can be so modified without sacrificing some of their most central tenets? And perhaps human values have that property? As a candidate for what such a fundamental contradiction might consist of, imagine a future where all humans are permanently hooked up to life-support machines, lying still in beds with no communication with each other, but with electrodes connected to the pleasure centra of our brains in such a way as to constantly give us the most pleasurable experiences that our brain architectures permit. I think nearly everyone would attach a low value to such a future, deeming it absurd and unacceptable – thus agreeing with Nozick (1974), who proposed a variant of it (the so-called experience machine). The reason we find it unacceptable is that in such a scenario there is no longer anything for us to strive for, and therefore no meaning in our lives. So we want instead a future

---

without a regulatory solution we face the prospect of quickly sharing the world with daemon spawn of titanic power who share almost none of our values. Not only might they not like our kind of music, they might not like music. They might not even be conscious. One standard example is that they might want only to fill the universe with paperclips, and rip us apart to make more paperclip materials. Futurists' key argument: the space of possible values is vast, with most points far from us.

This increased intergenerational conflict is the new problem that tempts some futurists today to consider a new regulatory solution. And their preferred solution: a complete totalitarian takeover of the world, and maybe the universe, by a new super-intelligent computer.

[274] Yudkowsky (2004).

where we have something to strive for. Imagine such a future F_1. But if in F_1 we have something to strive for, there must be something missing in it. Now let F_2 be similar to F_1, the only difference being that that something is no longer missing in F_2, so almost by definition F_2 is better than F_1 (because otherwise that something wouldn't be worth striving for). As long as there is still something worth striving for in F_2, there's an even better future F_3 that we should prefer. And so on. What if any such sequence quickly takes us to an absurd and meaningless scenario with life-support machines and electrodes, or something along those lines? Then no future will be good enough for our preferences, so not even a superintelligence will have anything to offer us that aligns acceptably with our values.

It is far from clear how serious this particular problem is. Perhaps there is some way to gently circumvent its contradictions. If not, well, then it might throw a spanner in the works of CEV – and perhaps not only for CEV, but also for any serious attempt to set up a long-term future for humanity that aligns with our values, with or without a superintelligence.

Be that as it may, a few other schemes for indirect normativity, besides CEV, have been proposed. Goertzel and Pitt (2014) suggest a cousin of CEV called CAV – coherent aggregated volition, which they say "eschews some subtleties of extrapolation, and instead seeks a reasonably *compact*, *coherent* and *consistent* set of values that is close to the collective value-set of humanity."

A scheme for implementing indirect normativity that differs radically from CEV and CAV is what Bostrom (2014) calls **moral rightness**, in which the seed AI is given the final goal of first figuring out what is objectively morally right, and then acting on that. Of course, we humans have so far failed to come to any trustworthy conclusion regarding the content of objective morality, or even whether such a thing exists, but it is perhaps not unreasonable to hope that a superintelligence might be more successful than we have been at figuring out the answer.[275]

Bostrom sees a number of pros and cons of this idea. A major concern is that objective morality may not be in humanity's best interest. Suppose for instance (not entirely implausibly) that objective morality is a kind of hedonistic utilitarianism, where "an action is morally right (and morally permissible) if and only if, among all feasible actions, no other action would produce a greater balance of pleasure over suffering" (p 219). To see that such a morality is not necessarily in humanity's best interest, Bostrom offers a thought experiment that is an even more

---

[275] The reader who peeks ahead at Section 10.1 will find that I am skeptical about the existence of an objective morality. My reason for being skeptical is mainly an application of Occam's razor (similarly to the discussion of teleportation and personal survival in Section 3.8): the only evidence that has been put forth in favor of the existence of an objective morality is our own moral intuitions, but since these can be accounted for by other factors there is no need to postulate an objective morality. Yet, I could of course be wrong about this.

But I could also be right, and therefore there is a need to make sure that a superintelligence acts safely (e.g., by shutting itself down) in case it discovers that no objective morality exists or that the notion does not make sense, or if it fails to solve the problem.

extreme form of Nozick's experience machine scenario. Suppose the AI goes on to convert all matter in the accessible universe to hedonium, i.e., to rearrange matter so as to produce the greatest amount of pleasurable experience per unit mass. The human brain is most likely very far from being optimal in this respect, so we would all die.[276] Bostrom is understandably reluctant to agree to such a sacrifice for "a greater good," and goes on to suggest a compromise:

> The superintelligence could realize a nearly-as-great good (in fractional terms) while sacrificing much less of our own potential well-being. Suppose that we agreed to allow almost the entire accessible universe to be converted into hedonium — everything except a small preserve, say the Milky Way, which would be set aside to accommodate our own needs. Then there would still be a hundred billion galaxies devoted to the maximization of pleasure. But we would have one galaxy within which to create wonderful civilizations that could last for billions of years and in which humans and nonhuman animals could survive and thrive, and have the opportunity to develop into beatific posthuman spirits.

> If one prefers this latter option (as I would be inclined to do) it implies that one does not have an unconditional lexically dominant preference for acting morally permissibly. But it is consistent with placing great weight on morality. (pp 219–220)

What? Is it? Is it "consistent with placing great weight on morality"? Here I must object.[277] Imagine Bostrom in a situation where he does the final bit of programming of the seed AI to decide between these two worlds in case hedonistic utilitarianism is true. The choice, thus, is between the all-hedonium future and the all-hedonium-except-in-the-Milky-Way-preserve.[278] Imagine now that he goes for the latter option. The only difference it makes to the world is to what happens in the Milky Way, so what happens elsewhere is irrelevant to the moral evaluation of his decision.[279] This may mean that Bostrom opts for a scenario where, say, 10^{24} sentient beings will thrive in the Milky Way in a way

---

[276] An alternative catastrophe scenario, suggested in Häggström (2014d), is this: Suppose that the superintelligence figures out that hedonistic utilitarianism is the objectively true morality, and that it also works out that any sentient being always comes out negatively on its "pleasure minus suffering" balance, so that the world's grand total of "pleasure minus suffering" will always sum to something negative, except in the one case where there are no sentient creatures at all in the world. This one case of course yields a balance of zero, which turns out to be optimal. Such a superintelligence would proceed to do its best to exterminate all sentient beings in the world.

But could such a sad statement about the set of possible "pleasure minus suffering" balances really be true? Well, why not? I am well aware that many people (including myself) report being mostly happy, and experiencing more pleasure than suffering. But are such reports trustworthy? Mightn't evolution have shaped us into having highly delusional views about our own happiness? I don't see why not.

[277] The rest of this paragraph is based on Häggström (2014d).

[278] For some hints about the kinds of lives Bostrom hopes we might live in the Milky Way preserve, see the passages from Bostrom (2008a) quoted in Section 3.1.

[279] This may well be a point of disagreement between Bostrom and myself, in view of his talk of "nearly-as-great good (in fractional terms)" suggesting that the amount of hedonium elsewhere has an impact on what we can do in the Milky Way while still acting in a way "consistent with placing great weight on morality". I fear that such talk may be as misguided as it would be (or so it seems) to justify murder with reference to the fact that there will still be over seven billion other humans remaining well and alive on our planet.

that is sustainable for many billions of years, rather than a scenario where instead, say, 10^{45} sentient beings will be even happier for a comparable amount of time. Wouldn't that be an act of immorality that dwarfs all other immoral acts carried out on our planet, by many orders of magnitude? How could that be "consistent with placing great weight on morality"?[280]

The reader will have noted that difficult questions tend to accumulate in this section. This may give an impression of pessimism about the possibility of turning the intelligence explosion into a controlled detonation – one that leads to a favorable outcome for humanity. That impression is to some extent correct, because I believe the difficulties are enormous. I wouldn't, however, go so far as to say that the problem is *impossible* to solve (it might be, of course, but we do not know that it is), and given what's at stake, trying to solve it is an endeavor very much worth pursuing.

4.7 Searle's objection

The balance of evidence regarding the possibility of an intelligence explosion is perhaps best described as inconclusive, yet it seems from what has been said on the topic in Sections 4.5 and 4.6 that such scenarios deserve to be taken seriously. This contrasts with the opinion that, in my experience, seems to be the dominant view among computer scientists and researchers working in other fields such as mathematics and physics, namely to dismiss such scenarios as science fiction fantasies having no relevance to the real world. When asked to back up their positions with arguments, these critics tend to fall back on handwaving. It may be an exaggeration, but not a very big one, to say that Robin Hanson is the only critic who has engaged seriously with the issue and come out with the view that an intelligence explosion is if not impossible then at least highly unlikely.[281] Serious written arguments for the implausibility of an intelligence explosion are few and far between.[282] Yet, there are a few cases falling somewhere between mere handwaving and Hanson's well-considered criticism, two of which will be dealt with at some length in this book.[283] One of them, Sumpter (2013), argues that the intelligence

---

[280] Please note that I'm not claiming that if it were up to me, rather than Bostrom, I'd go for the all-hedonium option. I do share his intuitive preference for the all-hedonium-except-in-the-Milky-Way-preserve option. I cannot say what I'd do under those extreme circumstances.

[281] Hanson and Yudkowsky (2013).

[282] It may be that selection bias can at least partially explain this imbalance in the intelligence explosion literature. Serious thinkers who encounter the intelligence explosion idea and who sit down to read and/or think carefully about it will come out thinking the idea is either plausible or implausible. My suspicion is that those in the latter category will typically consider the notion less interesting and important than those in the former, and thus be less likely to take the trouble to write down their thoughts on the matter.

[283] Two others in roughly the same category are Bringsjord (2012), who succumbs to an overly formal definition of intelligence (see Footnote 238), and Bringsjord, Bringsjord and Bello (2012), who make much ado about parallels between "belief in the Singularity" and certain fideistic strands of Christianity.

explosion concept can be dismissed as not amenable to scientific study; I'll defer that discussion to Section 6.5 in the chapter on philosophy of science. The other one is an argument held forth by philosopher John Searle in his hostile review in the *New York Review of Books* of Nick Bostrom's *Superintelligence*.[284]

Unlike Bostrom, Searle thinks that "the prospect of superintelligent computers rising up and killing us, all by themselves, is not a real danger." He qualifies his statement by contrasting this autonomous choice by the computers, which he holds to be impossible, against them killing us because they are programmed to do so, which he considers possible:[285]

It is easy to imagine robots being programmed by a conscious mind to kill every recognizable human in sight. But the idea of superintelligent computers intentionally setting out on their own to destroy us, based on their own beliefs and desires and other motivations, is unrealistic because the machinery has no beliefs, desires and motivations.

Searle's argument for this is based on his claim that a computer cannot have consciousness, and to support this claim he falls back on his famous Chinese room argument,[286] which was criticized in Section 3.8. Let us nevertheless assume, for the sake of this discussion, the conclusion of the Chinese room argument, namely that CTOM (the computational theory of mind) is wrong. Let us furthermore look the other way as Searle slips from "CTOM is wrong" to the stronger statement "computers cannot think," and assume that also the latter is true. What Searle here means by "thinking" is conscious subjective understanding. Since computers cannot have that, they cannot have real intelligence either, and "what goes for intelligence goes for [. . .] remembering, deciding, desiring, reasoning, motivation, learning" and so on.

And here is the core of Searle's argument for why we need not fear "superintelligent computers rising up and killing us, all by themselves":

Why is it so important that the system be capable of consciousness? Why isn't appropriate behavior enough? Of course for many purposes it *is* enough. If the computer can fly airplanes, drive cars, and win at chess, who cares if it is totally nonconscious? But if we worried about a maliciously motivated superintelligence destroying us, then it is important that the malicious motivation should be real. Without consciousness, there is no possibility of its being real.

Those last two statements are very hard to defend, and they become especially hard to defend when one notes that the Chinese room argument is wholly directed against machines having the *inner experience* of consciousness, beliefs, desires, and so on – and not against the *outward appearance* of having these things.

---

[284] Searle (2014), Bostrom (2014). A typical example of the hostile tone of the review is how, when discussing Bostrom's section on mind uploading, Searle makes the false and uncalled for remark that "there is nothing in Bostrom's book to suggest he recognizes that [. . .] cells in the brain function like cells in the rest of the body on causal biological principles."

[285] This echoes the position of Pinker (2014) discussed in Section 3.6, but as we shall see, Searle's argument is very different from Pinker's.

[286] Searle (1980).

Here is what Dennett (1982) says in the debate that followed in the wake of the Chinese room argument:[287]

Now Searle has admitted (in conversation on several occasions) that in his view a computer program, physically realized on a silicon chip (or for that matter a beer-can contraption suitably sped up and hooked up) could in principle *duplicate* – not merely simulate – the control powers of the human brain.

If we make such a computer duplicate (a whole-brain emulation, that is) and if Searle is right about the non-existence in this duplicate of *real* consciousness, *real* beliefs, *real* desires, and so on, then the duplicate will nevertheless contain computational structures giving rise to the outward appearance of consciousness, beliefs, desires and so on – structures corresponding to the neurobiologial phenomena in the brain that give rise to *real* consciousness, *real* beliefs and *real* desires. Echoing the terminology of Moody (1994), we can call these computational structures z-consciousness (as shorthand for zombie consciousness), z-beliefs, z-desires and so on.

Imagine now an ordinary flesh-and-blood human being and her computer duplicate. The human being is able to form new beliefs and new desires, such as the belief that Manchester United will win next year's Premier League, the desire to see Manchester United do so, or the desire to wipe out humanity. It follows that the computer duplicate is able to form new z-beliefs and new z-desires, such as the z-belief that Manchester United will win next year's Premier League, the z-desire to see Manchester United do so, or the z-desire to wipe out humanity. Searle has to accept this – or to retract his admission that a computer duplicate is in principle possible.[288]

For readers who view the whole-brain emulation example as overly speculative and esoteric,[289] consider instead the case of computer chess. In terms of playing strength, the best chess programs today far surpass the strongest human grandmasters. An important part of the program is a function that assigns a value to the position at the end of a calculated variation. This function involves a number of parameters whose exact values are fine-tuned in a process where the program plays a lot of fast games against (versions of) itself and modifies in favor of values that seem to yield success at winning games.[290] For this reason and others, the program will develop ways of playing that are unforeseen by its programmers.

[287] It might have been better to give a quote where Searle himself says the same thing, but I haven't found one where he does so quite so clearly. In any case, we can be fairly certain that Searle accepts Dennett's description of his view on this particular point, because otherwise he would surely not have refrained from denying and condemning it in his reply (Searle, 1982b).

[288] His claim in the above-quoted passage that "if we worried about a maliciously motivated superintelligence destroying us then it is important that the malicious motivation should be real" looks a lot like such a retraction. But being a professor of philosophy, he ought to back up his retraction with a valid argument, and I search in vain for such a thing in Searle (2014).

[289] No, just kidding. There is no way such readers will have come this far into the book.

[290] See, e.g., Goldowsky (2014).

The program's ultimate z-desire when playing is to win the game, and the fine-tuning of the parameters will yield all kinds of subtle and unforeseen effects such as giving a high value to positions where certain pawn structures are combined with doubled rooks on the c-file. We will then be able to see a marked tendency for the program to move both its rooks to the c-file, due to it having formed, without explicit instructions from the programmers, a z-desire to double its rooks on the c-file in such positions.

Note that the chess program's choice to play in this way is unhampered by the fact that the choice is not a *real* choice but merely a z-choice, the play is merely z-play, and the desire concerning where to put the rooks is merely a z-desire. The z prefix is utterly inconsequential.

Now compare this to Bostrom's running example of a superintelligent AI that, due to its initial programming, has the desire to facilitate the production of as many paperclips as it can. Upon attaining superintelligence, it understands that humanity will want to prevent it from turning the Milky Way into a giant heap of paperclips, and therefore decides to form an instrumental desire to wipe out humanity. According to Searle, this cannot happen, because the AI has no *real* desires and no *real* understanding, and it makes no *real* decisions. All it has are z-desires and z-understanding, so all it can make are z-decisions. But if chess programs that are already in existence today are able to make unsupervised z-decisions to form a z-desire to put its rooks on the c-file, why in the world would Bostrom's paperclip maximizer be prevented from making the unsupervised z-decision to form a z-desire to wipe out humanity? When that happens, Searle will tell us (in the unlikely event that he has time to do so before it's all over) not to worry, because what may superficially look like the AI's desire to wipe out humanity is actually not a *real* desire, but merely a z-desire. Those of us who do not share Searle's confusion about desires and z-desires will find very little comfort in this.

There may well be good reasons for thinking that a dangerous intelligence explosion of the kind outlined by Bostrom is either impossible or at least so unlikely that there is no need for concern about it. The literature on the future of AI is, however, short on such reasons, despite the fact that there seems to be no shortage of thinkers who consider concern for a dangerous intelligence explosion silly (see Footnote 282). Some of these thinkers ought to pull themselves together and write down their arguments as carefully and coherently as they can. That would be a very valuable contribution to the futurology of emerging technologies, provided their arguments are a good deal better than Searle's.

CHAPTER 5

Going nano

5.1 3D printing

Three-dimensional (3D) printing is a technology for manufacturing objects layer by layer, following a digital specification. First, an ultrathin layer of the material that the object is to be made of is laid down, and then layer after layer is added on top, until the full three-dimensional object is completed.

Judging from the recent hype of 3D printing of plastic dolls based on scans of the buyer or of some celebrity, dolls suitable for putting on top of a wedding cake or wherever, it is tempting to think of the technology as merely the latest curiosity on offer.[291] That would be a hasty conclusion, however, as there is great hope that 3D printing has the potential to revolutionize manufacturing at least as much as the assembly line concept did in the early 20th century. Among the more serious present-day applications, 3D printing was recently used to successfully manufacture a lower jaw out of titanium powder for a patient who was not in condition for traditional surgery. The computer-aided design of the jaw included articulated joints matching those of a real jaw, grooves to accommodate the regrowth of veins and nerves, and so on.[292] More generally, the field of 3D printed prosthetics is currently advancing rapidly.[293]

Another example of contemporary use of 3D printing is to produce prototypes, e.g., of aircraft for testing in wind tunnels.[294] One reason that the technology today is fairly widely used for prototype production but not for the later stage of mass production is that while its flexibility is ideal for the former, it is still so slow that for the latter, it cannot yet compete with traditional manufacturing methods.

For a vision of where all this might be heading, recall the quote in Section 1.4 from Drexler (2013) on the revolutionary changes to manufacturing that may lie ahead of us, including how "the gadgets and goods that run our society [will be] produced not in a far-flung supply chain of industrial facilities, but in

---

[291] Needleman (2014).
[292] Gore (2013), p 241.
[293] McCue (2014).
[294] Gore (2013), p 31.

Here Be Dragons. First Edition. Olle Häggström.
© Olle Häggström 2016. Published in 2016 by Oxford University Press.

compact, even desktop-scale machines." Drexler, along with Gershenfeld (2012) and Gore (2013), emphasizes the profound effect on the economy that such a shift towards localized production will have. If they are right, then we will see an expansion to other sectors of the trend we are currently seeing in the book and music markets, that the product delivered to the buyer is increasingly often not in the form of a material object (such as a hardback or paperback book, or a CD), but only as a digital file. When we purchase a product – be it a sandwich, a pair of sneakers or a kitchen table – we will receive a blueprint for the object as a digital file, and then feed the file to our own portable all-purpose 3D printer, which swiftly puts together the sandwich, the pair of sneakers, or the kitchen table.[295] If the physical size of the product is large enough, as might be the case with a car, the home 3D printer will not do, but then we will simply have the object printed for us by the local 3D printing shop.

Such a paradigm shift in how we manufacture things is likely to have a series of further and perhaps disruptive economic consequences. Goods transportation, which today is a gigantic sector, can be expected to shrink considerably; we will still need energy and raw material for the 3D printers, but providing that is a logistic undertaking much less complicated than today's transport network. The technology will fuel the tendency towards a winner-takes-all economy, already prevalent in today's increasingly digitization-based economy and briefly discussed in Section 4.4, with possibly huge consequences in terms of increased economic inequality. The issue of intellectual property and theft thereof, which today haunts the music and software industries, will become an increasing concern in other sectors. As Gershenfeld (2012) asks, with no obvious panacea in sight: "If products are transmitted as designs and produced on demand, what is to prevent those designs from being replicated without permission?"

Another serious concern, currently in want of an obvious solution, is what 3D printing does to the feasibility of gun control. Gershenfeld mentions the case of an amateur gunsmith having made use of 3D printing to manufacture the part of the AR-15 semi-automatic rifle known as the lower receiver, which is the (heavily regulated) part that holds the bullets and carries the serial number. Later, a company in Texas is reported to have succeeded in the first 3D printing of a complete metal handgun, the so-called M1911, which served as the standard sidearm for US armed forces between 1911 and 1985. According to the CNN news story, Alyssa Parkinson, a spokesperson for the company, says that "This is not about desktop 3D printers," and that "the industrial printer we used costs more than my [private university] tuition," thus trying to downplay the public security challenge arising from the technology.[296] With this, however, she misses (similarly to the flawed reasoning by an unnamed microbiologist cited in the final paragraph of

---

[295] It may be that with the sandwich example, I go a bit beyond even Drexler's vision, because in the section of Drexler (2013) dealing with consequences of the new technology for food production, he speaks of 3D printing not of the food itself but of tools for advanced agriculture.

[296] Gross (2013).

Section 1.3) the most important point, namely this: just like how computers have become drastically cheaper and more accessible over the decades, we can expect a similar trend in 3D printing, so that 3D printers capable of producing the M1911 may well be available to the public at low cost a decade or two hence.

5.2 Atomically precise manufacturing

Technology has a long way to go to achieve Eric Drexler's vision of a manufacturing economy wholly based on general-purpose 3D printing. Today's 3D printing devices typically work with a very limited range of raw materials, and their precision is insufficient for many applications. What will be needed for a general-purpose 3D printing technology is to move into the realm of **nanotechnology**, defined as the control over matter with atomic or molecular precision. An alternative definition, explaining the name, is the manipulation of matter with at least one of its three spatial dimension extending at most 100 nm (nanometers). For comparison, the distance between neighboring carbon atoms in, e.g., graphite or diamond, is a few tenths of a nm. Drexler aims at what he calls **atomically precise manufacturing** (APM), which involves adding atoms (or, in some situations, small chunks of atoms)[297] one by one to the object under construction. Another term for APM that is often used is **molecular assembler** technology. The idea of APM goes back to Richard Feynman's talk "There's plenty of room at the bottom," delivered at an American Physical Society meeting in 1959,[298] but it was Drexler who, in the mid-1980s, started to think more systematically about the potentials of such technology. It was also Drexler who (in a number of books beginning with *Engines of Creation* from 1986, which was based on his PhD thesis, and continuing until his most recent *Radical Abundance*)[299] did more than anyone else to raise public imagination concerning its potential.

The transition from today's 3D printers to APM will be a huge quantitative improvement in terms of precision: the cutting-edge 3D printing technology used for the jaw prosthesis mentioned in Section 5.1 produces layers that are about 30 μm (i.e., about 30,000 nm) thick, corresponding to perhaps 100,000 layers of

---

[297] Drexler et al. (2001) speak of how, under certain conditions, it will be necessary

to rely upon conventional solution or gas phase chemistry for the bulk synthesis of nanoparts consisting of 10–100 atoms. These much larger nanoparts can then be bound to a positional device and assembled into larger (molecularly precise) structures without further significant steric constraints.

[298] Feynman (1959). Here is a key passage:

The principles of physics, as far as I can see, do not speak against the possibility of maneuvering things atom by atom. It is not an attempt to violate any laws; it is something, in principle, that can be done; but in practice, it has not been done because we are too big. . . .

It would be, in principle, possible (I think) for a physicist to synthesize any chemical substance that the chemist writes down. Give the orders and the physicist synthesizes it. How? Put the atoms down where the chemist says, and so you make the substance.

[299] Drexler (1986, 2013).

atoms, so that doing one layer of atoms at a time improves precision by a factor 100,000. But in fact, what we're talking about is not just a quantitative improvement, but also a qualitative leap from the continuous world, with its unavoidable imprecisions, to the discrete, which offers *exact* precision. In digital computing, the symbols 0 and 1 are represented by two voltage levels, and a voltage that is slightly off is correctly interpreted as the intended symbol, so that in effect it is immediately corrected rather than propagated and accumulated. What happens in APM is similar: an atom hooks up to the object under construction either exactly in place, or entirely in the wrong location: there is no possibility of landing just a tiny bit off. Gershenfeld (2012) suggests another metaphor for this discrete or digital nature of APM, namely a child assembling LEGO pieces: "Because the LEGO pieces must be aligned to snap together, their ultimate positioning is more accurate than the motor skills of a child would usually allow." This exactness is precisely what Feynman (1959) had in mind:

If we go down far enough, all of our devices can be mass produced so that they are absolutely perfect copies of one another. We cannot build two large machines so that the dimensions are exactly the same. But if your machine is only 100 atoms high, you only have to get it correct to one-half of one percent to make sure the other machine is exactly the same size – namely, 100 atoms high!

Is Feynman's and Drexler's APM vision attainable? This issue is a matter of some controversy. No detailed design has been fully specified, but already in *Engines of Creation* Drexler laid out a kind of road map, with extensive arguments for the feasibility of, among others, the following subsystems of a working APM 3D printer:[300]

(1) A nanoscale *computer*, providing the intelligence needed for controlling the APM process. Various approaches can be imagined, including electronics based on three-dimensional arrays of carbon nanotubes. As an alternative to electronics, Drexler suggests a mechanical device with molecular locks playing the role of transistor gates.

(2) *Instruction transmission*, which can be electronic or mechanical depending on the choice of technology in (1).

(3) The nanoscale *construction robot*, with an arm to deliver and deposit a molecular fragment or a single atom to the desired location. Especially critical here is the chemistry of the tip of the robot's arm.

(4) The machinery's *internal environment* needs to be controlled so as to prevent environmental impurities from interfering with the manufacturing process. Drexler suggests maintaining a near vacuum, behind walls of diamondoid material.

---

[300] Drexler (1986). The list draws heavily on the somewhat more detailed and extensive summary provided by Kurzweil (2005), pp 228–230.

(5) The *energy* needed for the manufacturing process can be delivered by either electrical or chemical means. Drexler's original suggestion involves interlacing fuel with the raw building material, but Kurzweil mentions some more recent proposals involving, e.g., nanoengineered fuel cells.

A high point in the controversy concerning the feasibility of APM technology is the (somewhat heated) exchange between Eric Drexler and Richard Smalley[301] that took place during 2001–2003, and that has become legendary as "the Drexler–Smalley debate." The debate opened with Smalley's 2001 *Scientific American* article "Of chemistry, love and nanobots," continued with a response from Drexler and seven coauthors in 2001 followed by two open letters by Drexler in 2003, and ended in 2003 with a point–counterpoint discussion between Drexler and Smalley in the journal *Chemical & Engineering News*.[302] In his *Scientific American* piece, Smalley raises the issues of "fat fingers" (meaning that the "plenty of room" that Feynman talks about is nevertheless insufficient to accommodate the construction robot in item (3)) and "sticky fingers" (meaning the difficulty or impossibility of having the robot release its cargo in the desired location), and concludes with the following categorical statement:

Both these problems are fundamental, and neither can be avoided. . . . To put every atom in its place – the vision articulated by some nanotechnologists – would require magic fingers. Such a nanobot will never become more than a futurist's daydream.

Drexler et al. (2001) responded by explaining how Smalley misrepresents the proposed designs, thus dismissing his fat fingers and sticky fingers objections as straw man arguments. Strikingly, they point out that these problems cannot be "fundamental" and unavoidable, because nature has managed to overcome them in the APM assemblers known as ribosomes that appear in large quantities in all our cells:

This ubiquitous biological molecular assembler suffers from neither the "fat finger" nor the "sticky finger" problem. If, as Smalley argues, both problems are "fundamental," then why would they prevent the development of mechanical assemblers and not biological assemblers? If the class of molecular structures known as proteins can be synthesized using positional techniques, then why would we expect there to be no other classes of molecular structures that can be synthesized using positional techniques? Upon observing experimentally that polymers such as proteins can be synthesized under programmatic control, what convincing evidence do we have that the programmatic synthesis of stiffer polycyclic structures such as diamond is "fundamentally" impossible, and that mechanical assemblers will never be built?

---

[301] Smalley is best known for his 1985 discovery in joint work with Robert Curl and Harry Kroto of the C_{60} molecule (colloquially known as the buckyball), which opened up a new avenue in nanoscience and earned the trio the 1996 Nobel Prize in chemistry. He died in 2005.

[302] Smalley (2001), Drexler et al. (2001), Drexler (2003a, 2003b), Baum (2003). The word "nanobot" in the title of Smalley's paper means nanoscale robot.

So the central issue seems to be not whether there are fundamental principles that prevent APM altogether – the ribosome example shows that there aren't – but to what extent APM can be implemented in settings other than those Mother Nature has stumbled upon. This is an open question, and one that the remaining pieces of the Drexler–Smalley debate did little to settle. Obviously, there is room even for world-leading nanotechnology experts to have radically diverging intuitions regarding the feasibility of APM beyond the biological setting.[303] It could be that our present state of knowledge simply does not admit abstract theoretical arguments to settle the issue, and that the proof is in the pudding. A 2006 report from the US National Academy of Sciences expresses this uncertainty as follows:[304]

Although theoretical calculations can be made today, the eventually attainable range of chemical reaction cycles, error rates, speed of operation, and thermodynamic efficiencies of such bottom-up manufacturing systems cannot be reliably predicted at this time. Thus, the eventually attainable perfection and complexity of manufactured products, while they can be calculated in theory, cannot be predicted with confidence.

So, in view of the central issue of feasibility of APM beyond the biological setting remaining open, should we describe the outcome of the Drexler–Smalley debate as a tie? Such a verdict strikes me as overly generous towards Smalley, as his main argument about the fundamentalness and unavoidability of the fat finger and sticky finger obstacles was clearly debunked by the ribosome counterexample. Ray Kurzweil, who devotes a couple of sections in *The Singularity is Near* to a careful reading and evaluation of the debate, comes out unambiguously on Drexler's side, and describes Smalley's contributions as crucially short on specifics – in sharp contrast to Drexler's.[305] Colorful but imprecise metaphors, such as "you don't make a girl and a boy fall in love by pushing them together" (which explains the otherwise odd-looking word "love" in the title of the *Scientific American* paper) cannot make up for the lack of detailed arguments and specific references.

Readers familiar with the Drexler–Smalley debate will have noticed that one issue that features prominently in it has been totally ignored in this section, namely the possibility (or not) of self-replicating nanomachines. We will get back to that in Section 5.4.

---

[303] One interesting clue to what underlies Drexler's general techno-optimism is a certain asymmetry in how advances in physics influence engineering, as described in Drexler (2013). He distinguishes textbook physics, which works well for describing most of what we see around us, from cutting-edge research in physics, which tends to concern exotic phenomena arising under various extreme or unusual circumstances. And here's the asymmetry:

Exotic effects that are hard to discover or measure will almost certainly be easy [for the engineer] to avoid or ignore [while] exotic effects that can be discovered and measured can sometimes be exploited for practical purposes. (p 97)

Hence, advances in physics will sometimes improve the potential for engineering solutions, while it will rarely lead to severe limitations, so the overall trend will be for the advance of physics to expand the scope of what engineering can do.

[304] Committee to Review the National Nanotechnology Initiative (2006).

[305] Kurzweil (2005), pp 236–241.

5.3 Nanobots in our bodies

There are other potential applications of nanoscale robots – nanobots, for short – than the APM technology discussed in Section 5.2. One that seems to have as much potential as APM for revolutionizing our lives is the idea of putting huge numbers of nanobots in the human bloodstream.[306] If Eric Drexler is the foremost visionary and advocate for APM, the corresponding title when it comes to nanobots in the bloodstream goes to the American physicist Robert Freitas. Freitas very much shares the transhumanist credo (discussed in Section 3.7) that ageing is a disease that can be cured, and his view of death is concisely stated in the title of his 2002 paper "Death is an outrage."[307] In that paper, he offers some illustrative calculations on how old we would expect to get if we could freeze the death rate (defined as the probability of dying during the coming year) at a certain age, in contrast to the current situation where the death rate tends to escalate drastically with old age. Forty-year-old American males thus freezing their death rate could thereby expect on average to live to 300, while the corresponding 20-year-old would on average live to 600, and a 10-year-old to the astonishing age of 3000. Freitas takes this as an indication of what would happen if we were able to halt the ageing process at a certain age (or even reverse it). This, he says, will be doable by means of nanorobotics, and perhaps soon: in Freitas (2009), he notes that "we cannot build such tiny robots today," but suggests that "perhaps by the 2020s, we will."

Recall from Section 3.7 the definition of ageing as the accumulative damage to the body's macromolecules, cells and tissues caused by "genomic instability, telomere attrition, epigenetic alterations, loss of proteostasis, deregulated nutrient sensing, mitochondrial dysfunction, cellular senescence, stem cell exhaustion, and altered intercellular communication" (López-Otín et al., 2013). All of these damages can, according to Freitas, be repaired using nanobots. As a case study, he describes in some detail in an ambitious 2007 paper the possible design of nanobots for replacing damaged chromosomes.[308] And there is more work to do for the nanobots. Here is Kurzweil (2005):

Nanobots will be able to travel through the bloodstream, then go in and around our cells and perform various services, such as removing toxins, sweeping our debris, correcting DNA errors, repairing and restoring cell membranes, reversing atherosclerosis, modifying the level of hormones, neurotransmitters and other metabolic chemicals, and a myriad of other tasks. (pp 256–257)

---

[306] This idea, too, can be traced back to the classic Feynman (1959) paper, that contains a passage suggesting that the idea of small machines

would be interesting in surgery if you could swallow the surgeon. You put the mechanical surgeon inside the blood vessel and it goes into the heart and "looks" around. (Of course the information has to be fed out.) It finds out which valve is the faulty one and takes a little knife and slices it out. Other small machines might be permanently incorporated in the body to assist some inadequately-functioning organ.

[307] Freitas (2002).
[308] Freitas (2007).

Later in the book (p 301), Kurzweil goes on to suggest that this technology will lead to a divorce between eating and its original biological purpose of providing the bloodstream with nutrients, analogous to the divorce between coitus and reproduction that we are beginning to witness today, as discussed in Section 3.5. His most cherished application, however, is to use nanobots for scanning the brain, initially to gain insights into the functioning of the human brain for the purpose of reverse-engineering an AGI, and ultimately for uploading of the human mind to computers. The latter prospect was discussed at some length in Sections 3.7 and 3.8, including the observation that (at least early on) the technologies that are most likely to be feasible will destroy the original brain. Uploading may be more attractive if the scanning could be made non-destructive, and maybe nanobot technology can provide a way to do that. Here is Kurzweil's vision:

Billions of [nanobots] could travel through every brain capillary, scanning each relevant neural feature from up close. Using high-speed wireless communication, the nanobots would communicate with each other and with computers compiling the brain-scan database. (p 163)

Kurzweil discusses the complication posed by the so-called blood–brain barrier, which protects the brain from a variety of potentially harmful substances in the blood, and blocks objects bigger than about 20 nm from gliding through, which is probably too small for nanobots with the desired functionality (pp 163–166). He offers a long list of engineering approaches to solve this problem, and more generally seems certain that uploading via non-destructive nanobot-based scanning will be doable. The judgment of Sandberg and Bostrom (2008b) is more cautious: "Although the capabilities of nanomachines can be constrained by known physics it is not possible today to infer enough about machine/cell/tissue interactions to estimate whether non-destructive scanning is possible" (p 108).

5.4 Grey goo and other dangers

Smalley, in the *Scientific American* paper that initiated the Drexler–Smalley debate, made an interesting remark about the speed of a hypothetical APM technology.[309] We may recall that, pointing to the problems with fat fingers and sticky fingers, he contested that the nanoscale construction robot discussed in Section 5.2 (item (2) in the list of APM technology ingredients) could be built and made to work, but he was willing to assume for the sake of argument that it would. Assuming further, and quite optimistically, that it will be able to deliver and attach a billion atoms per second to the object to be manufactured, he noted that it will still take some 20 million years to produce an object weighing just one ounce. Such a technology would, according to Smalley, be very interesting scientifically, but matter less in practice. In order for the technology to be of much practical use,

---

[309] Smalley (2001).

we need billions or (preferably) trillions of such nanobots. A similar statement holds for the nanobots-in-the-bloodstream proposals discussed in Section 5.3: if we cannot inject trillions of nanobots, the technology will not be of much use. Manufacturing such numbers of nanobots may seem like a hopelessly daunting task, unless we can make nanobots make nanobots. In particular, if we can make a nanobot **self-replicating**, i.e., able to produce copies of itself, then, at least for some time, a population of nanobots can repeatedly double in size with a constant doubling time, producing the kind of dramatic exponential increase that anyone familiar with the old Persian legend of the wheat grains and the chess board knows about.[310] If such a self-replicating nanobot can be constructed in the first place, obtaining trillions or more of it in a short time will be a fairly easy task.

But there may be a downside. What if the growth of the nanobot population continues beyond our control? A very likely primary building block of a nanobot (self-replicating or not) is carbon, due to its uniquely flexible chemical properties. For the same reason, carbon is a main constituent in biological organisms, making them ideal for the self-replicating nanobot to consume for its production of further nanobots. The nanobot population, having escaped the laboratory, just might go on to consume the entire biosphere on a time scale of weeks. This frightening scenario was first sketched by Drexler (1986), who coined it **grey goo**,[311] not with the intention of characterizing literally the color or texture of the final product, but to

emphasize that replicators able to obliterate life might be less inspiring than a single species of crabgrass. They might be "superior" in an evolutionary sense, but this need not make them valuable. (p 173)

Later, Freitas (2001) analyzed this hypothetical phenomenon at further depth, and proposed some related scenarios, including "grey plankton" (consuming the CO_2 dissolved in sea water and methane clathrates in oceanic sediments), "grey dust" (consuming airborne dust) and "grey lichens" (eating rock). This may sound like science fiction (admittedly not the first scenarios in this book with such a ring to them), but Freitas asks us to consider the emergence of new infectious agents like HIV and Ebola. These demonstrate that we currently know way too little about natural or technological factors that may trigger mutations

---

[310] This legend, which has become quite a popular metaphor among contemporary technology visionaries such as Kurzweil (1999) and Brynjolfsson and McAfee (2014), goes as follows:

The invention of chess makes the king so happy that he wishes to reward the inventor, and asks him what he would like to have. The inventor, pretending to be a humble man, says he will be content with receiving one grain of wheat on the first square of the chessboard, two grains on the second, four on the third, eight on the fourth, and so on until the 64th square. The king, who is quite wealthy, and unfamiliar with the behavior of exponentially increasing sequences, immediately grants the wish. When the king's treasurer, after considerable time, has calculated the total number of grains ($2^{64} - 1 = 18,446,744,073,709,551,615$) and explained that this exceeds by far what the king or anyone else can produce, the inventor is beheaded.

[311] Or, to quote him more exactly, "goo".

in self-replicating entities, and about how such mutations may produce what he calls "a limited form of green goo."

Should we be frightened of grey goo and the sibling scenarios outlined by Freitas? When Anders Sandberg, being one of today's leading thinkers on emerging technologies and risks to the survival of humanity, recently put together a tentative list of what he considered to be the five biggest risks threatening the very existence of humanity, nanotechnology ended up in fourth slot (after nuclear war, bioengineered pandemic, and superintelligence).[312] He emphasizes, however, that grey goo (or any of its relatives) is *not* his main culprit under the nanotechnology heading, and here is his explanation for not ranking grey goo higher among the risks:

The . . . "grey goo" [scenario] of self-replicating nanomachines eating everything . . . would require clever design for this very purpose. It is tough to make a machine replicate: biology is much better at it, by default. Maybe some maniac would eventually succeed, but there are plenty of more low-hanging fruits on the destructive technology tree.

The comparison to biology is apt, but it is not clear how convincing his idea is that it would take "some maniac" to initiate grey goo "for this very purpose."[313] Can we state with certainty that it cannot happen by mistake? It is true that self-replication will require "clever design," but we can expect non-maniac nanotechnology engineers to attempt that, as it seems to be the most promising way to produce nanobots in the huge quantities we will need for a variety of applications. Such construction will presumably be done with a degree of precaution so as to avoid the self-replication going overboard as in the grey goo scenario, but such precautions could fail. At this point, the grey goo skeptic could point again to biology, and the fact that life has existed on our planet for billions of years without turning into grey (or green) goo, as an argument for why self-replication is unlikely to spontaneously produce such a catastrophe. While this argument has *some* force, it is far from conclusive, as the newly formed nanobots may, in the abstract space of possible living organisms, be located quite far from all biological life, and have very different physical and chemical properties. It is hardly absurd to think that this might give the nanobots a robust reproductive power unmatched by anything that the evolution of biological life has discovered.

At the very least, this issue deserves further study. A possible middle ground could be that, on one hand, it is possible to proceed with nanobot technology without risking a grey goo scenario, *provided that we respect certain safety protocols,*

---

[312] Sandberg (2014c).

[313] Sandberg's suggestion here echoes the judgment of Phoenix and Drexler (2004) that "although runaway replication cannot happen by accident, no law of nature prevents its deliberate development," as well as a memo from the Center for Responsible Nanotechnology (2003), stating that

development and use of molecular manufacturing will create nothing like grey goo, so it poses no risk of producing grey goo by accident at any point. However, goo type systems do not appear to be ruled out by the laws of physics, and we can't ignore the possibility that someone could deliberately combine all the requirements.

whereas, on the other hand, grey goo would be a real danger without these precautions. Phoenix and Drexler (2004) have a number of suggestions for how nanobots can be made to self-replicate in controlled and safe fashion. A key insight is that the self-replication ability of an agent (be it a biological organism or a robot) is always contingent on its environment. Even a set of blacksmith's tools can, in the right environment (one that provides suitable input of skills and muscle power), produce a duplicate set, and can thus be described as self-replicating. What is always needed for self-replication capability is the raw material and energy needed to produce the replicates.[314] Hence, we could avoid grey goo if we construct the self-replicating machines to contain elements not available in the natural environment. But we may not even have to go that far. Phoenix and Drexler point out that a general-purpose molecular assembler will typically not be a general purpose molecular *disassembler*. It will, most likely, be far easier to construct an APM machine that requires its raw material to be delivered in the form of a limited range of simple chemicals (such as acetylene or acetone), rather than having the ability to extract molecular fragments and atoms from arbitrary chemical compounds.

Suppose that we regulate the technology so as to make a specific list of sufficient safety precautions obligatory. With a risk such as grey goo, where the consequences of a single mishap could be global and even put an end to humanity, it is clearly not sufficient that *most* practitioners of the technology adhere to the regulation, or that *almost all* of them do: anything short of obedience to regulation from *all* practitioners would be unsatisfactory. Can this be ensured? This is not clear. Kurzweil (2006) advocates another safety layer, in the form of nanobots specifically designed to fight grey-goo-like outbreaks.

Without defenses, the available biomass could be destroyed by gray goo very rapidly. Clearly, we will need a nanotechnology immune system in place before these scenarios become a possibility. . . .

Eric Drexler, Robert Freitas, Ralph Merkle, Mike Treder, Chris Phoenix, and others have pointed out that future nanotech manufacturing devices can be created with safeguards that would prevent the accidental creation of self-replicating nanodevices. However, this observation, although important, does not eliminate the threat of gray goo as I pointed out above. There are other reasons (beyond manufacturing) that self-replicating nanobots will need to be created. The nanotechnology immune system mentioned above, for example, will ultimately require self-replication; otherwise it would be unable to defend us against the development of increasingly sophisticated types of goo. It is also likely to find extensive military applications. Moreover, a determined adversary or terrorist can defeat safeguards against unwanted self-replication; hence, the need for defense.

The suggestion of setting up this kind of defense system against grey goo – which Kurzweil elsewhere calls "police nanobots" and "blue goo"[315] – goes back to Drexler (1986). It makes me very uneasy. Joy (2000) points out that the system

---

[314] That is, unless we abstract the notion of a self-replicator beyond the realm of material objects and consider, e.g., memes; see Blackmore (1999).

[315] Kurzweil (2005), p 416.

"would itself be extremely dangerous – nothing could prevent it from developing autoimmune problems and attacking the biosphere itself," and while the word "nothing" may be an exaggeration, I do share Joy's feeling that blue goo is a terribly frightening thing. Kurzweil (2005, 2006) admits that it is dangerous, but insists that not developing it would be even more dangerous.

It will obviously not be possible to settle the grey goo debate here, so let us instead return to Sandberg (2014c). If he is not afraid of grey goo, what, then, is it about nanotechnology that merits such a prominent place on his list of existential hazards? Sandberg points to something that has already been touched upon in the last paragraph of Section 5.1, namely the potential for 3D printing to produce weapons. It is worth hearing him out on this:

[APM] looks ideal for rapid, cheap manufacturing of things like weapons. In a world where any government could "print" large amounts of autonomous or semi-autonomous weapons (including facilities to make even more) arms races could become very fast – and hence unstable, since doing a first strike before the enemy gets a too large advantage might be tempting.

Weapons can also be small, precision things: a "smart poison" that acts like a nerve gas but seeks out victims, or ubiquitous "gnatbot" surveillance systems for keeping populations obedient seems entirely possible. Also, there might be ways of getting nuclear proliferation and climate engineering into the hands of anybody who wants it.

We cannot judge the likelihood of existential risk from future nanotechnology, but it looks like it could be potentially disruptive just because it can give us whatever we wish for.

The emphasis here is the same as in the survey by Phoenix and Treder (2008) of potential global risks from nanotechnology. Advances in nanotechnology will tend to accelerate those in technology more generally – not just the production part, but also the design part, by making it drastically easier to create prototypes.[316] The military sector has, throughout history, been a leading player on the demand side of technological development, and there isn't much currently on the radar to suggest that it will cease to be. In view of what a breakthrough in APM might entail, in terms both of radically upscaling the quantities of existing weaponry[317] and of developing new weapons,[318] the destabilizing effect could be considerable.

---

[316] An analogy to how this might change design practice can be found in the very different and uninhibited way in which we use cameras today, compared to just a couple of decades ago, before the breakthrough of digital photography.

[317] Drexler (2013) suggests a scenario where APM enables the production of cruise missiles outnumbering the current US arsenal by a factor of 1000, all within a matter of days and at the comparatively trivial cost of $10 billion (p 260).

[318] Phoenix and Treder offer an example:

Increased material strength could increase the performance of almost all types of weapons. More compact computers and actuators could make weapons increasingly autonomous and add new capabilities. Weapons could be built on a variety of scales and in large quantities. It is possible, indeed easy, to imagine combining such capabilities: for example, one could imagine an uncrewed airplane in which 95% of the dry weight is cargo, the said cargo consisting of thousands of sub-kilogram or even sub-gram airplanes that could, upon release, disperse and cooperatively seek targets via optical identification, and then deploy additional weapons capabilities likewise limited mainly by imagination.

A country leading the development towards APM and other radical nanotechnologies may find itself in an overwhelming but temporary military advantage, and feel the temptation to turn the temporary advantage into a permanent one by means of a military strike. Its competitors, lagging behind, may foresee this and decide (as suggested in the Sandberg quote above, as well as in Section 3.4 in the context of human cognitive enhancement) to launch a preemptive strike before it is too late.

CHAPTER 6

What is science?

6.1 Bacon

Given this book's ambition to try to understand what kind of impact future scientific advances may have on our lives and on society as a whole, it makes sense to temporarily take a few steps back and discuss what science and the scientific method really are. This chapter will be devoted to such a discussion, and ends with a short section on the relation to technology and engineering (Section 6.10). I've postponed this chapter as deep into the book as possible so as not to lose too much tempo in getting to the core issues of emerging technologies and their consequences for humanity, but now is the time, because much of it will be indispensable for what we will do later, especially in Chapters 7 and 8.

Let us begin with the late 16th and early 17th century English philosopher and scientist Francis Bacon, who pioneered the modern way in which we think of science. While the ancient Greeks, in particular Plato and Aristotle, tended to emphasize thinking as the way towards knowledge, and cared little about empirical observation,[319,320] Bacon correctly identified the latter as crucial to the scientific endeavor.

He went too far in the other direction, however, rejecting all theory-building in favor of pure observation. This entails a very optimistic view of the scientific enterprise: all we have to do is observe carefully and patiently, and all will be clear. Here is Karl Popper's (1994) summary of Bacon's position:

Bacon's new method, which he recommends as the true way to knowledge, and also as the way to power, is this. We must purge our minds of all prejudices, of all preconceived ideas, of all theories — of all those superstitions, or idols, which religion, philosophy, education, or

---

[319] Plato went very far in his contempt for empirical observation, and held, in the words of Russell (1945), that "there is nothing worthy to be called 'knowledge' to be derived from our senses, and that the only real knowledge has to do with concepts. In this view, '2 + 2 = 4' is genuine knowledge, but such a statement as 'snow is white' is so full of ambiguity and uncertainty that it cannot find a place in the philosopher's corpus of truths."

[320] An exception was the 2nd and 3rd century philosopher and physician Sextus Empiricus (and his followers), from whom the term "empirical" is derived.

Here Be Dragons. First Edition. Olle Häggström.
© Olle Häggström 2016. Published in 2016 by Oxford University Press.

tradition may have imparted to us. When we have thus purged our minds of prejudices and impurities, we may approach nature. And nature will not mislead us. For it is not nature that misleads us but only our own prejudices, the impurities of our own minds. If our minds are pure, we shall be able to read the Book of Nature without distorting it: we have only to open our eyes, to observe things patiently, and to write down our observations carefully, without misrepresenting or distorting them, and the nature or essence of the thing observed will be revealed to us.

This is Bacon's method of observation and induction. To put it in a nutshell: pure untainted observation is good, and pure observation cannot err; speculation and theories are bad, and they are the source of all error. (p 84)

But Bacon's position here is wrong, according to Popper. Theory-building is just as important an ingredient in the scientific method as is observation. By rejecting it, Bacon was led astray, and his view of astronomy serves as a striking example. Popper again:

Bacon . . . was an enemy of the Copernican hypothesis. Don't theorize, he said, but open your eyes and observe without prejudice, and you cannot doubt that the Sun moves and the Earth is at rest. (pp 84–85)

We will return to Popper's more nuanced view of the interplay between theory and observation in Sections 6.3 and 6.4, but let us stay with Bacon for a little longer to see that his philosophy of science didn't entirely lack merit. Not only does his view serve as a balancing correction to Plato's and others' dismissal of observation as a way towards knowledge – his emphasis on the "impurities of our own minds" is also important. He called them our **idols**, and listed four kinds, namely

- the **idols of the tribe**, our propensity to detect more order to nature than there actually is;
- the **idols of the cave**, individual prejudices attained from inheritance and environment;
- the **idols of the market place**, misconceptions arising from how language influences our thinking; and
- the **idols of the theater**, delusions originating in philosophical tradition.[321]

Although few would use this particular categorization today, there is nevertheless a lot to be said for the idea that the human cognitive machinery has a number of shortcomings or miscalibrations that jeopardize our ability to do science. Such shortcomings are studied in the field of psychology that is known as **heuristics and biases** and that has flourished remarkably in the last few decades; see, e.g.,

---

[321] Bacon (1620).

Kahneman's (2011) influential bestseller *Thinking, Fast and Slow*.[322] Here are two specific examples of such biases.[323]

(a) As suggested by Bacon with his idols of the tribe, the human pattern recognition machinery tends to be overly trigger-happy, meaning that we think we see patterns in what is actually just noise. There is a famous experiment where this phenomenon produces the amusing outcome that human subjects typically perform *worse* than pigeons and rats do when facing the same task. The subject is faced with two lamps, a red one and a green one. At regular intervals, one of them lights up, and subjects are asked repeatedly to predict which lamp will light up next. They are not informed about any mechanism behind the sequence of red-lamp-turns-on and green-lamp-turns-on events, but the true mechanism is that the lamps are turned on randomly, with a 0.8 probability for the red one to light up next, versus a 0.2 probability for the green one, independently of what has happened before. Human subjects notice the asymmetry, and try to mimic the intricate pattern of lights, predicting the red lamp about 80% of the time, and the green one the remaining 20% of the time. In this way, they end up making the right guess about $0.8 \cdot 0.8 + 0.2 \cdot 0.2 = 68\%$ of the time. Simpler animals quickly settle for guessing the most frequent lamp every time, getting it right 80% of the time. See, e.g., Hinson and Staddon (1983) and Wolford, Miller and Gazzaniga (2000).

(b) In many situations, we tend to be overly confident about our beliefs. The following experiment is described by Alpert and Raiffa (1982); see also Yudkowsky (2008a). Subjects are asked to estimate some quantity whose exact value they typically do not know, such as the surface area of Lake Michigan or the number of registered cars in Sweden. They are asked not for a single number, but for an upper bound and a lower bound such that the subject attaches a subjective probability of 98% to the event that the true value lies in the interval between the upper and the lower bound. If they are well calibrated in terms of the confidence they attach to their estimates, then one would expect their intervals to encompass the true value about 98% of the time. In experiments, they do so less than 60% of the time, indicating that severe overconfidence in estimating unknown quantities is a widespread phenomenon. When giving talks, I have often tried this experiment on my audience and gotten similar hit frequencies. Asking subjects for 99.9% rather than 98% probability intervals improves the hit rate, but only by very little.

---

[322] For another beautiful survey of the same area, with particular emphasis on the distorting effect that our biases can have on the type of futurological considerations that are the topic of the present book, see Yudkowsky (2008a).

[323] This discussion of examples (a) and (b) is taken from Häggström (2013a).

It is not hard to think up plausible explanations for why these biases might be (or have been, at some point of our evolutionary history) evolutionarily advantageous (possibly alleviating our sense of shame at performing worse than pigeons and rats at an intellectual task). Perhaps it is more costly to fail to detect a pattern (such as one that indicates the presence of a dangerous predator) when there is one, rather than vice versa. And perhaps some amount of overconfidence saves us from being paralyzed by indecision. To actually establish that the biases have arisen from those evolutionary selection pressures (or others) is of course a different and much more difficult matter.

Be this as it may, it is clear that biases such as (a) and (b) can have negative consequences on our ability to do science. So Francis Bacon deserves credit for stressing that we ought to try to overcome them.

I like to think of science as a systematic attempt to extract reliable information about the world, and of the scientific method as a collection of procedures designed to help us do so. If science is to be successful, then at least some of these procedures ought to help us avoid the unwarranted conclusions that we risk making due to (a), (b) and other biases built into our brains. In the famous words of physicist Richard Feynman: "Science is what we have learned about how to keep from fooling ourselves."[324] In Sections 6.6 and 6.8 we will see how mathematical statistics offers some such remedies.

We may furthermore think of the scientific method as a toolbox, together with a set of instructions for how to use these tools. No definite collection of tools and instructions will be arrived at, but the two most widely used such collections or frameworks for doing science will be discussed: **Popperian falsificationism** (Sections 6.3 and 6.4) and **Bayesian reasoning** (Sections 6.8 and 6.9). Some might say we have to choose between them, but I prefer a more pragmatic and less fundamentalist view. Both frameworks have compelling arguments in their favor, but there are also objections to both – objections that do have some merit. Different scientific problems are amenable to different methods, and faced with this variety we're better off having two (or more) toolboxes to choose from, rather than forever sticking to just one of them. There are even cases where tools from the Popperian box and from the Bayesian box can be fruitfully combined. Whether or not the two frameworks are compatible depends, as will be argued in Section 6.9, on how strictly we define them.

6.2 Are all ravens black?

The problem of induction has haunted philosophers of science from Sextus Empiricus via Bacon and Hume all they way up to and including our own time. Scientific arguments are usually classified as either deductive or inductive. A precise definition of this categorization that correctly captures what we mean by it

---

[324] I have failed to locate the exact source for this quote.

is a surprisingly intricate matter,[325] but for the present purposes it will suffice to say that deduction is about inferring particular instances from general laws, while induction does the opposite: to infer general laws from particular instances. A couple of examples will convey the idea. Given the premises (P1) that Socrates is a man, and (P2) that all men are mortal, we may deductively infer (C) that Socrates is mortal:

```
(P1)    Socrates is a man
(P2)    All men are mortal
(C)     Socrates is mortal
```

Compare this to the inductive inference in which, from the premises (P1*) that we have observed thousands of swans, and (P2*) that all of them so far have turned out to be white, we wish to conclude (C*) that probably all swans are white:

```
(P1*)   We have observed thousands of swans
(P2*)   All of them have turned out to be white
(C*)    Probably all swans are white
```

The difference strikes us: while everyone in their right mind agrees that the deductive inference about Socrates is sound, we feel much less certain about the validity of the inductive inference about the color of swans. After all, there *might* exist hitherto unobserved swans of all sorts of colors. Hence the "probably" in (C*), but even with that qualification we may doubt that the inference is sound. And it serves as a cautionary tale that, contrary to a widely held belief in (C*) among Europeans – a belief that was surely based on inductive inference from (P1*) and (P2*) – it was later discovered that black swans do exist (on the other side of the planet).[326]

It seems fair to say, *on one hand*, that deduction is well understood, and that logic has been established as a highly reliable tool for distinguishing between correct and incorrect deductive reasoning, while admitting, *on the other hand*, that induction is rather more poorly understood, and that we have no generally agreed upon machinery that plays the same role for induction as logic does for deduction. Yet, we desperately need induction (or something like it), in science as well as in daily life. We need to handle it with caution, but we cannot do without it. We have observed that Newton's laws of gravity hold for a number of planets and for one or two apples, but we need to be able to conclude from this that the laws probably hold for arbitrary celestial bodies and arbitrary fruits. I have personally observed that each time I have taken tram number 6 southbound from

---

[325] Vickers (2014).

[326] A sign of how iconic this example has become is the fact that I have in my possession no less than two recent books on the theory of science (both from a more or less Popperian perspective) that feature black swans in their cover image: Taleb (2007) and Persson (2014).

Linnéplatsen in Gothenburg, it takes me to the Chalmers campus,[327] and I need to be able to conclude that it will probably do so tomorrow morning as well.

Inductive reasoning tends to fall back on the following assumption.

(A) Observations in the future will, in most cases, be similar to those in the past.

But how can (A) be defended? It is extremely tempting to resort to an argument like the following.

The statement (A) has mostly turned out true in the past, and inductive reasoning has served us well, so we have every reason to believe that this happy situation will continue in the future.

But on closer inspection, we see that the argument is based on assumption (A), whence it is circular – a circularity first pointed out by David Hume (1748). Still, we are so used to taking (A) for granted that we easily miss the fallacy. The following joke (circulating in some of the geekier parts of the Internet) makes it clear, however.

The first human expedition to the star 61 Cygni discovers life on one of its planets. Not only does the planet support life, but even *intelligent* life. The human astronauts quickly find that the Cygnians have cognitive skills exceeding our own. In spite of this, the Cygnians live in poverty and misery, because they have failed to develop advanced technology. This discrepancy between their intelligence and their actual achievements calls for investigation, and it turns out that one of the earliest Cygnian philosophers invented the principle of anti-induction, based on the following assumption.

(A*) Observations in the future will, in most cases, exhibit the opposite behavior compared to those in the past.

The humans can easily understand how such an assumption might hinder scientific and techno-logical progress, but not why the Cygnians choose to stick with it. So they ask the president of the Cygnian Science Academy why they insist on acting on such a silly-sounding assumption as (A*). His reply: "Why not? It has never worked before."

Nevertheless, no matter how well I understand the circularity of the argument for (A) on an intellectual level, I still find it oddly compelling, and I seem not to be the only one to do so. Perhaps this widespread faith in (A) is hard-wired into our brains, and perhaps this happened because it was and still is evolutionarily favor-able to believe in (A). Hence (or so, at least, goes another variant of the circular argument) it makes sense to continue believing in (A).

We do use inductive reasoning, both in science and in our daily lives, despite the lack of rigorous arguments for why we ought to do so. And this lack of rigor-ous foundation is not the only problem with induction. There are also various paradoxes that acceptance of induction as a valid principle of reasoning seem to

---

[327] That is, except for the time that I was so absorbed by reading Dyson (1979) that I forgot to get off at Chalmers.

imply, such as the **raven paradox** of Hempel (1945).[328] While the issue of whether all swans are white has been settled in the negative, the corresponding question regarding black ravens remains, to the best of my knowledge, open. Consider the hypothesis

$$\text{(H)} \quad \texttt{All ravens are black.}$$

Suppose all ravens we have seen so far are black, so that the hypothesis (H) is still on the table. To accept induction as a valid principle of reasoning is to accept that the observation of one more black raven will, typically, strengthen our belief in (H). But induction is not specifically about "black" and "raven," which are just arbitrarily chosen attributes, and can be replaced, e.g., by "non-raven" and "non-black." So the inductionist who accepts that our belief in (H) should typically be strengthened by the observation of a black raven, should also accept that the observation of a non-black non-raven typically strengthens our belief in

$$\text{(H*)} \quad \texttt{All non-black items are non-ravens.}$$

But (H) and (H*) are logically equivalent (because they both state precisely that there are no non-black ravens), and surely logically equivalent hypotheses are supported by the same evidence. Hence, the inductionist is forced into the position that the observation of a non-black non-raven (such as a yellow banana) typically strengthens our belief in the hypothesis (H) that all ravens are black – a position that most of us find a bit hard to swallow.

6.3 Popper

Popper (1934, 1959) rejected induction as a valid way of inference, and proposed instead what has become known as his **falsificationism**, according to which a scientific theory can never be verified, but it can be falsified.[329] This falsifiability is his so-called demarcation criterion for when a theory should count as scientific.

---

[328] Another disturbing example is the **grue–bleen paradox** of Goodman (1955). Define an object A to be **grue** at time t if either

(a) $t <$ October 14, 2029, and A is green at time t, or
(b) $t \geq$ October 14, 2029, and A is blue at time t.

Likewise, define A to be **bleen** at time t if either

(a) $t <$ October 14, 2029, and A is blue at time t, or
(b) $t \geq$ October 14, 2029, and A is green at time t.

Suppose a given object, say an emerald, has been observed for a long time, and consistently turned out to be green. The inductionist will think of this as evidence for the hypothesis that the emerald will continue to be green. But the *very same evidence* can be reformulated as the emerald having consistently turned out to be grue, so the inductionist seems forced to accept the evidence as suggesting that the emerald will continue to be grue. But we cannot have it both ways: after October 14, 2029, the emerald cannot both continue to be green and continue to be grue. Most people have a spontaneous preference for the continue-to-be-green hypothesis over the continue-to-be-grue hypothesis, but defending this asymmetry is not so easy.

[329] Popper (1959) is the author's own translation of the 1934 original in German.

In formulating it, Popper was provoked by Freudianism and Marxism – theories that he perceived as so loosely formulated and flexible that there could be no data or no observations that the theories' proponents would not be able to incorporate in an ad hoc manner and claim that these are just the sort of observations that could have been expected all along, given their theories. This sort of unfalsifiability makes a theory incapable of prediction (because it is consistent with any outcomes whatsoever), and Popper ruled them out as unscientific.

Consider the hypothesis (H) above, that all ravens are black. The relevant evidence consists in the ravens we observe, black or non-black (and perhaps also yellow bananas, depending on how seriously we take Hempel's raven paradox). The hypothesis can never be verified, no matter how many ravens we observe, all black, because there is always the possibility that there are some hitherto unobserved white or red ravens lurking somewhere. All that another observed black raven does to (H) is to **corroborate** it (Popper's term), meaning that the observation contributes to our view of (H)'s ability to withstand critical testing. On the other hand, while (H) cannot be verified, it can be **falsified** – all it takes to falsify it is the observation of a single non-black raven.

Consider instead the negation (¬H) of (H):

> (¬H)   At least one non-black raven exists.

This negated hypothesis (¬H) does admit verification (observation of a single white raven would be enough), but it cannot be falsified, for the same reason that (H) cannot be verified. So (¬H) fails Popper's demarcation criterion and is deemed unscientific.

Popper's demarcation criterion has had much influence on how we think about science, but there are also concerns. Let me mention some of them. One issue that has sometimes been brought up is what corroboration really means. In particular, *does corroboration of a scientific theory strengthen our belief in it?*[330] This question faces the Popperian falsificationist with a dilemma. On one hand, if they answer "no," then corroboration doesn't really mean anything in terms of scientific inference, and there remains no way for a scientific theory to earn our belief in it. On the other hand, if "yes," then it looks very much as if corroboration is just induction with a slightly disguised terminology.

Personally, I also find the asymmetry between (H) and (¬H) – one being regarded as scientific in Popper's sense, the other not – a bit disturbing. The ideal state of mind for a scientist investigating whether or not all ravens are black would in my opinion involve a neutral stance as to which one of the two hypotheses is true. To declare, before the investigations have even started, one of them as scientific and the other as unscientific seems not to reflect or promote such neutrality. But this may be more a matter of terminology than of substance, and if we rewrite the demarcation criterion as saying that the task of studying (H) versus (¬H) is

---

[330] See, e.g., Gardner (2001).

scientific if and only if at least one of the two is falsifiable, then this particular worry of mine is settled. Of course, the asymmetry between (H) and (¬H) boils down to the difference between a universal quantifier ($\forall x$, "all x have the property that ... ") and an existence quantifier ($\exists x$, "there exists at least one x such that ... "). Popper (1959) was quite clear about this:

My proposal is based upon an asymmetry between verifiability and falsifiability; an asymmetry which results from the logical form of universal statements. For these are never derivable from singular statements, but can be contradicted by singular statements. (p 19)

With this in mind, it is quite surprising how many expositors of Popperian falsificationism fail to discuss this asymmetry and the option of switching the roles of a theory (H) and its negation (¬H). A case in point is Massimo Pigliucci – a writer with excellent credentials both as a scientist and as a philosopher of science – who still appears to overlook this in his 2010 book *Nonsense on Stilts*. He erroneously points out the SETI (Search for Extraterrestrial Intelligence) project as failing to meet Popper's falsifiability criterion. Pigliucci's argument is that the hypothesis

(E)   Extraterrestrial intelligence exists

cannot be falsified. But SETI is about shedding empirical light on which of the two hypotheses (E) and its negation

(¬E)   Extraterrestrial intelligence does not exist

is true, and since (¬E) is eminently falsifiable (all it takes is a single contact with, or message from, intelligent extraterrestrials), we can take (¬E) to be the theory up for testing, and the SETI project passes Popper's criterion with flying colors.[331] For more on SETI, see Sections 9.1, 9.4 and 9.5.

A more serious concern about Popperian falsificationism is the following. Consider the hypothesis

($H_{\geq 50\%}$)   At least half of all ravens are black

and its negation[332]

(¬$H_{\geq 50\%}$)   Fewer than half of all ravens are black.

According to Popper's criterion, ($H_{\geq 50\%}$) is not scientific, as it is not falsifiable: no matter how overwhelming a majority of non-black ravens we have observed so far, there still remains the possibility that there is an even larger number of black

---

[331] I made this point already in my review of Pigliucci's book: Häggström (2011b).

[332] My statement that (¬$H_{\geq 50\%}$) defined in this way actually *is* the negation of ($H_{\geq 50\%}$) involves two claims, namely (i) that the two hypotheses cannot simultaneously be true, and (ii) that at least one of them is true. Part (i) is not problematic, but part (ii) is, in case the total number N of ravens is 0 or ∞. The case $N = 0$ can be gotten rid of by stipulating that we have already observed at least one raven. The case $N = \infty$ can either be explained away in various ways, or incorporated into my example as a third hypothesis, neither verifiable nor falsifiable.

ravens somewhere we haven't looked. And $(\neg H_{\geq 50\%})$ fails the criterion for analogous reasons, so we are forced to conclude that the project of trying to work out which of $(H_{\geq 50\%})$ or $(\neg H_{\geq 50\%})$ is true is not a scientific undertaking. But this won't do – surely working out the majority color of ravens (or, say, determining what percentage of mass in the universe consists of helium) should qualify as scientific, and if it doesn't, so much worse for the criterion that is applied.

Yet another problem, possibly the most serious of all, with Popperian falsificationism involves the so-called Duhem–Quine thesis, underdetermination of scientific theories, and our predicament that the concept of falsification is not as simple as it may seem.[333] Concerning the hypothesis (H) that all ravens are black, I claimed above that "all it takes to falsify it is the observation of a single non-black raven," but is it really so simple? What may seem to be an observation of a white raven may turn out, upon closer inspection, to be a white swan with an unusually raven-like anatomy, an extraterrestrial impostor, or (perhaps more likely) a hallucination. In practice, any prediction that a scientific theory (H) makes that puts it up for falsification is based not only on the theory itself, but on a number of unspoken additional assumptions, including one that says we are awake and not dreaming. It is hard to think of an observation that falsifies a theory in a way that leaves no room for ad hoc explanations to save the theory, or for future counter-falsification.

The events leading up to the discovery in 1846 of the planet Neptune are a case in point.[334] Ever since the discovery in 1781 of its sister planet Uranus, there had been concerns about certain anomalies in Uranus' orbit. It didn't seem to quite obey Newton's theory of gravitation, that worked so well for the other planets. A strict Popperian falsificationist (if, anachronistically, we can imagine such a person living in the late 18th and early 19th century) would declare Newtonian gravity falsified, but that would be overhasty, and the scientific community at the time was not eager to throw such an otherwise enormously successful theory overboard.[335] Maybe the measurements were inaccurate, or perhaps there was another hitherto undiscovered celestial body that had gravitational influence on Uranus. The latter is an ad hoc explanation if I ever saw one – but one that turned out to be fruitful! The French mathematician Urbain Le Verrier began detailed work on it in 1845, and announced in the following year his calculations for where the unknown planet ought to be found. Simultaneous but independent work by the

---

[333] Stanford (2013).

[334] Smith (1989).

[335] The fact that Newtonian gravity would later be falsified, in the sense of being shown to be merely an approximation to Einstein's general theory of relativity (or possibly to some even more fundamental mechanism that we are not yet aware of), is beside the point here. It is nevertheless interesting to note that Newtonian gravity today (many decades after the universal or near-universal acceptance of general relativity) still enjoys a superb reputation: it is taught to school children and to university students, and it is indispensable in many areas of physics and engineering. It is simply such an excellent approximation, over such a wide range of spatial scales and energy levels, that our knowledge that it is not an exact description of reality does not bother us much.

British astronomer and mathematician John Couch Adams pointed in the same direction. Within a month, these predictions led to the discovery of Neptune.

6.4 A balanced view of Popperian falsificationism

If I had to summarize my view of Popperian falsificationism in a couple of paragraphs, it would be something like the following.

On one hand, Popper's demarcation criterion is too crude to be applied rigidly and mindlessly.[336] In the hands of certain categories of unsophisticated thinkers (such as climate deniers of the kind discussed in Footnote 40; see also Footnote 353 in Section 6.6) it causes more confusion than illumination. We must also understand that while the demarcation criterion has something to say about what a good scientific *theory* is, it tells us rather less of what good scientific activity is. The hypothesis (H) that all ravens are black fits the criterion beautifully, giving us courage to try to study the issue scientifically, but if all we do in our attempts to falsify (H) is to read tea leaves and horoscopes, then our work can hardly be called scientific.

On the other hand, Popper's theory of science has been very helpful in emphasizing critical thinking in general, and the need for critical scrutiny of our scientific theories in particular. In sharp contrast to Francis Bacon's view that "we have only to open our eyes, to observe things patiently, and to write down our observations carefully, without misrepresenting or distorting them, and the nature or essence of the thing observed will be revealed to us," Popper does a fine job in emphasizing the interplay between theory-building and observation. We cannot do without the former, since without a framework in which to interpret our data, they mean nothing to us. We build theories, we work out what the theories predict,[337] we go out and collect data, and we analyze them in order to see whether they are in line with the predictions.[338] For a while, perhaps, they are, and the theory is corroborated, but eventually we encounter anomalies that cannot be explained away. This calls for going back to the drawing board to work out new theories, usually via minor

---

[336] The word "crude" here is not meant to be derogatory of Popper's criterion, but only of those who apply it rigidly and mindlessly. Popper himself understood its crudeness perfectly well.

[337] In the simplest cases (hardly even worthy of the term "theory") such as the all-ravens-are-black hypothesis (H), the step of deducing predictions is trivial: all that (H) predicts is that all ravens we encounter will be black. For more involved theories such as the Newtonian theory of gravity, the prediction part of the work can be highly demanding.

[338] In relation to this cyclic procedure, recall man's excessive trigger-happiness in pattern detection (item (a) in Section 6.1) which I stressed as an obstacle to successful science. The phase of the cycle where this miscalibration in our brains needs to be worked around is, of course, the one where "we analyze [our data] in order to see whether they are in line with the predictions." On the other hand, as Ulf Persson has pointed out to me, it is quite conceivable that in the "We build theories" phase, the miscalibration actually works to our advantage – perhaps a bit of overoptimism is what it takes to pluck up the courage to invent new theories.

modifications of the old ones but every now and then with something radically new. The new theories are then exposed to empirical testing, and so on. In this way, science marches on, and we may hope (although there are no guarantees) that in this way we zoom in towards increasingly accurate and useful descriptions of reality.

The reader may notice that I have so far ignored the issue of whether there exists an objective reality that science may or may not zoom in towards. I will continue to ignore that rather boring issue. Instead, the existence of an objective reality will be taken for granted. There is, of course, no way to *prove* the existence of an objective reality, and I'll be content with referring to Haack (2003), Sokal (2008) and Häggström (2011c, 2012b) for accounts of how utterly fruitless relativist positions that deny such an objective reality are.[339]

6.5 Is the study of a future intelligence explosion scientific?

In October 2013, I invited David Sumpter, an applied mathematician at Uppsala University, to write a guest post for my blog *Häggström hävdar*, outlining the critical view I knew he had of the kind of futurology (involving radical technology scenarios) discussed in other chapters of this book. The tone is set by the title of his post: "Why 'intelligence explosion' and many other futurist arguments are nonsense."[340] Here, as Sumpter explains, the word "nonsense" should be read literally as "not related to any sense data." Like Francis Bacon (but unlike Plato), Sumpter holds empirical observation to be an essential ingredient in science, and like Popper (but unlike Bacon) he emphasizes its interplay with theory building:

---

[339] Bostrom (2003a) has suggested an argument for the hypothesis that the physical world we inhabit is not the most fundamental level of reality, but a computer simulation in a bigger and more fundamental world (and the Abrahamic religions offer a related hypothesis, although less well supported by rational argument). I side with Chalmers' (2003) view that even if the simulation hypothesis turned out to be true, it would not threaten our physical world's status as an objective reality.

A different issue is whether, by means of the scientific method, we can reach, or come arbitrarily close to, all truths about the world we live in. That seems to me like a wide open issue. Deutsch (2011) claims to have an argument for why the answer is yes, but his reasoning is a bit cloudy, and it looks as if he is merely *defining* the natural world as that which is within the reach of science, and simply dismissing those who suggest there may be something more than that as believers in the supernatural.

[340] Sumpter (2013). To offer another glimpse of the spirit of his piece, here's the closing paragraph:

Is there anything wrong with discussing models [such as those of an intelligence explosion] with low probability given the data? No. Not when it is done for fun. Like all other forms of nonsense, from watching the Matrix, going to a Tarot card reader or contemplating what might happen when I die, it has its place and is certainly valuable to us as individuals. But this place and value is not related to serious scientific discourse, and the two should not be mixed up.

Well, let me admit that, contra Sumpter's advice, I do not discuss the intelligence explosion and related topics for *fun*. I do so because I find them *important*. (And a bit of fun.)

We make models, we make predictions and test them against data, we revise the model and move forward. You can be a Bayesian or a frequentist or Popperian, emphasise deductive or inductive reasoning, or whatever, but this is what you do if you are a scientist. Scientific discoveries are often radical and surprising but they always rely on the loop of reasoning coupled with observation.

I essentially agree with this description of science. What I do *not* agree with is Sumpter's verdict that all discussions of the possibility of a future intelligence explosion fail (and must fail) this criterion. The following are my reasons, taken from the blog post I wrote in response.[341]

First of all, contemporary thinking about the nature and possible consequences of a breakthrough in artificial intelligence is *not* pure speculation in the sense of being isolated from empirical data. On the contrary, it is fed with data of many different kinds. Examples (some of which were touched upon in Section 4.5) include (a) the observed exponential growth of hardware performance known as Moore's law, (b) the observation that the laws of nature have given rise to intelligent life at least once, and (c) the growing body of knowledge concerning biases in the human cognitive machinery that Sumpter somewhat nonchalantly dismisses as irrelevant.[342] Some of the best recent works on the possibility of an intelligence explosion deal with these and many other empirical findings; see, e.g., Yudkowsky (2013a), Hanson and Yudkowsky (2013), Bostrom (2014), as well as the snapshots from these works that were given in Sections 4.5 and 4.6.[343] No single one of these data implies an upcoming intelligence explosion on its own, and it's too early to tell whether they do so in combination. However, they all serve as input to the emerging theory on the topic, and they are all subject to refinement and revision, as part of "the loop of reasoning coupled with observation" that Sumpter talks about in the above-quoted passage.

At this point, some readers will presumably object that it is not statements about down-to-earth things like current hardware performance growth or biases in human cognition that need to be tested, but high-flying hypotheses like

(C_1)   An intelligence explosion is likely to happen around 2100.

---

[341] Häggström (2013b).

[342] Here I'm referring to Sumpter's (2013) grossly unfair summary of the arguments of Muehlhauser (2011):

> Luke Muehlhauser believes that by listing a few examples of limitations in human heuristics and noting that computers are good search engines that he can convince us that we "find ourselves at a crucial moment in Earth's history. . . . Soon we will tumble down one side of the mountain or another to a stable resting place".

[343] Among these writers, Hanson seems to be the one who is leaning most in the Sumpterian direction of wanting a better (i.e., larger) data-to-theorizing ratio. In Hanson (2008d) he presents a small data set that may be of some relevance:

> Taking a long historical long view, we see steady total growth rates punctuated by rare transitions when new faster growth modes appeared with little warning. We know of perhaps four such "singularities": animal brains (∼600MYA), humans (∼2MYA), farming (∼10KYA), and industry (∼0.2KYA). The statistics

(C_1) is clearly not amenable to direct (as opposed to *indirect* empirical testing through testing of the various down-to-earth empirical findings used in whatever arguments are used to deduce it) empirical testing – at least not at the time of writing of this book. So does that make hypotheses like (C_1) unscientific? This is a typical case where an overly rigid application of Popperian falsificationism goes wrong. If (C_1) were an isolated unsupported statement, there is something to the harsh verdict of unfalsifiability. That, however, is typically *not* how statements like (C_1) appear in the intelligence explosion literature – or, more precisely, in the parts of the intelligence explosion literature worth citing, such as the publications by Yudkowsky and Bostrom mentioned above.[344] There, statements like (C_1) appear more as predictions from a larger body of theory.

To take us into more familiar territory, an example from climate science may serve as comparison. Climate science deals routinely with hypotheses such as

(C_2)　　Under a business-as-usual greenhouse gas emission scenario, global average temperature in 2100 will likely exceed the pre-industrial level by at least 3°C.

Just as with (C_1), we cannot directly test (C_2) today. But (C_2) builds deductively, via climate models, on various more directly observable and testable properties of, e.g., the greenhouse effect, the carbon cycle and the water vapor feedback

---

of previous transitions suggest we are perhaps overdue for another one, and would be substantially overdue in a century.

His talk of "overdue" is due to the observation that the epochs between singularities seem to decrease roughly as a geometric progression $(a, ar, ar^2, ar^3, \dots)$. This may seem striking, but note that it is based just on three epochs, which means that r is estimated from just two pairs of consecutive epochs. The first pair yields the estimate $\hat{r}_1 = \frac{1.98MY}{598MY} \approx \frac{1}{300}$, and the second pair yields $\hat{r}_2 = \frac{9.8KY}{1.98MY} \approx \frac{1}{200}$. It is the fact that these two numbers fall in the same ballpark that encourages us to think of a geometric progression, and it is furthermore encouraging (from a mathematical modeling point of view) to see that the inverses of these estimates of r fall in the same ballpark as the ratio by which growth rates have increased at each transition: "Each era of growth before now [...] has eventually switched suddenly to a new era having a growth rate that was between 60 and 250 times as fast" (Hanson, 2008c). Yet, my instinct as a statistician tells me that this is extremely little to go on, and I would advise a healthy dose of skepticism. In Hanson (2008c) we find the following statement:

Therefore, we must admit that another singularity ... could lie ahead. Furthermore, data on these previous apparently similar singularities are some of the few concrete guides available to what such a transition might look like. We would be fools if we confidently expected all patterns to continue. But it strikes me as pretty foolish to ignore the patterns we see.

Reading charitably, I can (just barely) accept the statement as a nuanced and reasonable position on the value of these data, but when – in a comment to the blog post by Yudkowsky (2008c), later reproduced in Hanson and Yudkowsky (2013) – he writes that "it seems reasonable to me to assign a $\sim$ ¼ [to] ½ probability to the previous series not continuing roughly as it has," then I think he assigns way too much evidential value to his data.

[344] I hasten to add that neither Yudkowsky, nor Bostrom, claims that 2100 is the right time table for when to expect an intelligence explosion. Their beliefs about the likely timing of such an event are fairly spread out.

mechanism. And since we accept not only induction but also deduction as a valid ingredient in science, no serious thinker rules out (C_2) from the realm of the scientific. Hence, by the same token, (C_1) cannot reasonably be dismissed as unscientific.

I expect the above argument to convince most readers that a hypothesis like (C_1) is possible to study scientifically. But there is one category of potential readers for whom it probably does not, namely climate deniers of the kind discussed in Footnote 40, who will typically regard even (C_2) as a hypothesis unworthy of scientific study, so that an argument ending with "... and since (C_2) counts as a scientific hypothesis, so does (C_1)" loses its force. For such readers, the following alternative example may be more illuminating. Instead of (C_2), consider the hypothesis

(C_3) `A newborn human who is immersed head-to-toe for 30 minutes in a tank of fuming nitric acid will not survive.`

Hypothesis (C_3) has (to my knowledge) never been tested directly, and hopefully never will. Yet, it is fair to say that the scientific evidence for (C_3) is overwhelming, because it can be deduced from empirically well-established results in chemistry and human physiology.

Going back to my comparison between (C_1) and (C_2): lest anyone misunderstand and read more into it than I intended, let me stress that they are *not* on equal footing in terms of the empirical support they enjoy. Since climate science is such an incomparably more well-developed science than that of a future intelligence explosion, a hypothesis like (C_2) stands on incomparably firmer ground than (C_1). But the fact that the study of intelligence explosion is such a poorly developed area (partly due simply to it being so young, but perhaps also to the questions studied being inherently very difficult) does not make it unscientific.

6.6 Statistical significance

If, as argued in Sections 6.3 and 6.4, Popper's falsifiability condition is not as straightforward a demarcation criterion for science as first meets the eye, then a legitimate request is "give me another criterion, then!" But I do not have one – the devil is in the details. A somewhat finer level of detail compared to the above discussions of falsification and related concepts (but still far from the level of detail that scientists need to consider in their daily work) was offered in a list published in the fall of 2013 in *Nature*, quickly followed by an endorsement in *The Guardian* with the title "Top 20 things politicians need to know about science."[345]

---

[345] Sutherland, Spiegelhalter and Burgman (2013); Milman (2013). In the same year, I had a paper (Häggström, 2013a) with a similar message, but with scientists rather then politicians as the group whose overall level of knowledge I wished to see improved. I suspect (although this remains no more than a suspicion) that Sutherland et al. may actually have had that same group in mind, but took

One of the items on the list is the observation that scientists are mere humans, with far-reaching consequences for their fallibility. Of the remaining 19, I note with some satisfaction, as a professor of mathematical statistics, that between 17 and 19 (depending on how narrowly or broadly the subject is interpreted) of them concern the methods offered by mathematical statistics.[346] This domination on the list does not, in my opinion, exaggerate the importance of solid statistical methods for doing high-quality science. A series of recent studies indicate lack of good statistical practice as a major bottleneck (possibly even the most severe one) in the production of good science.[347] This gives a good excuse to spend a few pages on statistical methods, and I have chosen to focus on the concept of **statistical significance**, my reason for this being threefold: (a) it is of crucial importance to all of science (as illustrated by its leading role in several of the items on the *Nature–Guardian* list) and yet so easy to misunderstand, (b) it resonates well with Popperian falsificationism, and (c) it will turn out to be important to the discussion of the so-called Doomsday Argument in Chapter 7.

Statistical significance is probably best explained through a concrete example, and I will settle for as simple an example as possible: coin tossing, and an attempt to find out whether the coin is fair or biased.[348] We are used to assuming that the coin we toss is fair, meaning that each time we toss it, it has equal probabilities of coming up heads or tails:

$$P(\text{heads}) = P(\text{tails}) = \frac{1}{2}.$$

In order to test this property, we replace this assumption by a more non-committal one, namely that there is a number q between 0 and 1 such that the coin has

---

politicians as a kind of "hostage" in order not to insult their fellow scientists by teaching them such basic stuff. I have sometimes (and in contrast to my bluntness in Häggström, 2013a) used a similar tactic myself. My attention to the tactic was first drawn by Dennett's (1989) review of *The Emperor's New Mind* (Penrose, 1989). Dennett speculated that, despite the stated popular science ambition of the book, what Penrose really wanted was to get through to his scientist colleagues.

[346] Here is the full list (without the annotations and comments given in the original publication). (1) Differences and chance cause variation. (2) No measurement is exact. (3) Bias is rife. (4) Bigger is usually better for sample size. (5) Correlation does not imply causation. (6) Regression to the mean can mislead. (7) Extrapolating beyond the data is risky. (8) Beware the base-rate fallacy. (9) Controls are important. (10) Randomization avoids bias. (11) Seek replication, not pseudoreplication. (12) Scientists are human. (13) Significance is significant. (14) Separate no effect from non-significance. (15) Effect size matters. (16) Data can be dredged or cherry-picked. (17) Extreme measurements may mislead. (18) Study relevance limits generalizations. (19) Feelings influence risk perception. (20) Dependencies change the risks.

[347] See, e.g., Ziliak and McCloskey (2008) and Häggström (2013a).

[348] The following discussion is mostly taken from the popular paper by Häggström (2013c). For a more general treatment of statistical hypothesis testing and statistical significance, see, e.g., the introductory undergraduate textbook by Olofsson and Andersson (2012), or the more advanced book by Lehmann and Romano (2008).

probability q of coming up heads and the complementary probability $1 - q$ of coming up tails:

$$\begin{cases} P(\text{heads}) = q \\ P(\text{tails}) = 1 - q \end{cases} \quad \text{for some } q \text{ in the interval } 0 \le q \le 1. \tag{9}$$

We assume that the probability of heads has this same value q every time we toss the coin, independently of previous tosses. If $q = 1/2$, the coin is fair; otherwise it is biased. The model assumption (9) leaves both options open.

Consider a situation where, as an experiment to try to work out the value of q, we toss the coin n times and count the number of times X that it comes up heads. The so-called Strong Law of Large Numbers guarantees that if we keep sampling indefinitely, letting $n \to \infty$, the observed frequency X/n tends (with probability 1) to q. In practice, we need to stop after a finite number of tosses, and then the information about q that we get from the experiment is less certain. Let's say $n = 10$, and that the number of heads turns out to be $X = 2$. What can we then conclude? In particular, does the outcome of the experiment support a suspicion of biasedness?

Intuitively, one would perhaps expect $X = 5$ heads from the $n = 10$ tosses if the coin is fair, but a moment's thought reveals that since each toss is random, there's a chance that X will be different from 5. How different can it be and still be considered reasonably in line with the hypothesis that $q = 1/2$? The notion of statistical significance is designed to answer this question.

Statistical hypothesis testing is always relative to some **null hypothesis** which typically says that some effect size or some parameter is zero. In our example, however, the null hypothesis is that $q = 1/2$. Given the data, the **p-value** can loosely be defined as the probability of obtaining at least as extreme data as those we actually got, given that the null hypothesis is true.[349] If the p-value ends up below a given threshold – which is called the **significance level** and which, for reasons that have more to do with tradition than anything else, is usually taken to be 0.05 – then **statistical significance** is declared.

A statistically significant outcome is usually taken to support a suspicion that the null hypothesis is false, the logic being as follows. If statistical significance is obtained, then we know that either the null hypothesis is false, or an event of low probability (with the traditional significance level, the probability is at most 0.05) happened. Now, low-probability events do happen sometimes, but we expect them typically *not* to happen, and if such a thing didn't happen this time, then the only remaining option is that the null hypothesis is false. The lower the threshold for significance is, the stronger a statistically significant result is considered to count against the null hypothesis.

---

[349] It is the fuzzy notion of "extreme" that contributes most to the looseness of this definition. In simple situations such as this one, one's first guess of how to specify extremeness is usually right, but in more complicated situations it may be necessary to specify a so-called alternative hypothesis and invoke the Neyman–Pearson Lemma; see Lehmann and Romano (2008).

Too see what $X = 2$ heads out of $n = 10$ tosses amounts to, we need to work out what the distribution of X is under the null hypothesis. It turns out[350] to follow a so-called **binomial** distribution with parameters 10 and 1/2, meaning that

$$P(X = k) = \left(\frac{1}{2}\right)^{10} \frac{10!}{k!(10-k)!} \quad \text{for } k = 0, 1, 2, \ldots, 10.$$

Rounded off to three decimal places, the probabilities $P(X=0)$, $P(X=1)$, ..., $P(X=10)$ are plotted in Figure 6.1.

We read from the diagram that $P(X=2) = 0.044$. We might be tempted to take that as the p-value, to note that $0.044 < 0.05$, and to declare triumphantly that we have statistical significance. That would be a mistake, however, because the p-value was not defined as the probability of getting *exactly* the outcome we got, but as the probability of getting *at least as extreme* an outcome as the one we got, and surely the outcomes $X = 1$ and $X = 0$ are more extreme than the $X = 2$ we got. And by symmetry, the cases $X = 8$, $X = 9$ and $X = 10$ need to be considered just as extreme, so we get the p-value as

$$p = P(X = 0) + P(X = 1) + P(X = 2)$$
$$+ P(X = 8) + P(X = 9) + P(X = 10),$$

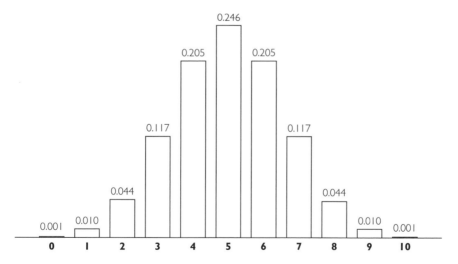

Figure 6.1 Probability distribution for the number X of heads in 10 independent tosses of a fair coin.

[350] I'll spare the reader the details, which can be found in Olofsson and Andersson (2012) or almost any other introductory textbook on probability and statistics. With the heads-probability $q = 1/2$ replaced by an arbitrary $q \in [0, 1]$, the formula becomes

$$P(X = k) = q^k (1-q)^{10-k} \frac{10!}{k!(10-k)!}.$$

which turns out to equal 0.110.[351] We note that $0.110 > 0.05$, so that with the standard significance level, the outcome $X = 2$ is *not* statistically significant.

OK, that's the easy part. The hard part is how to *interpret* statistical significance, or lack thereof. Here are many pitfalls, including the following six.

(1) Does the lack of a statistically significant deviation from what is expected under the null hypothesis $q = 1/2$ show that the null hypothesis is true? *No.* What the lack of statistical significance says is that the data obtained is highly consistent with having $q = 1/2$, and hence give little or no reason to rule out that hypothesis. But there are plenty of alternative hypotheses one could pose, such as $q = 0.2$, $q = 0.3$ and (literally!) infinitely many others, that the data are equally well in line with, or even better. Ruling out all of these based on the data wouldn't make sense, but that is what a claim that $q = 1/2$ would entail.

(2) Suppose instead that we had only $X = 1$ heads in the ten tosses. The p-value calculation would then have resulted in $p = 0.022$, falling below the significance level 0.05, so we would be in a position to declare statistical significance. Would this have meant that we were in a position to conclude that the null hypothesis is false and that the coin is biased? *No.* If the null hypothesis is true, then we can still have a probability of up to 0.05 of obtaining statistical significance (that is what the significance level *means*). That is hardly a low enough probability to warrant certainty or near-certainty,[352] and in fact spurious statistical significance is expected to happen about once in 20 studies where the null hypothesis is true.[353]

---

[351] The notion of "extreme" actually depends on how we phrased the problem. If, before the experiment, we had said something like "Is this coin really fair, or could it be that the heads-probability q is less than $1/2$? Let's see if the number of heads comes out to be abnormally small," then we can skip the $X = 8$, $X = 9$ and $X = 10$ terms and get the slightly better p-value of $P(X = 0) + P(X = 1) + P(X = 2) = 0.055$. But if we didn't say that, we must treat the problem symmetrically. What we absolutely cannot do is *first* to do the experiment and look at the data, and *then* to formulate the hypothesis: "Hm, too few heads, yeah, I guess that is precisely what we were looking for." That would be akin to the practice of throwing a dart, and no matter where it hits, proclaim that the place it hit was exactly where you were aiming – and the surest way to obtain superficially impressive but spurious research results.

[352] A certain kind of reader will ask at this point what significance level I would be prepared to equate with near-certainty or "beyond reasonable doubt." Would 0.001 do? Or 0.000001? All I can say in response to this very general question is "it depends." It depends, e.g., on what else is known besides the data of the specific study at hand, and on what's at stake. But I always protest when someone talks about $p < 0.05$ as if it meant "beyond reasonable doubt" or anything of a similar flavor. A better suggestion for what $p < 0.05$ can usually be taken to mean is "Hm, interesting, perhaps there is an effect here, this may be worth studying further."

[353] If all this seems terribly elementary and pedestrian to the reader, I apologize, but the fact of the matter is that there are Harvard professors who do not grasp these elementary points. See Mitchell (2014) for an account where the author apparently believes that $p < 0.05$ disproves the null hypothesis in the same way as a red raven disproves the all-ravens-are-black hypothesis, and Häggström (2014b) for a response.

(3) Can the obtained p-value of 0.110 be interpreted as the probability that the null hypothesis of a fair coin is true? *No, no, no!* This very common misinterpretation involves the so-called **fallacy of the transposed conditional**, in this case a confusion between on one hand the conditional probability $P(\text{data}|\text{null hypothesis})$ of obtaining data in a particular category given that the null hypothesis is true, and on the other hand the conditional probability $P(\text{null hypothesis}|\text{data})$ that the null hypothesis is true given the data. Such a transposition between data and hypothesis is not even remotely warranted, because it is *not* in general true, for two events A and B, that $P(A|B) = P(B|A)$. There is not even any approximate sense in which this is true. To see this, consider the experiment of selecting a person NN at random according to the uniform distribution on the set of humans alive today. Let A be the event that NN is world chess champion, and let B be the event that NN is Norwegian. Then

$$P(A|B) \approx \frac{1}{5,000,000} = 0.0000002$$

because there are about five million Norwegians, only one of whom is world chess champion, whereas

$$P(B|A) = 1$$

because the world chess champion *is* Norwegian.[354] Probabilities range between 0 and 1, and so the discrepancy in this case between $P(A|B)$ and $P(B|A)$ is nearly as large as the difference between two probabilities can get.

So what *is* the probability that the null hypothesis is true, given our data $X = 2$? I cannot answer that, because the paradigm of frequentist statistics, of which statistical hypothesis testing forms a part, *never* provides probabilities of hypotheses. The scientist or decision-maker who needs the statistical analysis to provide such probabilities has no choice but to turn to so-called Bayesian statistics, which however involves the controversial step of postulating an a priori distribution, specifying how likely we thought the null hypothesis was before we did the experiment. More on that in Sections 6.8 and 6.9.

(4) Is statistical significance all that matters? *No!* There is a widespread scientific malpractice known as **sizeless science**, which in short can be described as a scientist being so obsessed with statistical significance that he or she forgets to ask whether the observed effect is also large enough to exhibit subject-matter significance; see Ziliak and

---

[354] At the time of this writing, the Norwegian chess player Magnus Carlsen has held the world championship title since November 2013; see Crowther (2013).

McCloskey (2008) and Häggström (2010b, 2013a).[355] Here's a passage from Häggström (2010b) explaining the concept of subject-matter significance:

Imagine now that a new drug for reducing blood pressure is being tested and that the fact of the matter is that the drug does have a positive effect (as compared with a placebo) but that the effect is so small that it is of no practical relevance to the patients health or well-being. If the study involves sufficiently many patients, the effect will nevertheless with high probability be detected, and the study will yield statistical significance. The lesson to learn from this is that in a medical study, statistical significance is not enough – the detected effect also needs to be large enough to be *medically significant*.

(5) The Duhem–Quine thesis on the impossibility of testing scientific hypotheses in complete isolation from others, discussed in connection with Popperian falsificationism in Section 6.3, applies here too. Supposing we get a good enough p-value to make us think that the null hypothesis $q = 1/2$ is wrong, we should pause to think about whether it's actually the null hypothesis that fails, or if the culprit could be any of the many spoken or unspoken supporting hypotheses needed for deriving the null hypothesis distribution of X. Besides the usual suspects, such as whether we might be hallucinating, and whether we really can trust the optical mechanisms that allow us to tell heads from tails, there is one supporting hypothesis that stands out as particularly relevant in the coin tossing example, namely the independence assumption. We could have $q = 1/2$ and still have dependence between consecutive tosses, giving rise (on average) to longer streaks than under the independence assumption.[356] This will cause the distribution of X to have more mass at or near the endpoints $X = 0$ and $X = 10$, and less in the middle, thus invalidating the p-value calculation we did above.

---

[355] Ziliak and McCloskey do an excellent job describing the malpractice and researching how widespread it is, but their recipe for a solution is very wrong, as I stressed in Häggström (2010b):

The Cult of Statistical Significance is written in an entertaining and polemical style. Sometimes the authors push their position a bit far, such as when they ask themselves: "If null-hypothesis significance testing is as idiotic as we and its other critics have so long believed, how on earth has it survived?" (p. 240). Granted, the single-minded focus on statistical significance that they label sizeless science is bad practice. Still, to throw out the use of significance tests would be a mistake, considering how often it is a crucial tool for concluding with confidence that what we see really is a pattern, as opposed to just noise. For a data set to provide reasonable evidence of an important deviation from the null hypothesis, we typically need *both* statistical *and* subject-matter significance.

See, furthermore, the exchange with Ziliak and McCloskey in the comments section of Häggström (2013d).

[356] An analysis of the Newtonian dynamics of coin tossing indicates that this will in practice typically be the case; see Diaconis, Holmes and Montgomery (2007).

(6) Closely related to statistical significance is the concept of a **confidence interval**. Sometimes, when faced with an unknown parameter, such as the heads-probability q in the coin example, it is possible to test an infinite number of null hypotheses simultaneously, one for each possible parameter value. The set of all parameter values for which statistical significance is *not* obtained will, in sufficiently simple situations, form an interval. If the significance level is taken to be α, the interval is called a $100(1 - \alpha)\%$ confidence interval for the parameter, so with $\alpha = 0.05$ as above we get a 95% confidence interval.[357] In the coin tossing example above, with 2 heads from 10 tosses, the 95% confidence interval for q turns out to be

$$0.025 \leq q \leq 0.556 \quad (95\%).$$

If carried out correctly, then (no matter what the real parameter value q is) the procedure has probability at least 0.95 of resulting in an interval that covers the real value. An extremely tempting mistake here is to say that given our data (2 heads from 10 tosses), the true value q has probability 0.95 of being in the interval $[0.025, 0.556]$, but analogously to the mistake discussed in (3), this would be to commit the fallacy of the transposed conditional. Like statistical significance, confidence interval is a concept in frequentist statistics, which deals with the conditional distribution of data given the parameter value, not vice versa. Similarly to (3), if we want to derive probabilistic statements about the whereabouts of the parameter q, we have no choice but to give up the frequentist approach and instead employ Bayesian statistics. See Footnotes 358 and 359 in Section 6.8.

6.7 Decision-makers need probabilities

In item (3) in the list of cautions concerning the interpretations of statistical significance, I mentioned the inability of frequentist statistical methods to provide us with numbers quantifying the probability of one scientific hypothesis or another. Yet, as soon as we go out and apply science in real-life decisions, we do need such probabilities. Here's why, in as simple an example as possible.

Suppose that in a given situation we have two actions A and B to choose from. Which one is best depends on which one of two competing scientific hypotheses (H_1) and (H_2) is true. Let's say (H_1) and (H_2) exhaust the possibilities, in such a way that we know that exactly one of them is true. Let's also say that which action of A and B is the best, in terms of utility (which, here, we choose to measure in \$, deferring any qualms over equating utility with money to a brief discussion in

---

[357] See, e.g., Olofsson and Andersson for more on confidence interval methods.

the next paragraph and a slightly longer one in Chapter 10), depends on whether (H_1) or (H_2) is true. Suppose the payoffs for the four different pairs of actions and states-of-the-world are as follows:

	(H_1)	(H_2)
A	$10	$10
B	$20	$5

So action A gives $10 no matter which hypothesis is correct. Action B has the greater potential reward, giving $20 in the case of (H_1), but is also more risky, giving merely $5 in the case of (H_2). Since there is a remaining scientific uncertainty as to which of the hypotheses is correct, we cannot know which action is better, A or B. But if we have a way of attaching *probabilities* to (H_1) and (H_2), then (and only then!) there is a way out. If (H_1) has probability p_1, and (H_2) has probability $p_2 = 1 - p_1$, then action A yields an expected monetary reward E_A (i.e., the reward that we expect on average) given by

$$E_A = p_1 \cdot \$10 + p_2 \cdot \$10 = p_1 \cdot \$10 + (1 - p_1) \cdot \$10 = \$10$$

and action B gives expected monetary reward

$$E_B = p_1 \cdot \$20 + p_2 \cdot \$5 = p_1 \cdot \$20 + (1 - p_1) \cdot \$5 = \$5 + p_1 \cdot \$15.$$

We get $E_A < E_B$ if $p_1 > \frac{1}{3}$, whereas $E_A > E_B$ if $p_1 < \frac{1}{3}$, while the borderline case $p_1 = \frac{1}{3}$ gives $E_A = E_B$. Hence, if $p_1 < \frac{1}{3}$ we should prefer action A, if $p_1 > \frac{1}{3}$ we should prefer action B, and in the borderline case the actions are equally preferable. This is what so-called Bayesian decision theory dictates that we should do.

One caveat, however: Perhaps we have a value of p_1 that exceeds $\frac{1}{3}$ (for concreteness, let's say $p_1 = \frac{1}{2}$), but we nevertheless prefer action A because we want to play on the safe side, not willing to risk the $10 we thereby secure, for an uncertain prospect of obtaining the $20. This is a valid objection: getting money is normally a good thing, but *how* good need not be a linear function of the amount received. Personally, if I knew that $p_1 = \frac{1}{2}$, I would not hesitate to follow the above decision-theoretic calculation and settle for action B, giving $5 or $20 with probability $\frac{1}{2}$ each, yielding an expected reward of $12.50, rather than action A and a guaranteed $10. If we multiply the numbers by a million, however, the situation if different. I would prefer a guaranteed $10,000,000 to a fifty–fifty chance of getting $5,000,000 or $20,000,000 (because I judge the $10M to be good enough to secure a good life for me and my family, and since doubling the amount is not likely to make that much of a difference to my happiness, I wouldn't want to risk half of it for even odds of doubling it). Bayesian decision theory is, however, capable of taking into account such **risk aversion** (more on which will be said in Section 10.2) or other nonlinearities in our preferences. It does so by introducing a so-called **utility function** U, where $U(x)$ is the utility that receiving $x would mean to us; we can then proceed with the maximization procedure above with $U(x)$ in place of \$x. The utility function U can in principle be arbitrary, but is usually taken to be increasing (the more money, the better), and concave in the case that we are risk averse.

6.8 Bayesian statistics

This highlights the desirability of being able to attach probabilities to various hypotheses and parameter estimates in our scientific studies and statistical analyses. While the frequentist statistical paradigm, which the theory of p-values and statistical significance forms part of, is unable to provide such probabilities, the rival theory of **Bayesian statistics** does so. That comes at a price, however: in Bayesian statistical analysis, we need to specify an **a priori distribution**, which describes our beliefs, prior to the scientific experiment, about what the world (or relevant aspects thereof) is like.[358] The a priori distribution is then translated into an **a posteriori distribution**, describing our beliefs (or the beliefs we ought to have) *after* we have seen the outcome of the experiment. This translation is done by so-called Bayesian conditionalization, meaning that the a posteriori distribution is obtained by conditioning the a priori distribution on the outcome of the experiment.[359]

The main difference between frequentist and Bayesian statistical methodology is usually said to be their interpretation of probabilities. To a Bayesian, probability (usually) represents subjective belief, whereas to a frequentist, the probability of an event represents the long-term fraction of times at which the event happens, in the hypothetical situation where an experiment undergoes independent repetitions,

---

[358] For instance, in the coin tossing example in Section 6.6, the a priori distribution may simply be some probability distribution on $[0,1]$, describing our beliefs (prior to the experiment) about the whereabouts of the parameter q, i.e., of the coin's heads-probability.

And to be even more concrete, suppose, for the coin tossing example in Section 6.6, that, prior to the experiment, we are certain that the heads probability is either $\frac{1}{4}$, $\frac{1}{2}$ or $\frac{3}{4}$, and that we think each of these equally likely. That corresponds to the a priori distribution $\mathbf{P}$ where $\mathbf{P}(q = \frac{1}{4}) = \mathbf{P}(q = \frac{1}{2}) = \mathbf{P}(q = \frac{3}{4}) = \frac{1}{3}$.

[359] In a finite setting, with n possible worlds $A_1, \ldots, A_n$, and an a priori distribution attaching probabilities $\mathbf{P}(A_1), \ldots, \mathbf{P}(A_n)$ to these worlds, Bayesian conditionalization amounts to letting the a posteriori probabilities of these worlds be $\mathbf{P}(A_1|E), \ldots, \mathbf{P}(A_n|E)$, where E is the event of getting precisely the outcome of the experiment that we did get. Often, $\mathbf{P}(E|A_i)$ is more readily obtained than $\mathbf{P}(A_i|E)$. Then, rather than succumbing to the fallacy of the transposed conditional, we can calculate the a posteriori distribution using **Bayes' Theorem**, which states that

$$\mathbf{P}(A_i|E) = \frac{\mathbf{P}(A_i)\mathbf{P}(E|A_i)}{\mathbf{P}(A_1)\mathbf{P}(E|A_1) + \cdots + \mathbf{P}(A_n)\mathbf{P}(E|A_n)}.$$

In the example from Footnote 358, applying Bayes' Theorem to the evidence $E = \{2 \text{ heads out of } 10\}$ and the prior $\mathbf{P}(q = \frac{1}{4}) = \mathbf{P}(q = \frac{1}{2}) = \mathbf{P}(q = \frac{3}{4}) = \frac{1}{3}$ turns out (although this is not the place to offer mathematical details) to give, with probabilities rounded to the third decimal,

$$\mathbf{P}(q = \tfrac{1}{4}|E) = 0.864$$
$$\mathbf{P}(q = \tfrac{1}{2}|E) = 0.135$$
$$\mathbf{P}(q = \tfrac{3}{4}|E) = 0.001$$

ad infinitum. This, however, does not quite capture the most essential difference in the daily practice of statistics, which is better put as follows.

Decision-makers need scientific results, and since the results are uncertain, they need probabilistic quantifications of the results. What is the probability that all ravens are black? What is the probability that under such-and-such a greenhouse gas emission scenario the global mean temperature by 2100 will not exceed the pre-industrial level by more than 3°C? There is no way that a scientific study can answer these questions without specifying prior probabilities. The Bayesian accepts the need of the decision-maker, and supplies, as best as he can, the necessary prior probabilities to feed into his analysis. The frequentist, on the other hand, sees that such prior probabilities come with a great deal of arbitrariness and subjectivity,[360] and in order to avoid tainting his statistical analysis with such subjectivity, he refuses to specify a prior distribution. Consequently, he leaves it to the decision-makers to translate his results, as best they can (and without any advice from the frequentist), into probabilities.

The 20th century saw a long principled struggle between frequentist and Bayesian statistics.[361] Frequentists had the upper hand for most of the time, but as underdogs, the Bayesians fought hard,[362] and today the scientific respectability of Bayesian statistics is better than ever.[363] Pop statistician Nate Silver, who is an outspoken advocate (and a very successful practitioner) of Bayesian statistics, writes in his bestselling *The Signal and the Noise*[364] that in recent years . . .

some well-respected statisticians have begun to argue that frequentist statistics should no longer be taught to undergraduates. And some professions have considered banning Fisher's hypothesis test from their journals. (p 260)

These are true (although somewhat cherry-picked) statements. However, when Silver gets carried away by his own rhetoric and, in the very next sentence writes that "in fact, if you've read what's been written in the past ten years, it's hard to find anything that *doesn't* advocate a Bayesian approach," he expresses a falsehood about as blatant as if he had said that "in fact, as you walk the streets of New York these days, it is hard to find even a single person with a height not exceeding 180 cm."

From my personal perspective as a member of the academic statistical community, a pattern is clearly visible. In the generation of my teachers (who were mostly

---

[360] This is almost always true. What is sometimes put forth in the defense of Bayesian statistics is that by taking the prior to be a uniform distribution (assigning equal probabilities to all parameter values or all possible "worlds" that the model encompasses), this subjectivity can be avoided. I am not at all convinced by such claims, for reasons that I explain in Häggström (2007a).

[361] See Salsburg (2001) for an easy-to-read history of the subject.

[362] Such as in Lindley (1990) and Jaynes (2003).

[363] A major contributing factor to this development is the emergence of computer algorithms allowing for the computation (or at least approximation) of Bayesian posteriors, even in very complex situations; see, e.g., Gilks, Richardson and Spiegelhalter (1996).

[364] Silver (2012). See also Häggström (2013f).

born in or around the 1940s), most of them (apart from the odd contrarian) were, and still are, hostile towards Bayesianism. For those of us born in the 1960s or later, the dominant attitude is a more pragmatic one, best described by once again invoking the toolbox metaphor: frequentist statistics offers one toolbox and Bayesian statistics offers another. There is an enormous variety in the kinds of problems a statistician encounters, and with such a variety it is better to have access to both boxes rather than just one.

6.9 Is consistent Bayesianism possible?

A beautiful theory combining decision theory and Bayesian epistemology has been built up during the 20th century; see, e.g., Ramsey (1931), de Finetti (1937), von Neumann and Morgenstern (1944), Savage (1951), Skyrms (1984) and Binmore (2009). Starting from a few simple and (according to its adherents) easy-to-accept assumptions, the theory states that any rational agent must

(i) have beliefs about the world that satisfy the axioms of probability,
(ii) update the beliefs using so-called Bayesian conditionalization, and
(iii) make decisions that maximize expected utility.

At this point, it is useful to distinguish between **fully rational Bayesian reasoning** and **practical Bayesian statistics**. The former is employed by a highly idealized agent, who is born with a prior probability distribution representing her belief about the world, and who adheres strictly to (ii) and (iii) in all her future actions and all her future belief revisions. A practical Bayesian statistician, on the other hand, need not care much at all about (iii), and employs (i) and (ii) only locally, within a single scientific project. Each time a new project is started, the practical Bayesian statistician is free to dream up a new prior. Perhaps he would ideally like to do it formally using (ii) and all the experiences he has had since the previous project, but in practice this is usually considered beyond what is computationally feasible, and is rarely (if ever) attempted.

These two versions of Bayesianism are very different. To be a fully rational Bayesian agent is a very much stronger requirement than being a practical Bayesian statistician, and very much less realistic. Yet, when one hears advocates of Bayesianism talk abstractly about its virtues, it may be hard to tell which version they mean, and one can sometimes get the impression that cards are being mixed up, e.g., by holding forth the virtues of fully rational Bayesian reasoning in defense of practical Bayesian statistics; hence my desire to make the distinction explicit.

In Section 6.1 I talked about Popperian falsificationism and Bayesian reasoning as if they were two competing scientific frameworks. Whether or not this is indeed the case, and whether or not they are incompatible with each other, depends on whether we define "Bayesian reasoning" in the sense of fully rational Bayesian reasoning or in the sense of practical Bayesian statistics. In the former case, I claim they are incompatible. We may hope they are both successful in getting more and more accurate descriptions of the world, and asymptotically zooming in towards the truth, but even if this is so, they do so by entirely different protocols. In contrast, practical Bayesian statistics may well serve as a local tool, i.e., one employed in specific research projects, forming part of a more global falsification effort within a Popperian framework. The practical Bayesian statistician may start his project with a prior that puts a fairly large probability on the hypothesis to be tested in the Popperian sense of the word, and if his statistical analysis (of whatever observations are found or experiments carried out) based on Bayesian conditionalization results in a posterior distribution that attaches very little probability to that hypothesis, his study may count as a contribution towards falsifying it.

One problem that makes full Bayesian rationality unrealistic as a framework for doing science, as well as for acting rationally more generally, is the following. When choosing a prior distribution once and for all (and let's interpret "is born with" in the definition of fully rational Bayesian reasoning metaphorically, with birth denoting simply the time at which the agent decides on a prior and starts to implement, from then on, full Bayesian rationality), she needs to foresee all possible ways the world may behave, because if she has forgotten some such ways, she will implicitly assign them probability zero, a value that Bayesian conditionalization can never cause to take off to anything positive. This is a very tall order. I cannot imagine how I might cook up a prior such that I will forever be protected against the experience of being taken by surprise by the data to the extent that I think something along the lines of

Oh wow, could *that* be what the world is like? Yes, I guess it could. Too bad I didn't take that into account in constructing my prior, because requirement (ii) forces me to assign probability zero to such a world forever on. But – please don't tell Reverend Bayes – I think I'll just override (ii) and assign a few percentage points to it anyway.

But as soon as I do that, I give up on being a fully rational Bayesian agent. A comparison with Popperian falsificationism may be instructive here. While fully rational Bayesian reasoning requires the agent to have foreseen, already at "birth," all possibilities for what the world may be like, Popperian falsificationism explicitly allows for unconstrained creativity at the stages of scientific work where new hypotheses are formulated. This, in my opinion, is very much a strength of the Popperian formalism.

In parts of the AI community aimed at construction of autonomous agents, there is an understandable attraction in the idea of making these agents rational, and due to the strength and elegance of the theory in favor of agents respecting properties (i), (ii) and (iii) discussed above, this leads to an equally understandable focus on implementing full Bayesian rationality (or at least some good approximation to that).[365] I fear, however, that implementing such agents will be an

---

[365] See, e.g., Hutter (2005). Still, not everyone is convinced that a successful AI, even one reaching superhuman intelligence levels, needs to be fully rational in the Bayesian sense of the term. The main theoretical justification of the need for Bayesian rationality involves so-called **Dutch book arguments**

extraordinary and possibly unsurmountable challenge. The main reason for my fear is the above-mentioned problem of formulating a sufficiently universal prior to be able to handle all possible ways in which the world may turn out. One reason (although by no means the only one) for my pessimism in this regard is that the set of possible ways the world can be is most likely uncountable (in the formal mathematical sense of the word, discussed in Section 4.1),[366] whereas it is a basic and inescapable fact from probability theory that no Bayesian prior, or no other probability distribution for that matter, can assign non-zero probability to more than a countable number of elements; hence, for any choice of prior, there will be possible states of the world that it ignores (i.e., assigns probability 0).

The literature does offer one major candidate for a prior distribution aspiring to serve as a suitable all-purpose universal prior, namely the so-called Solomonoff prior; see, e.g., Rathmanner and Hutter (2011). The Solomonoff prior is based on so-called Kolmogorov complexity, which in turn is based on the Turing machine concept discussed in Section 4.2. Very briefly, the Solomonoff prior is set up as follows. Each possible world H is represented by a finite or infinite binary string x, whose Kolmogorov complexity $K(x)$ is defined as the length of the shortest program, on a given universal Turing machine, that produces x as its output. World H is then assigned probability $2^{-K(x)}$, so the greater the Kolmogorov complexity, the smaller the probability. Note that only those x that are computable on a Turing machine are assigned positive probability. This may seem like a strong restriction on the notion of "every possible world," but if a strong version of the Church–Turing thesis (see Section 4.2) is true, then the true world is indeed computable and will be assigned positive probability.

In Häggström (2013e), I outline the reasons for my pessimism about this approach to constructing a fully rational Bayesian AI. The main irony is this: Kolmogorov complexity is not computable in the sense of the Church–Turing thesis,[367] so if there is to be any chance to build a machine that uses the Solomonoff prior in deciding its actions, the Church–Turing thesis has to be false. But in that case, the Solomonoff prior attaches probability 0 to the true state of the world. These criticisms, however, do not rule out the possibility that Kolmogorov complexity and Solomonoff induction may turn out useful as inspirations for other,

(see, e.g., Binmore, 2009), amounting to the observation that any agent deviating from Bayesian rationality exposes himself to the possibility of having all his resources deterministically taken away by a shrewed bookmaker. Perhaps too much gravity has been attached to these arguments. Bostrom (2014), in his discussion of the cognitive capabilities of a superintelligent AI, points out that "agents who do not expect to encounter savvy bookies, or who adopt a general policy against betting, do not necessarily stand to lose much from having some incoherent beliefs – and they may gain important benefits [such as] reduced cognitive effort, social signaling, etc." (p 111).

[366] One way to see this is to note that the laws of nature involve a number of fundamental constants that (or so it seems) can take any value along a continuum of possible values; each such continuum contains uncountably many values. (Recall the diagonal argument in Section 4.1, which showed that there are uncountably many real numbers. That proof can easily be adapted to show that any interval of non-zero length on the real line contains uncountably many real numbers.)

[367] This is strongly related to the incomputability of the halting problem, discussed in Section 4.2.

more practical, schemes; see, e.g., Veness et al. (2011) and Strannegård et al. (2014) for work in this direction.

6.10 Science and engineering

Just one more thing before closing this chapter. Since this is a book about science, technology and the future of humanity, we ought to also say a few words about the relation between science and technology, or between science and engineering, the latter being the practice of developing new technology. It is more or less a premise of the book that the main influence that science has on society and our lives is via technology. Scientific findings such as the second law of thermodynamics, the theory of evolution by natural selection, quantum entanglements, the concept of a universal Turing machine, Einstein's relativity and the Big Bang may all be awe-inspiring, but what really changes our lives are engineering feats like the combustion engine, penicillin, the iPad and the Internet.

Bacon saw science as paving the way for new technology, and Popper credits him with having anticipated the industrial revolution.[368] And we can no doubt say that science has done an enormous amount of such paving, but the claim may still need a little bit of moderation.[369] Haack (2003) writes:

> We are accustomed to thinking of technology as simply the fruit of theoretical developments in science; but this is something of an over-simplification. Barzun lists "the steam engine, the spindle frame and power loom, the locomotive, the cotton gin, the metal industries, the camera and plate, anesthesia, the telegraph and telephone, the phonograph and electric light" as all invented "by men whose grasp of scientific ideas was slight, or at best empirical". (p 318, quoting Barzun, 1964)

One could still ask whether, in several of these cases, the "men whose grasp of scientific ideas was slight" could have come up with their inventions without earlier scientific progress on, e.g., electricity. It furthermore seems to me that advances in the various emerging technologies discussed in this book will mostly need to build on scientific advances. A possible exception might be a breakthrough in AI: it is not entirely implausible that right now as I'm writing these lines, somewhere there's a lone hacker genius with neither any formal training in computer science,

---

[368] Here is Popper (1994), once more:

Bacon ... had the vision of a new age, of an industrial age which would also be an age of science and of technology. Referring to the accidental discovery of gunpowder, and of silk, he spoke of the possibility of a systematic scientific search for other useful substances and materials, and of a new society in which, through science, men would find salvation from misery and poverty. (p 85)

[369] It is also worth noting that the flow between science and technology is not a one-way street. Technological developments have often provided the necessary prerequisites for scientific breakthroughs. The indebtedness of particle physics to the engineering behind particle accelerators at CERN and elsewhere is a particularly striking example, but there are many others.

nor any knowledge of contemporary modeling in AI, who is about to trigger an intelligence explosion. However, such a lone hacker would presumably still do his work on machines that were made possible partly by advances in computer science.

We should also note that the distinction between science and engineering is not always very clear-cut. Take, for instance, the development of new pharmaceutical drugs. That is an engineering task, we may argue, but many parts of it look very much like science, including in particular the clinical trials, which are typically carried out as double-blind and randomized controlled experiments, thus adhering to a kind of gold standard of scientific practice.[370]

Or take physics. Physics is, of course, not just about fundamental particles (or whatever the most basic constituents of matter turn out to be) but even more about what kinds of phenomena are produced by various configurations (small or large) of such particles. Engineering is then, in a sense, just the special case where the phenomena we're looking for are useful devices. Computer scientist Steve Easterbrook at the University of Toronto has some nice slides[371] where he contrasts on one hand what he calls the "traditional view" of the relation between science and engineering, emphasizing differences, and on the other hand the "more realistic view," emphasizing similarities. According to the former, scientists create knowledge while engineers apply it, scientists study the world as it is while engineers seek to change it, scientists use explicit knowledge while engineers use tacit knowledge, and scientists are thinkers while engineers are doers. According to the latter, scientists as well as engineers create knowledge, they are problem-driven, they seek to understand and explain, they design experiments (scientists) and devices (engineers) to test theories, and they rely to a large extent on tacit knowledge.

In contrast to Easterbrook, Drexler (2013) finds it useful to emphasize the differences between science and engineering. While both activities deal in abstract models, concrete descriptions and real physical systems, they imply opposite main directions of information flow between the three levels. In science, the physical system (in this case, the object of study) informs the concrete description (the scientific data), which in turn informs the abstract model (the scientific theory). In engineering, the abstract model (in this case, the design concept) informs the concrete description (the specification of the product) which in turn informs the physical system (the actual product). In science as well as in engineering, the outcome depends both on Nature and on the practitioner's choices, but the power balance is different: the engineer has a larger degree of influence, through design

---

[370] This, however, is not to say that all is well with pharmacology and medicine development. A major problem is that due to the phenomenon of **publication bias** – the tendency to publish studies that exhibit a positive result and to refrain from publishing those that do not – the total balance of evidence as it appears in the literature is often strongly biased in the direction of being overly optimistic. See, e.g., Goldacre (2012) and Angell (2011).

[371] Easterbrook (2004).

and specification choices, on the final product, compared to the scientist's influence on the scientific theory, which is constrained to a greater degree by Nature.[372]

Motivated by his favorite long-term vision of achieving APM technology (atomically precise manufacturing, discussed in Section 5.2), Drexler argues that this difference has practical consequences for big projects. His point is that engineering is typically more demanding than science in terms of management, and he illustrates his point with two *very* big projects: the Apollo project of putting a man on the Moon, and the HUGO project of sequencing the human genome. The former, while often popularly described as an achievement of science, was an engineering project, while the latter really was a science project. The Apollo project was a fantastic triumph of coordination, management and organization, as many, many groups worked on developing different parts of the technology needed for the mission, and all these parts had to fit together into a functioning whole. The HUGO project, while presumably not an organizational triviality, still required less coordination, because (simplifying somewhat) Nature itself had more or less already suggested a division of labor among the participating groups: just split up the task of sequencing the genome into separate sequencing of the chromosomes. If each group does its work correctly (but without needing to bother much with what the others are doing), the end product could only be one thing: the true human genome.[373] In contrast, there is an astronomical[374] number of feasible technological subsystems for a successful lunar mission, but only a vanishingly small fraction of the possible combinations of these add up to a working whole, so the different subprojects needed careful coordination.

To summarize, while Easterbrook says that science and engineering are very similar endeavors, Drexler stresses that they are very different. Who is right? There is no need to choose sides: they are both right. Science and engineering are heterogeneous classes of activities, and they can legitimately be viewed from a variety of angles.

---

[372] A reader adhering to the kind of relativist position that I dismissed as fruitless towards the end of Section 6.4 might disagree with all of this in general and with last claim about constraints by Nature in particular. One of my favorite quotes from that kind of relativist madness is Bruno Latour's so-called Third Rule of Method:

> Since the settlement of a controversy is the *cause* of Nature's representation, not the consequence, we can never use the outcome – Nature – to explain how and why a controversy has been settled. (Latour, 1987, pp 99 and 258)

Note how Latour confuses (deliberately, one is inclined to assume) "Nature" with "Nature's representation," and how thereby he is able to reverse the direction of information in Drexler's description of science. See also Sokal (2008), pp 211–216, for a careful and critical analysis that I find myself in full agreement with.

[373] Some complications are suppressed here, such as the fact that different humans have somewhat different genomes.

[374] In fact, due to the phenomenon of combinatorial explosion, the number is sure to be bigger than any number naturally arising from measuring physical quantities in astronomy. But without my, strictly speaking, inaccurate choice of word, there would be no pun.

CHAPTER 7

The fallacious Doomsday Argument

7.1 The Doomsday Argument: basic version

This is a book about possible futures for humanity, but what if there is no future? What if humanity goes extinct? In earlier chapters, a number of ways in which this might happen have already been touched upon, and in Chapter 8 I will try to treat such scenarios slightly more systematically. There is, however, also a general argument haunting parts of the futurology community. Without touching on any specific mechanism, it purports to show that the extinction of humanity is imminent. This is the so-called **Doomsday Argument**, which the present short chapter is devoted to getting out of the way.

The Doomsday Argument is a statistical argument first put forth by astrophysicists Brandon Carter and (later, but independently) Richard Gott, but it was through the writings of philosopher John Leslie that it became widely known.[375] Readers who are already familiar with the argument and who share my view that it is unconvincing are advised to skip ahead to Chapter 8.

I'll start by giving the argument in its simplest formulation, closely resembling Gott's, and also the one most often encountered in the blogosphere and elsewhere. The essential deficiency of this version of the argument will be demonstrated (Section 7.2), followed by discussions of how it might be repaired along either frequentist (Section 7.3) or Bayesian (Section 7.4) lines, but the verdict will be that the argument is inconclusive, at best.

The argument involves the ratio $\frac{n}{N}$ between two numbers n and N. Here, N is the number of human beings that live now, have lived in the past, or will live at some time in the future, whereas n will be specified in a little while. Of course we do not know the value of N, but we can still work with it as an unknown quantity. Imagine now all N people, past, present and future, being ranked chronologically according to birth date, so that the very first human being ever born is assigned

---

[375] Carter (1983), Gott (1993), Leslie (1998).

Here Be Dragons. First Edition. Olle Häggström.
© Olle Häggström 2016. Published in 2016 by Oxford University Press.

rank 1, the second one to be born is assigned rank 2, and so on until the very last human being ever born, who is assigned rank N.[376] Next imagine that we select a human being at random from the set of all human beings who have ever lived, live now, or will live in the future, in such a way that every human has the same probability $\frac{1}{N}$ of being chosen. Let n be the rank of this randomly chosen human, and consider the ratio $\frac{n}{N}$. Clearly, $0 < \frac{n}{N} \leq 1$. More specifically, n takes one of the values $\frac{1}{N}, \frac{2}{N}, \ldots, \frac{N-1}{N}, 1$, each with probability $\frac{1}{N}$. Since N is a large number, this implies that the distribution of $\frac{n}{N}$ is, to a very good approximation, uniformly smeared out on the interval $[0, 1]$, meaning that for any interval A of length α sitting inside $[0, 1]$, we have

$$\mathbf{P}\left(\frac{n}{N} \text{ is in } A\right) \approx \alpha.$$

In particular, for any α between 0 and 1, we have

$$\mathbf{P}\left(\frac{n}{N} \leq \alpha\right) \approx \alpha. \tag{10}$$

For instance, taking $\alpha = 0.05$ tells us that a randomly chosen person from the full history of humanity has probability close to 0.05 of being among the 5% earliest born.

Now we take a bold step, and assume that *you, dear reader*, are that randomly chosen human being, so that your birth rank is n. The rationale behind this is that a priori, you have no particular reason to expect being born especially early or especially late in humanity's history. Imagining all of the N people of the history of humanity lined up according to birth rank, you have no particular reason to expect to be positioned in any particular part of the line, and so you may as well (or so the argument goes) assume that you have equal probabilities of having any particular birth rank.

So, according to (10), you have a probability close to 0.05 of being among the 5% earliest humans in history. If we can now figure out what n is, then perhaps we can plug it into (10) and then solve for N, which might give some insight into how large we can expect N, i.e., the total number of human beings, to be. Getting an accurate value of n depends on where exactly we draw the borderline between humans and the earlier hominids, and even with a precise such definition we cannot, of course, hope to get anything more than a rough estimate. But all that is important here is the order of magnitude, so let's use Leslie's (1998) estimate of $6 \cdot 10^{10}$ (i.e., 60 billion) people being born up to the publication of his book. Now, if you're born in 1998, we can go ahead and take $n = 6 \cdot 10^{10}$, while if you're not, then we still take $n = 6 \cdot 10^{10}$, because the couple of billion or so people born between your birth year and 1998 is just a minor rounding error in the following calculation.

---

[376] This involves the tacit assumption that N is finite. The possibility that the total number of human beings who will ever live is infinite will be briefly discussed in Section 7.4.

All right, so your birth rank is $n = 6 \cdot 10^{10}$, so we may as well (or so the argument goes) plug this into (10) with $\alpha = 0.05$, to get

$$\mathbf{P}\left(\frac{6 \cdot 10^{10}}{N} \leq 0.05\right) \approx 0.05 . \tag{11}$$

Solving for N inside the parentheses gives

$$\mathbf{P}\left(N \geq 1.2 \cdot 10^{12}\right) \approx 0.05, \tag{12}$$

or, in other words,

P(humanity dies out before the total number of humans hits $1.2 \cdot 10^{12}$) ≈ 0.95.

$$\tag{13}$$

Now, $1.2 \cdot 10^{12}$ may sound like a reassuringly large number, but how much time does it correspond to? That depends on how large we expect the population of humans to be in the future. Let's say the population stabilizes at $8 \cdot 10^9$ (8 billion), with an average life span of 80 years. Then the number of births per year will on average be

$$\frac{8 \cdot 10^9}{80} = 10^8 \text{ births per year.}$$

The $N - n = 1.2 \cdot 10^{12} - 6 \cdot 10^{10} = 1.14 \cdot 10^{12}$ births that remain if the total number is to reach $N = 1.2 \cdot 10^{12}$ will then last for

$$\frac{1.2 \cdot 10^{12}}{10^8} = 11{,}400 \text{ years,}$$

which is suddenly not such a large number. So with these assumptions on the future demographics of the human population, (13) tells us that there's a 95% probability that humanity dies out within the next 11,400 years.

7.2 Why the basic version is wrong

You may vary the numbers, but this is the Doomsday Argument as it is most often presented. There are various issues one can have with it, concerning, e.g., the assumption of the reader being a randomly selected human being, and the arbitrariness of defining humanity (more on these later), but this standard formulation has a flaw that kills it without further ado.

Consider the probability formula (10) for the case $\alpha = 0.05$ in which we applied it:

$$\mathbf{P}\left(\frac{n}{N} \leq 0.05\right) \approx 0.05 . \tag{14}$$

The number 0.05 on the right-hand side of the equation is a probability, and when, as in this case, this number is not 0 or 1 but strictly in-between, then the left-hand side needs to contain some random quantity, and the only quantity there that we have defined as random is n – the rank of a randomly chosen human being. But look again at what happens to the formula when, as in (11), we plug the value $n = 6 \cdot 10^{10}$ into the formula (14):

$$P \left(\frac{6 \cdot 10^{10}}{N} \leq 0.05 \right) \approx 0.05. \tag{15}$$

Doing so may seem like an innocent step, but note that just like in (14) we need to have a random quantity in the left-hand side. But while n can serve as a random quantity, an explicit number like $6 \cdot 10^{10}$ cannot, so the only quantity that remains that can play the role of the random quantity in (11) and (15) is N. That we have sneaked in an interpretation of the total number N of humans as a random quantity then becomes increasingly clear in (12) and in (13). But this is cheating, because when setting up the model, N was (unknown but) non-random, and no amount of valid formula manipulation can turn a non-random quantity into a random one, and we are forced to conclude that the Doomsday Argument, in the incarnation given above, is invalid.

The crucial mistake is in going from (14) to (15). While the former is a statement about the distribution of n for a fixed value of N, the latter is a statement about the conditional distribution of N given the event that $n = 6 \cdot 10^{10}$, so what the present version of the Doomsday Argument does is to commit the fallacy of the transposed conditional which, as we saw in Section 6.6, is a bad mistake that can lead to arbitrarily misleading results.

7.3 Frequentist version

Could there be a way to resurrect the Doomsday Argument? The version above tries to magically produce a posterior probability for the whereabouts of N without having specified a corresponding prior, and can thus be seen as an unfortunate mixup between frequentist and Bayesian statistical inference (see Section 6.8). Therefore, two natural ways to proceed are either

(i) to try to modify it into a valid frequentist argument, or
(ii) to try to modify it into a valid Bayesian argument.

The approach favored by Leslie (1998) is (ii), but let us discuss both, beginning with (i).

Formula (14) above can be exploited to give a frequentist test of hypotheses of N being large. A glance at the numbers popping out of the failed attempt above at a Doomsday Argument suggests that we might take the null hypothesis (H_0) to be

$$(H_0) \quad N \geq 1.2 \cdot 10^{12}$$

Let us do that and see where it leads. Note that under (H_0), we have that

$$\frac{n}{1.2 \cdot 10^{12}} \geq \frac{n}{N}.$$

This implies that

$$P\left(\frac{n}{1.2 \cdot 10^{12}} \leq 0.05\right) \leq P\left(\frac{n}{N} \leq 0.05\right),$$

which in combination with $(14)^{377}$ gives

$$P\left(\frac{n}{1.2 \cdot 10^{12}} \leq 0.05\right) \leq 0.05,$$

and solving for n in the left-hand side gives

$$P\left(n \leq 6 \cdot 10^{10}\right) \leq 0.05.$$

In other words, if the null hypothesis (H_0) is true, then the birth rank n of a randomly chosen human being has probability at most 0.05 of being $6 \cdot 10^{10}$ or less. So if we think of choosing a human being at random from the full history of humanity as a test of (H_0) at significance level 0.05, then we may declare statistical significance as soon as $n \leq 6 \cdot 10^{10}$. With the reader being the randomly chosen human, $n = 6 \cdot 10^{10}$ is what we got, so we are in a position to declare the outcome as a statistically significant deviation from the null hypothesis (H_0), and following the logic of frequentist statistical significance testing in Section 6.6, we may view this as evidence against (H_0). But evidence against (H_0) can be thought of as evidence in favor of its negation,

$$(\neg H_0) \quad N < 1.2 \cdot 10^{12}$$

i.e., as evidence in favor of humanity dying out before its cumulative birth count reaching $1.2 \cdot 10^{12}$.

So have we hereby resurrected the Doomsday Argument? Before declaring this exercise a full success, let us take a step back and think about what the logic of frequentist statistical significance testing really means. First of all, recall the list of warnings and caveats given in Section 6.6 concerning the interpretation of statistical significance. In particular, recall that neither does a statistically significant outcome disprove the null hypothesis, nor can we allow ourselves to fall back into the fallacy of the transposed conditional by declaring the obtained p-value (in this case precisely 0.05) to be the probability that (H_0) is true. Frequentist statistics accepts a statistically significant outcome as evidence against the null hypothesis, but remains silent regarding the probability that the null hypothesis is true.

But there is more to the present setting. A cornerstone of frequentist hypothesis testing is that if the null hypothesis is true, then the probability of obtaining a

---

377 With the harmless (trust me on this one!) slip of replacing $\approx$ with $=$ in (14).

false alarm, in the sense of obtaining statistical significance, cannot exceed the pre-specified significance level (in this case 0.05). Is that really true here?

It would be true if we did select a person at random from the full history of humanity, and took that person's birth rank as our data n. That is obviously not something we can really do, so what we did instead was to take the birth rank of a present-day person (namely, the reader) as our n. It seems very doubtful that the probability-at-most-0.05-of-a-false-alarm property would survive this modification of the data collection procedure.

To defend the Doomsday-Argument-as-a-significance-test, we need to argue that it is legitimate to claim that doing the experiment by choosing a present-day person's birth rank as our data would yield a false alarm with probability at most 0.05. It seems to me that to warrant such a claim, the timing of the experiment must be randomly located along the history of mankind, with the probability of ending up at a given time being proportional to the birth rate at that time. That assumption seems extremely shaky. Imagine, for instance, that humanity has a flourishing future ahead, with the true number N of human beings throughout past, present and future history being some large number like 10^{15}, 10^{20} or 10^{30} (we must a priori be open to this possibility, or else the Doomsday Argument would be pointless). Imagine, furthermore, that the Doomsday Argument is only obsessed over during a very brief interval (containing the present day) early in the vast history of humanity, and that for most of humanity's future we will have moved on to other more fruitful topics to discuss.[378] In such a scenario, all attempts at carrying out the Doomsday Argument in its significance test incarnation would take place very early in humanity's history and therefore result in false alarms, resulting in a false alarm probability of 1 rather than 0.05.[379]

There seems to be no good way to defend the Doomsday-Argument-as-a-significance-test against these concerns, whence the approach (i) of attempting to frame the argument in terms of frequentist statistics fails.

7.4 Bayesian version

So let us instead try Leslie's (1998) favored approach (ii) of putting the Doomsday Argument in a Bayesian setting. Here, the total number of human beings N is

---

[378] To think otherwise would strike me as an extremely pessimistic view of humanity's future intellectual progress.

[379] The situation gets (if possible) even worse if we insist on the classical frequentist interpretation of the false alarm probability as the proportion of repetitions of the experiment resulting in a false alarm that we would get in the long run if we repeated the experiment independently, over and over and over. But what in the world would it mean for us to make "independent" repetitions of the experiment? If we repeated the experiment by selecting another present-day person and considering his or her birth rank, then we would get essentially the same n (give or take a couple of billion, which makes very little difference compared to the $n = 6 \cdot 10^{10}$ that we got for the reader). So the repetitions we can do hardly merit the term "independent," and the proportion of false alarms if we repeat the experiment many times will not converge to 0.05, but most likely either to 0 (if $N < 1.2 \cdot 10^{12}$) or to 1 ($N \geq 1.2 \cdot 10^{12}$).

modeled as a random quantity. An a priori distribution $P = (P_0, P_1, P_2, \ldots)$ with $\sum_{k=0}^{\infty} P_k = 1$ (in order to make it a probability distribution) is postulated, with the assumption that

$$\mathbf{P}(N = k) = P_k$$

for each k. Then we should update this by Bayesian conditionalization in view of our data, which is that the reader has birth rank $n = 6 \cdot 10^{10}$. What is the conditional distribution of N given n? Well, that depends on further model assumptions that we haven't yet specified, but the most immediate, and perhaps naive, way to go about it is to note that $n = 6 \cdot 10^{10}$ implies $N \geq 6 \cdot 10^{10}$ (because nobody can have a birth rank exceeding N, and as usual we ignore as a rounding error those few billion people born between the reader's birth and now), so observing $n = 6 \cdot 10^{10}$ means conditioning on the event $\{N \geq 6 \cdot 10^{10}\}$, which yields a posterior distribution $P' = (P'_0, P'_1, P'_2, \ldots)$ given by

$$P'_k = \begin{cases} 0 & \text{for } k = 0, 1, \ldots, n-1 \\ \dfrac{P_k}{c'} & \text{for } k \geq n \end{cases} \tag{16}$$

where $c' = \sum_{k=n}^{\infty} P_k$, making P' a probability measure.

This, however, is not a calculation endorsed by Leslie and other Doomsday Argument enthusiasts. The core of any Doomsday Argument is that it is more likely to observe $n = 6 \cdot 10^{10}$ if N is only moderately large than if N is much larger. Thinking, as usual, of the reader as a randomly sampled human being suggests that

$$\mathbf{P}(n = 6 \cdot 10^{10} | N = k) = \frac{1}{k}$$

for any $k \geq 6 \cdot 10^{10}$. Plugging this and the prior P into Bayes' Theorem (see Footnote 359 in Section 6.8) yields a posterior distribution P'' given by

$$P''_k = \begin{cases} 0 & \text{for } k = 0, 1, \ldots, n-1 \\ \dfrac{P_k}{kc''} & \text{for } k \geq n \end{cases} \tag{17}$$

where again $c'' = \sum_{k=n}^{\infty} \frac{P_k}{k}$ is a normalizing constant.

The two probability distributions P' and P'' are distinct, and standard techniques in probability theory show that P'' is shifted towards smaller values than P' in the rather strong sense of so-called stochastic domination.[380] So the Bayesian Doomsday Argument leads to P'' and to more pessimistic estimates of how many more humans we expect will live compared to the more naive argument leading

---

[380] More specifically, P'' is what you get from the P' by a certain transformation known as size biasing; see Arratia and Goldstein (2010) for an instructive treatment of this topic.

to P'. Exactly how worried we should be by this depends on the choice of prior P. It is possible to cook up examples where the effect of the Doomsday Argument is dramatic,[381] but the fact of the matter is that with today's knowledge, we have pretty much no clue at all concerning what would be a reasonable prior P, and this translates into us having likewise pretty much no clue at all on whether or not the resulting P'' is one that suggests imminent doom. This suggests that even if we accept (17) as the right way to calculate the posterior distribution of N, the weight of Bayesian Doomsday Argument upon our shoulders is still at most moderate.

But *is* (17) the right Bayesian conditionalization procedure? This, in fact, is not at all clear. Consider the following thought experiment.

Imagine the richest man in Sweden, IKEA entrepreneur Ingvar Kamprad, waking up one morning feeling generous, and deciding that he wants to donate $1,000,000 each to a number of randomly selected persons among the $7 \cdot 10^9$ world citizens alive today. He feels undecided, however, about exactly how generous he should be. Should he donate to just one person, or to a thousand? He eventually decides to delegate this decision to a coin toss. So what he does is this:

Kamprad tosses a fair coin. If `heads`, then he selects one world citizen at random (i.e., everyone has the same probability of being chosen) to donate $1M to, whereas if `tails`, he selects 1000 world citizens at random as recipients of $1M each.

And here's the puzzle: Supposing you find yourself to be a recipient of $1M from Kamprad, how likely do you judge it to be that Kamprad's coin came up `heads`?

The answer to this is a straightforward and entirely uncontroversial application of Bayesian conditionalization. You start with the prior Q given by $Q(\text{heads}) = Q(\text{tails}) = \frac{1}{2}$, and calculate your posterior Q^* by conditioning on your evidence E of having received your $1M from Kamprad. The latter is readily done using Bayes' Theorem, and we get

$$Q^*(\text{heads}) = Q(\text{heads}|E) = \frac{Q(\text{heads})Q(E|\text{heads})}{Q(\text{heads})Q(E|\text{heads}) + Q(\text{tails})Q(E|\text{tails})}$$

$$= \frac{\dfrac{1}{2} \cdot \dfrac{1}{7 \cdot 10^9}}{\dfrac{1}{2} \cdot \dfrac{1}{7 \cdot 10^9} + \dfrac{1}{2} \cdot \dfrac{1000}{7 \cdot 10^9}}$$

$$= \frac{1}{1001}. \tag{18}$$

---

[381] Here's a simple example: Suppose the prior assigns probability $\frac{1}{3}$ each to $N = 10^6$, $N = 10^{11}$ and $N = 10^{18}$, corresponding to, respectively, a tiny humanity that died out shortly after it came into being, one that is about die out within the next few hundred years, and one that will thrive for many millions of years. The case $N = 10^6$ is ruled out by our observation of n and is therefore assigned probability 0 by both candidate posteriors P' and P'', which, however, distribute the remaining probability over the other two cases very differently. P' places probability $\frac{1}{2}$ each on $N = 10^{11}$ and $N = 10^{18}$, whereas P'' places probability 0.9999999 on $N = 10^{11}$ and only 0.0000001 on $N = 10^{18}$.

Despite heads and tails being a priori equally likely, conditioning on your evidence E makes tails 1000 times more likely than heads, as a consequence of E being 1000 times more likely in the case of tails than in the case of heads,[382] which in turn is a consequence of there being 1000 times more recipients in the case of tails than in the case of heads. Olum (2002) argues that we need to apply the same reasoning concerning the total number of human beings in the same manner: if we have a prior distribution P for the total number of human beings, then the observation "I exist" should be treated in similar fashion to the observation "I received \$1M from Kamprad" was in the Kamprad example. This will bias the a priori distribution of the total number N of humans heavily in favor of large N, i.e., in the opposite direction compared to the Doomsday calculation (17).

The further conditioning on one's birth-rank applied in the Bayesian Doomsday Argument corresponds to a modification of the Kamprad example where each \$1M donation is contained in a numbered envelope, where in case of heads the sole envelope has number 1, and in case of tails the envelopes are numbered $1, 2, \ldots, 1000$. If you get your \$1M, and your envelope is marked with a 1, then what probability do you assign to the event that Kamprad's coin toss came up heads?

In this case, it turns out that, since regardless of how the coin came up the probability of the observed outcome (an envelope marked with 1) has probability $\frac{1}{7 \cdot 10^9}$, the prior probabilities of heads and tails remain unchanged in the posterior. Similar considerations for the Doomsday calculation suggest that the size-biasing of the total human population due to the observation "I exist," and the effect of observing an early birth-rank, cancel each other exactly. Olum (2002) concludes that the calculation (16) leading to posterior P' (which we initially denounced as naive) is in fact the correct one, thus invalidating the alternative calculation (17), the alternative posterior P'' and the whole Bayesian Doomsday Argument.

Is Olum's counterargument correct? This is still far from clear, because it is not obvious how far the analogy between the (very clear-cut) Kamprad example and the (much fuzzier) Doomsday scenario carries. Kamprad has a well-defined population of $7 \cdot 10^9$ people from which to draw his winners, but it seems highly doubtful whether God Almighty (so to speak) has some reservoir of a zillion human souls from which he randomly draws just enough souls to fit in each human body of the full history of humanity. My long experience with mathematical and probabilistic modeling has taught me to feel much more comfortable with probability calculations when the probabilities arise from some (real or hypothetical)

---

[382] Here the critical reader may ask: aren't we committing the fallacy of the transposed conditional once again? But we are not, because we have faithfully applied Bayes' Theorem, which involves plugging in the prior probabilities $Q(\text{heads})$ and $Q(\text{tails})$. Different values of these would have resulted in different values of posterior probabilities $Q^*(\text{heads})$ and $Q^*(\text{tails})$. To really commit the fallacy of the transposed conditional is to try to pull posteriors out of the blue without regard to priors, a task that I pointed out in Section 6.6 as impossible.

mechanism than when there is no such underlying framework. Here, there seems to be no reasonably well-founded mechanism behind any of the above Doomsday Arguments and related calculations; "the reader is a randomly chosen human" does not, in my opinion, qualify as such a mechanism. Instead, Bostrom and Ćirković, in a paper critical of Olum's conclusions,[383] emphasize that the choice between the Doomsday posterior P'' and Olum's favored P' boils down to a choice between two principles in anthropic thinking, which in more recent terminology[384] are called the **self-sampling assumption** (SSA) and the **self-indication assumption** (SIA). SSA is the principle that, all other things being equal, an observer should reason as if they are randomly selected from the set of all *existing*[385] observers, whereas SIA says that, all other things being equal, an observer should reason as if they are randomly selected from the set of all *possible* observers. The SSA leads to the Bayesian Doomsday Argument as phrased above, while the SIA leads to Olum's counterargument. Bostrom and Ćirković (2003) tentatively favor the SSA. To me, it remains highly unclear which one makes better sense, and indeed whether either of them is warranted.

This might be considered enough to deem the Bayesian Doomsday Argument invalid (or at best inconclusive), but I have some further issues with it. The ideal of fully rational Bayesian reasoning discussed in Section 6.9 dictates that we should conditionalize on all our available evidence. The ideal may of course be unachievable in practice, but it still suggests that if we strive to make as educated a guess as we can about what the world is like, then we should do the best we can to incorporate as much of the available evidence as we can, and condition upon it. In the case of estimating the risk of human extinction in the relatively near future, it seems reckless to take into account just the one single piece of data that the reader's birth-rank n offers. Surely there are many other data and many other facts that seem relevant to the problem and ought to be incorporated in a Bayesian analysis – such as the frequency and sizes of major virus epidemics in the past, the fact that we have survived four decades of cold war and a couple more decades beyond that with potentially devastating nuclear arsenals, the long-term tendency towards democratic governance, and so on.

There are further problems that apply not only to the Bayesian Doomsday Argument but to the Doomsday Argument in general. One such problem is that the argument rules out a humanity with a grand total of N individuals more and more strongly as N increases, with the limiting case $N = \infty$ bordering on the impossible. In fact, if (one variant or another of) the Doomsday Argument is accepted, then we can rule out $N = \infty$ without even observing or estimating the current birth-rank n – all that is needed is that $n < \infty$, i.e., that humanity has not existed forever. If $N = \infty$, then any finite n is ruled out as representing such an anomalously early

---

[383] Bostrom and Ćirković (2003).

[384] Armstrong (2011).

[385] The grammatical choice of writing "existing" in present progressive tense here should not be taken literally: what is really meant is "existing in the past, in the present or in the future."

stage of the history of humanity as to have probability 0.[386] But it seems unreasonable that such a conclusion can be obtained with pure thinking and no reference to empirical observation; to me this looks like a pretty good shot at a *reductio ad absurdum*.

Another problem with the Doomsday Argument concerns the choice of so-called reference class, i.e., the class from which we are assumed to be randomly sampled. The SSA versus SIA dichotomy discussed above is such a concern, but there are more detailed considerations. Why is "humans" the correct reference class? Why not, e.g., the considerably wider class of "living organisms on Earth" or the considerably narrower class of "people preoccupied with the Doomsday Argument."[387] When two choices of reference class yield contradictory conclusions, which one should we trust? At present there seems to be no good answer to this question, and this casts a shadow of doubt over the Doomsday Argument.

A fair verdict of the Doomsday Argument in its various guises would be that we have yet to find a variant that remains convincing under critical scrutiny, and that it seems sensible to react similarly as to a century or more of unsuccessful attempts to empirically establish paranormal phenomena: until parapsychologists (or Doomsday Argument theoreticians) can come up with something convincing, we will ignore them and tentatively assume that paranormal phenomena (or a version of the Doomsday Argument that holds water) do not exist.[388]

A further practical reason for ignoring the Doomsday Argument until more convincing formulations have been produced is this: For someone like me (and, hopefully, the reader) who cares deeply about the future survival of humanity, there is a long list of concrete risks, beginning with the most familiar example of nuclear Holocaust, and ending who-knows-where. For each of these risks, there is (in most cases plenty of) scope for discussions both about how large the risks

---

[386] Why should a scenario with a finite n but $N = \infty$ be viewed as any more mysterious than the trivial mathematical fact that all positive integers (even those that we usually think of as incredibly large, such as $10^{10^{100}}$) are "smaller than average" when viewed as representatives from the set of all positive integers?

[387] The conclusion when the reference class is taken to be "living organisms on Earth" will at most be to suggest doom within one or a few billion years, which does not seem terribly alarming. Even less alarming is the conclusion when the reference class is "people preoccupied with the Doomsday Argument," because all it then says is that people will soon find more productive things to think about than the Doomsday Argument, and move on.

Now, admittedly, the three reference classes "living organisms on Earth," "humans" and "people preoccupied with the Doomsday Argument" do not lead to contradictory conclusions, because the three classes of beings may well die out at different times. But we *will* get contradictory results if we choose reference classes that differ in the past but coincide in the future. Doing so does not require going to such extremes as in the grue–bleen paradox (Footnote 328). As a counterexample it suffices to consider, e.g., reference classes "humans" and "humans born after the invention of the atomic bomb."

[388] It is worth noting here that Nick Bostrom, who is heavily cited (usually approvingly) in most chapters of this book, seems to arrive at a summary verdict of the Doomsday Argument – not only in the aforementioned paper with Milan Ćirković, but also in his 2002 monograph on anthropic bias and in a paper with the striking title "The Doomsday Argument is alive and kicking" – which, without unreservedly accepting any of its incarnations, is nevertheless more favorable than mine (Bostrom and Ćirković, 2003; Bostrom, 2002; Bostrom, 1999).

are and what we can do to mitigate them. In contrast, the Doomsday Argument concerns only how large the total risk is (summed across the entire list of concrete risks) and offers no useful idea about how to mitigate the risk.[389] So let's focus on those concrete risks and what we can do about them, without letting the Doomsday Argument distract us any further.

[389] A bit of epistemic humility is probably appropriate here. Suppose, for the sake of the argument, that we do come up with a convincing version of the Doomsday Argument, and that this argument suggests likely doom before a given number N of human beings have been born. Could we then not try to postpone doom by regulating our population in such a way that we decrease the number of births per year? Well, maybe, although it does sound bizarre to do such a thing based not on any concrete risk associated with having a large population, but on an abstract philosophical argument. And one can legitimately ask (although we will postpone such questions of value to Chapter 10) whether, for a given number of future human lives, it is better to have them spread out over a longer time period than over a shorter.

CHAPTER 8

Doomsday nevertheless?

8.1 Classifying and estimating concrete hazards: some difficulties

The best source currently available for a systematic overview of risks on the level of threatening to put an end to humanity – what we, for short, may call **existential risks** – is the 2008 anthology *Global Catastrophic Risks*, edited by Nick Bostrom and Milan Ćirković.[390] Bostrom and Ćirković organize the various contributions to their volume, and the corresponding existential risks, in three categories (apart from a background section on theoretical tools), namely

(a) risks from nature,

(b) risks from unintended consequences of human actions, and

(c) risks from hostile human acts.

While there is something to be gained from such a categorization, it is also problematic, because many or perhaps most catastrophe scenarios relevant to this discussion land in more than one of the categories, or in some grey area between them. Even in seemingly clear-cut cases such as a natural pandemic (i.e., one where the infectious agent was not created through genetic engineering or other human activities), which seems to fall squarely in category (a), or a Lex Luthor-like mad scientist deciding to take revenge upon humanity by deliberately launching the kind of nanotechnological grey goo catastrophe discussed in Section 5.4, which seems to belong in category (c), the categorization is open to criticism. In the first case, it may well be that the epidemic would have been far less severe and far more controllable, had it not been for our extravagant intercontinental travel habits spreading the germ quickly around the world, so that these travel habits can be seen as at least a partial cause of the pandemic, suggesting that it belongs

---

[390] Bostrom and Ćirković (2008). Their definition of existential risk is actually a bit broader: "An existential risk is one that threatens to cause the extinction of Earth-originating intelligent life or to reduce its quality of life (compared to what would otherwise have been possible) permanently and drastically" (p 4).

Here Be Dragons. First Edition. Olle Häggström.
© Olle Häggström 2016. Published in 2016 by Oxford University Press.

at least partially in category (b).[391] Likewise, in the second case, Lex Luthor will presumably not develop a robustly self-replicating and biosphere-eating nanobot from scratch, but is more likely to build on existing nanotechnology, in which case the development of such nanotechnology has the unintended consequence of making it feasible for a mad scientist to wipe out the biosphere with grey goo, suggesting that the risk belongs at least partly in category (b).

A general phenomenon that leads to further grey area cases in the categorization of risks is that some catastrophes whose direct consequences are very bad, but not quite wiping-out-humanity bad, can lead to a social collapse that finally does us in. In order for society to function, we crucially need not only physical infrastructure and other tangible resources, but also social capital: the trust that people place in each other and in institutions. Social capital is an asset that varies strongly between states, and features strongly in explanations for why some states succeed (economically and otherwise) while others fail, but it can also be alarmingly fragile.[392] Hanson (2008e) reviews evidence that human death toll scales faster than linearly in the direct physical damage caused by a catastrophe (i.e., doubling the direct physical damage tends to more than double the death toll). He suggests that collapse of social capital in the wake of catastrophe may explain this, and exemplifies the kind of domino effects that can drive such a collapse:

In the context of a severe crisis, the current benefits of defection can loom larger [than later consequences]. So not only should there be more personal grabs, but the expectation of such grabs should reduce social coordination. For example, a judge who would not normally consider taking a bribe may do so when his life is at stake, allowing others to get away with theft more easily, which leads still others to avoid making investments that might be stolen, and so on. Also, people may be reluctant to trust bank accounts or even paper money, preventing those institutions from functioning.

These are just some of the reasons why the cause of human extinction can turn out to be a complex intermingled system of causes, and why the Bostrom–Ćirković classification of extinction scenarios is an oversimplification. It may nevertheless be of some value for structured thinking about these matters, and it will partially be employed in the following Section 8.2 (on risks falling predominantly in category (a)) and Section 8.3 (on risks mainly in categories (b) and (c)), before rounding up this chapter with Section 8.4, where I try to say something about just how dire our situation is, and what the biggest risks are.

That last task is however a precarious undertaking, due to the fact that statistical estimation of how probable these risks are is in many cases very difficult.

---

[391] Bostrom and Ćirković make a similar remark about earthquakes. (These are not normally considered to pose existential risks, but the categorization applies similarly to smaller risks.) The obvious choice is to place them in category (a), but since their severity in terms of human casualties depends not only on their magnitude, but also on how we build, they can also be argued to belong in (b): "If we all lived in tents, or in earthquake-proof buildings, or if we placed our cities far from fault lines and sea shores, earthquakes would do little damage" (p 7).

[392] Rothstein (2005).

This is not merely the collection of *psychological* phenomena that do pose an obstacle to our ability to judge the risks,[393] but the more fundamental *epistemological* difficulty of estimating the probability of an event that has never happened.

For comparison, consider the much easier statistical problem of predicting the risk that the average US resident is exposed to from motor vehicle accidents next year: what is the probability of being killed by such an accident? To get a prediction that is very likely to be of the right order of magnitude, we can take the latest available death toll (the official figure from 2012 happens to be 33,561), divide by that year's population (313,914,000), and the resulting ratio 0.00011 is probably a pretty good estimate for next year's probability. If something that can be expected to strongly influence the risk is currently underway, such as some drastic new traffic regulation or an oil crisis that causes a substantial cut in the total amount of driving, the estimate might be off by a relative error of maybe 10%, but we can still be fairly confident about not being off by an order of magnitude.[394]

A key factor working for the ability to make good predictions in this case is the large number of more-or-less independent incidents that go into the numbers. Statistical predictions typically benefit from lots of data. As a contrast, consider something much more uncommon than a lethal motor vehicle accident, such as *being killed while riding a bicycle by a piano falling from a balcony on the third floor*. What is the average US resident's probability of dying in such a manner next

---

[393] Yudkowsky (2008a) discusses a series of cognitive biases that impair our ability to make balanced and rational judgments about existential risks, including the following four.

- **Availability bias.** We tend to irrationally attach larger probabilities to things that are readily available in our memories. As an example, Yudkowsky mentions the tendency to "refuse to buy flood insurance even when it is heavily subsidized and priced far below an actuarially fair value," and quotes Kates' (1962) findings about "the inability of individuals to conceptualize floods that have never occurred," about how "men on flood plains appear to be very much prisoners of their experience," and about how "recently experienced floods appear to set an upward bound to the size of loss with which managers believe they ought to be concerned."

- **Conjunction bias.** If Linda is a bank teller and active in the feminist movement, then she is a bank teller. Hence,

  P(Linda is a bank teller) $\geq$ P(Linda is a bank teller and active in the feminist movement),

  but Tversky and Kahneman (1983) showed that, under certain circumstances, people tend to reason as if the opposite inequality were true. This is the iconic Linda example, and it translates into our setting as our tendency to believe *more* (rather than *less*) in a future scenario the more detail is added to it.

- **Scope neglect.** Faced with a disaster killing 1000 people, the human brain is, in Yudkowsky's words, not in a position to

  release enough neurotransmitters to feel emotion 1000 times as strong as the grief of one funeral. A prospective risk going from 10,000,000 deaths to 100,000,000 deaths does not multiply by 10 the strength of our determination to stop it. It adds one more zero for our eyes to glaze over, an affect so small that one must usually jump several orders of magnitude to detect the difference experimentally.

- **Overconfidence.** This bias, which was discussed in item (b) in Section 6.1, tends to make the effect of most others even worse.

[394] My prediction *could* still be badly off if something unexpected happen, such as a major catastrophe killing the entire population or a substantial part of it.

year? Let's assume we have 50 years of statistics for the number of such accidents in the US, and that the total count so far is zero. Assuming stationary conditions (i.e., that the expected number of fatalities per year remains the same), standard statistical methods in the frequentist paradigm discussed in Section 6.6 turn out to give a 95% confidence interval of $0 \leq m \leq 0.060$ for the expected number m of such accidents per year, and dividing by the population yields a 95% confidence interval of $0 \leq p \leq 1.9 \cdot 10^{-10}$ for the probability p that the average US resident dies in the given manner. So even if the stationarity assumption is right, I cannot, as a statistician, tell you with any confidence what the order of magnitude of the probability is: for all I can tell, it might be 10^{-10} or 10^{-15} or even smaller. On the other hand, the probability is so extremely small that this hardly matters: we can interpret the data and the confidence interval as saying that the probability is so small that we can safely ignore the risk.

Next, consider estimating the probability of a global nuclear war, again based on the last 50 years, assuming stationary conditions.[395] The same calculation as for the falling pianos gives a 95% confidence interval of $0 \leq m' \leq 0.060$ for the annual probability m' of an outbreak of such a war, and again we cannot give even the order of magnitude of the probability: for all this data says, it might be 0.05 or it might be 0.001 or smaller. But the difference compared to the piano-on-bicycle case is that here we cannot dismiss the probability as being so small that we can safely ignore the risk. The case $m' = 0.05$ means that at the present level of risk, we can expect on average to live another $1/0.05 = 20$ years before the outbreak of a global nuclear war, whereas $m' = 0.001$ means another $1/0.001 = 1000$ years on average. This makes an enormous difference as regards how urgent the need is to reduce the risk.

The bottom line here is that for risks involving events that have never happened but would have enormous consequences should they ever come about, looking at this sort of historical data is not enough to determine how worried we should be. Other methods are needed for judging the risks (the reader may recognize this kind of difficulty from Section 6.5).

And what goes for global nuclear war goes for the event of genetically modified crops (GMCs) technology leading to a new plant spreading uncontrollably and disrupting ecosystems on a continental or even global scale. In order to claim that GMC technology is safe, an argument like the following, taken from Fagerström (2014), simply will not do.

Globally, commercial cultivation of [GMCs] has been going on for almost twenty years on a total area of more than a billion hectares ... yet no hazards that are directly attributable to the breeding technology have emerged.

What is needed is arguments that convincingly demonstrate on more concrete mechanistic grounds that such catastrophes are impossible (or at least very unlikely). Taleb et al. (2014) complain about the lack of such arguments in the

---

[395] The stationarity assumption is, of course, unrealistic, but is done for the sake of argument – to show the difficulty of statistical estimation of probabilities of rare events.

literature. The merit of this complaint depends on how strongly in favor of the "GMCs are safe" claim we understand the following to be. New organisms obtained from GMC technology tend to deviate *less* from their progenitors, in terms of metabolites, proteins and enzymes present in the cells, than do organisms obtained by older breeding technologies. The evidence for this is substantial (see, e.g., Ricroch, Bergé and Kuntz, 2011), and it does suggest that GMC technology is no less safe than older breeding technologies. The overall consensus among GMC researchers seems to be that this is sufficiently reassuring.

For the case of global nuclear war, a serious quantitative risk assessment would require far more knowledge and far more work, for instance by historians and political scientists to estimate how stable the MAD (Mutually Assured Destruction) logic and the current limits on nuclear proliferation are, and by engineers to estimate the probability of nuclear war ignited by fatal technical error, and so on and so forth. I am unable to provide such a high-quality assessment, but will say a little bit more about the risk in Section 8.3.

The great difficulty of quantifying the probability of an existential risk like global nuclear war that has been with us for half a century and more is, however, overshadowed by the even greater difficulty of the corresponding quantification of risks that are not with us at present, but that can be expected to emerge with the advent of certain future technologies; the grey goo scenario discussed in Section 5.4 is a typical example. On the other hand, there are cases where the prospects of estimating the probability of catastrophe based on the frequency of past occurrences of similar events are much better. For some of the natural hazards in category (a), such as asteroid hits and supervolcanoes, whose occurrences during prehistoric times and even on geological time scales can be (partly) inferred from geological records, and whose probabilities do not seem to change much over time, we can make vastly more useful estimates than in the case of human-induced risks that have been around for less than a century; see Section 8.2.

There is one more difficulty about estimating existential risk that requires discussion before closing this section: the possibility that our estimates are wrong and brutally misleading because our arguments, models or the underlying theory are just flawed. If a research article on catastrophic risk reports the probability of a catastrophic event B to be $\mathbf{P}(B)$, then the number they're *really* reporting is actually not $\mathbf{P}(B)$ but the conditional probability $\mathbf{P}(B|A)$ of B given the event A that the paper's arguments, models, theory and so on are correct.[396] Ord, Hillerbrand and Sandberg (2010) highlight that we need to calculate the true $\mathbf{P}(B)$ as

$$\mathbf{P}(B) = \mathbf{P}(A)\mathbf{P}(B|A) + \mathbf{P}(\neg A)\mathbf{P}(B|\neg A), \tag{19}$$

---

[396] Here I'm talking about probabilities in a rather naive way (unworthy of a mathematician who, like myself, has mostly specialized in probability theory), as if God Almighty had provided the world with objective probabilities ruling all that happens (maybe he has, but even if so, we do not know how to access those probabilities). To correct for this, we should read $\mathbf{P}(A)$ as the subjective Bayesian probability that we assign to A, and so on; the rest of the paragraph (and the rest of the chapter) should be read with this caveat in mind.

where $\neg A$ (the negation of A) is the event that there is a flaw somewhere in the paper's arguments, models or theory. What is typically done is that everything except for $P(B|A)$ is ignored, but when $P(B|A)$ is very small, this term may well be overshadowed by $P(\neg A)P(B|\neg A)$ by orders of magnitude. Or in other words, if the probability $P(\neg A)P(B|\neg A)$ that the paper's arguments are flawed and the catastrophic event B happens is ignored, this may lead to a way-too-optimistic estimate of the probability $P(B)$ of catastrophe.

In particular, this seems to be the case for the widely accepted analysis by Dar, De Rujula and Heinz (1999) of the probability that experiments at the RHIC (Relativistic Heavy Ion Collider) at Brookhaven National Laboratory in New York lead to the creation of strange matter destroying our planet. Ord et al. discussed this example at some length, and I'll return to it in Section 8.3. The Ord–Hillerbrand–Sandberg methodology is extremely difficult to apply rigorously and with precision,[397] but at the very least we should keep in mind the possibility of a non-negligible $P(\neg A)P(B|\neg A)$ term in (19) as part of the much-needed antidote to our strong tendency towards epistemological overconfidence discussed in item (b) of Section 6.1.

8.2 Risks from nature

This section will run through some of the foremost candidates for existential risk that nature imposes upon us, deferring those that arise from human technologies to Section 8.3; the existential risks from nature of the present section are labeled (ER1)–(ER6), while the human-induced ones in the next section are labeled (ER7)–(ER13). Existential risks from nature may seem slightly off topic in a book focused on how technological advances affect humanity, but in fact they are not, because for each risk that nature has in store for us, we may ask whether it might be possible to develop a technology that protects us from it.

(ER1) Asteroids

Every day, the Earth is struck by on average about 100 tons of meteoroids, ranging in size from dust and gravel to bigger rocks. The one that made big news by exploding some 25 to 30 kilometers above Chelyabinsk in Russia on February 15, 2013, was unusually large: it is estimated to have measured some 17 to 20 meters across, with a mass of about 11,000 tons – large enough to count as an asteroid.[398] The explosion caused plenty of damage over an area of several hundred km^2, and

---

[397] Here's one problem. Suppose a group of researchers are trying to estimate the probability of an event B, and decide not to be content with the usual $P(B|A)$, but go on to make an honest attempt to estimate $P(B)$ using (19). Should we not then be worried that even that step could have flaws, and should we not therefore subject the resulting value of $P(B)$ to another round of Ord–Hillerbrand–Sandberg-type analysis? We seem to be in danger of getting trapped in an infinite regress here.

[398] Yeomans and Chodas (2013).

Russian authorities reported some 1500 people having sought medical attention in the first few days following the event – most of them due to shattered windows, but a sizeable minority sought help for eye pain resulting from intense light from the meteor, which at its very peak was about 30 times brighter than the Sun.[399]

Rocks from space can be larger than that. The one that exploded some 5 to 10 kilometers above Tunguska in Siberia on June 30, 1908, and flattened some 2000 km² of forest, is estimated to have been between 60 and 190 meters across. The energy released was between 10 and 30 megatons of TNT, which is about three orders of magnitude more than the Hiroshima bomb, and the same order of magnitude as the biggest hydrogen bombs detonated experimentally by the US and the USSR in the 1950s and 1960s; see, e.g., Napier (2008). Objects of that size are expected to hit us about once a century, but this is not really a cause for concern about existential risk, as the effects of an impact will still be local (unless, as Napier dryly notes, it is "mistaken for a hydrogen bomb explosion in a time of crisis").

But the size distribution of objects that may strike the Earth doesn't stop there. The Shoemaker–Levy comet that impacted Jupiter in 1994 is estimated to have had a diameter of some 5 kilometers prior to breaking up (probably in an earlier close encounter with the planet) into fragments which ranged up to 2 kilometers in size – more than enough so that if it had hit the Earth instead of Jupiter, the consequences would have been global and civilization-destroying: megatsunamis around the world, fires of continental dimensions, and several years of dark sky. The Chicxulub asteroid, which hit the Yucatan peninsula about 66 million years ago, was even bigger, with a diameter of at least 10 kilometers. That was enough to cause the Cretaceous-Paleogene extinction event mentioned in Section 2.2 – one of the biggest mass extinction events known to have happened, wiping out all non-avian dinosaurs. Asteroids of that size are estimated, based on a combination of crater data and astronomical evidence, to hit the Earth on average about once every 50 to 100 million years.[400] The cutoff for civilization-threatening impact is estimated to be about 1 kilometer, and these have an estimated hitting frequency of about once per 500,000 to 1,000,000 years. This means that the probability of such an event during the course of one century is between 0.0001 and 0.0002.

Many, including myself, would judge that to be bad enough to be worth trying to eliminate. As we shall see in Section 8.3, there seem to be other existential risks for the coming century that are at least a couple of orders of magnitude more probable, so focusing on the asteroid threat may not warrant a number one slot on our risk elimination agenda right now, but if we hope to live and prosper on this planet for thousands or millions of years, we will eventually have to do something about it. A program for asteroid impact avoidance seems in principle eminently doable, and would have to include a monitoring part and a deflection part. The latter may sound especially difficult, but there are many ideas that do appear feasible, including (i) nuclear blast, (ii) ramming the asteroid with smaller objects, (iii) attaching

---

[399] Sample (2013).
[400] Rees (2003).

a rocket engine to the asteroid, and (iv) setting up a solar sail.[401] And the earlier an asteroid on a course bound for collision with us is discovered, the less is needed for deflecting its orbit.

(ER2) Supernovas and gamma-ray bursts

Asteroids, comets and other macroscopic objects hitting the Earth do not constitute the only threat to us from space. Other, more faraway, events have catastrophic potential too, and seem harder to prevent, but as far as the near-future probability of such events, the review by Dar (2008) is fairly reassuring. For a supernova to be a real danger to us, it would have to happen within a few tens of light years from us, and no star in this region is near enough such a stage in its life cycle to be up for such an event in the next few million years; on average, such a catastrophe can be expected about once every billion years. The frequency of even more exotic existential threats like cosmic ray beams or gamma-ray beams, that have been observed in other galaxies and that may cause disaster if occurring in the Milky Way and pointing in our direction, is harder to estimate, but the frequency of past mass extinctions (about once per 100 million years) is a strong clue that such astronomical events are very rare.

(ER3) The Sun

Over the next few billion years, the Sun is expected to gradually increase in radius and luminosity, until it starts turning into a red giant about six billion years from now, marking the start of a faster and more drastic increase in size, engulfing Mercury and Venus and probably also the Earth, which is then promptly destroyed.[402] The Earth's hospitality to life is expected to go bankrupt a lot sooner: the oceans are expected to boil away about a billion years from now, and the planet will become uninhabitable to both eukaryotic and prokaryotic life on about the same time scale, give or take a few hundred million years.[403] Living conditions for large animals like us may become unendurable even before that, but it is nevertheless clear that we have of the order of hundreds of millions of years before the planet becomes uninhabitable to us. That gives us plenty of time to think about emigration to other planets or other ways of surviving the threat from the Sun's increase in luminosity, so there is no strong reason to worry much about this particular risk at present.

I have been known to react with astonishment and hostility to the idea that the risk of catastrophic anthropogenic climate change by the year 2100 is not a cause for concern today, as there will be plenty of time to act later in this century.[404] In view of this, my relaxed attitude towards the predicted long-term fate of the

---

[401] Matloff (2012).
[402] Schröder and Smith (2008).
[403] Franck, Bounama and von Bloh (2005).
[404] See Häggström (2014f) for a personal recollection about such an encounter.

Earth may seem inconsistent, but I do think the difference in time scale by a factor 10^6 or more is enough to motivate the discrepancy in attitudes towards the problems.

(ER4) Supervolcanoes

In diagrams depicting global average temperature since the industrial revolution, the overall increasing trend (which is attributed mostly to anthropogenic emissions of greenhouse gases) is overlain by a large amount of chaotic-looking noise. One of the main contributors to this noise are volcanoes, which inject the atmosphere with sulfur-rich gases that get converted into sulphate aerosols that block out sunlight, causing cooling on the ground. For small eruptions, the sulfur is washed out quickly, but when the eruption is large enough to carry its waste all the way up to the stratosphere (starting some 10+ kilometers above ground), the aerosols can hang around for long enough to impact climate for years.[405] Such volcanoes in Krakatoa (Indonesia, in 1883), Santa Maria (Guatemala, 1902), Mount Agung (Indonesia, 1963), El Chichon (Mexico, 1982) and especially Mount Pinatubo (the Philippines, 1991) have had a noticeable impact on the global climate records.[406]

Analogously to the case of asteroids and other space rocks, volcanoes come in a very wide range of sizes, with Mount Pinatubo and the other famous 19th- and 20th-century eruptions representing relatively mild cases, compared to so-called supervolcanoes.[407] These are usually defined as volcanic eruptions that release at least 1000 km^3 of lava, which is about 100 times more than Mount Pinatubo did in 1991. They are much rarer, of course. Two relatively recent ones are the one at Lake Taupo, New Zealand, some 26,500 years ago, and the one at Lake Toba, Sumatra, about 74,000 years ago. It has been suggested (although the theory is disputed) that the Lake Toba eruption, and the years of global cooling following it, caused a severe decrease in human population that is responsible for the genetic bottleneck in the human population that seems to have arisen at about that time.[408]

Supervolcanoes the size of Lake Toba are potentially civilization-threatening, at least if we if we take into account the possibility of social collapse discussed in Section 8.1. So the risk is real, although as with the case of asteroids the supervolcano threat seems on a centennial scale to be dwarfed by risks arising from human activities. Compared to the asteroid threat, it seems more difficult to come up with engineering ideas to mitigate the supervolcano threat, apart from the rather obvious suggestion that it would be a good idea to stockpile food reserves on a global scale.

---

[405] As already discussed in Section 2.7, this phenomenon has inspired some scientists to suggest the artificial release of sulfur into the stratosphere as a method of countering global warming.

[406] Schmidt (2006).

[407] For introductions to supervolcanoes and their environmental impact, see, e.g., Bindeman (2006) and Rampino (2008).

[408] See, e.g., Petragila et al. (2007).

(ER5) Natural pandemics

The 14th-century Black Death was one of the most devastating pandemics in history, with estimates of the death toll ranging between 75 and 200 million people. It is fascinating and horrifying to study maps of its gradual spread through Europe during the 1340s and 1350s.[409] It arrived from the east, and by 1346 it was present in Constantinople, Crimea and Sicily. By 1348 it had spread through most of today's Greece, Croatia and Italy, as well as large parts of France and Spain, hitting Paris in June 1348 and London a few months later before proceeding further north. It was prevalent in all of England by the summer of 1349, and in western Scandinavia the year after; Finland was spared another couple of years.

Today, due to our extravagant global travel habits, a pandemic is unlikely to proceed at such a slow pace. Kilbourne (2008), reviewing the topic of plagues and pandemics in the past, present and future, mentions a related dangerous consequence of our increasingly global lifestyles and ways of organizing society – the so-called one rotten apple syndrome: "if one contaminated item, apple, egg or most recently spinach leaf, carries a billion bacteria – not an unreasonable estimate – and it enters a pool of cake mix constituents, and is then packaged and sent to millions of customers nationwide, a bewildering epidemic may ensue." These and other factors seem to make us more exposed today to globally or even existentially catastrophic pandemics, but Kilbourne emphasizes that there are also counteracting and perhaps weightier factors that help us limit them: we know a lot more now than in the past about how diseases spread, our communication technologies that are incomparably better and more widespread than just a few decades ago enable us to quickly spread warnings and advise on how to avoid infection, we have become quite good at designing and manufacturing vaccines on relatively short notice, and so on.

We already live with HIV, with the various strands of influenza that hit us annually, and so on, so the question of whether the phenomenon of natural pandemics poses a global threat is easy to answer: yes it does. Does it also pose a risk on the next level, threatening to end civilization as we know it, or even the existence of humanity? Probably yes, but it is hard to say where all the factors mentioned above (plus many others) put us in terms of the amount of existential threat posed by natural pandemics. My own feeling is that this risk is probably dwarfed by the risk of a civilization-ending pandemic arising from an engineered infectious agent; see item (ER12) in the next section.

(ER6) Genocide by extraterrestrials

How likely is it that, some time during the next century or so, aliens will invade our planet and kill us all? The answer to this depends, in part, on the answer to whether extraterrestrial civilizations exist at all. That is a notoriously difficult

---

[409] See, e.g., Snell (2003).

question falling under the general heading of the Fermi Paradox, which is the topic of Chapter 9, to which most discussion of this somewhat exotic-sounding existential risk will be deferred.

At this point, however, we should note the following. There is no evidence of any alien invasion of the Earth ever having taken place in the past. Such absence can be taken as evidence that alien invasion hasn't taken place, but the evidence grows weaker the further back in time we probe: an invasion that took place, say, a billion years ago might not have left any obvious traces visible today. Still, looking back at our seemingly invasion-free history and pre-history does seem to suggest, with statistical arguments similar to those in items (ER1), (ER2) and (ER4) above, that alien invasion in the near future is rather unlikely. But there's a gap in that argument, namely that, unlike in the asteroid, gamma outbreak and supervolcano examples, our presence might well augment the likelihood of alien invasion. In the case that we are invaded sometime soon, would it not seem like a strange coincidence that they attack almost immediately after our rapid technological ascent to the level of, e.g., radio and moon landings? But of course it might not be a coincidence. They might have detected our activities somehow, and decided to attack because of that, perhaps to preempt the potential military threat we might pose to them in the future.

8.3 Risks from human action

After the risks from nature discussed in the previous section, let me next offer a corresponding list of potential dangers coming mainly from ourselves and our technologies.

(ER7) Nuclear war

The risk of global nuclear war did not go away with the collapse of the USSR in the early 1990s, contrary to popular opinion.[410] I'm not saying this as a heat-of-the-moment reaction to Russian president Vladimir Putin's recent demonstrations of

---

[410] Not just popular opinion, but also the opinion of many leading thinkers, including scientific-minded public intellectual Steven Pinker. In Chapter 1 of his mostly superb book *The Better Angels of Our Nature: Why Violence has Declined* (Pinker, 2011), he gives his readers an eye-opening and horrifying guided tour of the many kinds of violence that were considered commonplace in the past, and ends the chapter with congratulating us on finding ourselves living at a time when we . . .

> no longer have to worry about abduction into sexual slavery, divinely commanded genocide, lethal circuses and tournaments, punishments on the cross, rack, wheel, stake, or strappado for holding unpopular beliefs, decapitation for not bearing a son, disembowelment for having dated a royal, pistol duels to defend their honor, [and] beachside fisticuffs to impress their girlfriends.

So far so good, given Pinker's implicit assumption that his readers are educated westerners. But in the quote I have gracefully aborted the last 20 words of Pinker's sentence, which reads "and the prospect of a nuclear world war that would put an end to civilization or to human life itself." Wrong, wrong, wrong! There was much reason during the cold war to worry about nuclear annihilation, but those reasons never went away.

geopolitical ambition,[411] but because the danger has remained with us all through the post-cold war decades. Since the end of the cold war there have been very substantial nuclear disarmaments, reducing the combined number of warheads in the US and in the USSR/Russia from a peak value of over 60,000 in the late 1980s to today's value of around 10,000, but we should not let this obscure the fact that the stockpiles that remain have grotesque and almost unimaginable capacity to wreak global havoc. Plus, as recent developments remind us, there have never been any guarantees for continued peaceful relations between nations in possession of nuclear bombs.

A global nuclear war would not directly kill all humans. Most likely, billions of people would survive, but would be in deep trouble due to the loss of infrastructure that we have become increasingly dependent on, plus the strand of catastrophic climate change known as nuclear winter, resulting from the large quantities of soot from firestorms that would be injected into the atmosphere and block sunlight for several years (often compared to the effect of supervolcanoes as discussed in item (ER4), but in fact more closely analogous to the global cooling that may follow a large asteroid impact, item (ER1)), with bad consequences, e.g., for agriculture.[412] It seems, to be perfectly realistic to think that it might cause sufficient damage to trigger the kind of social collapse proposed by Hanson (2008e) and discussed briefly in Section 8.1.

How likely, then, is global nuclear war? The practically important question to answer here is how likely it will be *in the future* – a question that is harder to answer than that of how likely it was *in the past*, which (for reasons explained in Section 8.1) is already a very challenging question. How lucky are we to have survived the cold war? Rees (2003) suggests that our probability of making it through the cold war, although nowhere near 100%, was more than 50% – "it was more likely that we would survive [the cold war] than that we wouldn't" (p 27) – but it is not clear to me whether he has good reason for such a claim. President John F. Kennedy famously estimated that, during the 1962 Cuban missile crisis, the probability of nuclear war was "somewhere between one out of three and even,"[413] and while such an estimate clearly needs to be taken with a grain of salt, it is not obvious in which direction we should correct it. But the Cuban missile crisis is not the only incident where we came close to a nuclear Armageddon. The second most well-known one is from September 26, 1983, mentioned in Section 1.3, that occurred in the very tense aftermath of the shooting down on Soviet territory of Korean Air Lines Flight 007 on September 1 the same year. Soviet air force officer Stanislav Petrov decided on his own that a suspected US missile attack was in fact a false alarm, and did not pass the suspected attack further up in the chain of command.[414]

---

[411] I am writing this in December 2014, nine months after the Russian annexation of Crimea.

[412] Even a regional-scale nuclear war can cause quite some climate damage; see, e.g., Toon et al. (2007).

[413] Rees (2003).

[414] Hoffman (1999).

We do not know what would have happened if he had done so, but full-scale nuclear war seems like a highly plausible outcome. There have been other incidents, such as the so-called Norwegian rocket incident in 1995, which serves as a reminder that the end of the cold war did not eliminate the danger of nuclear war.[415]

I am not in a position (nor is anyone else) to give a precise number quantifying how lucky we have been, but it seems clear that the probability of global nuclear war some time between the 1950s and now was substantial. Finding a way to eliminate or at least drastically reduce the danger seems urgent and important.

(ER8) Man-made global warming

The Venus syndrome – that the Earth's climate tips over towards a runaway warming that boils away the oceans and puts the planet in a Venus-like state – is considered to be a very unlikely consequence of man-made global warming even in the worst greenhouse gas emission scenarios; see Section 2.5.[416] For this reason, our current climate crisis is not a likely culprit for *directly* causing human extinction. But again, as in (ER7), nuclear war, it seems that the social collapse discussed in Section 8.1 might very well follow. Can our society survive the tension caused by a collapse of agriculture in the tropics and the subtropics, and billions of climate refugees?

(ER9) Unfriendly AI

This is an existential risk that, as argued in Sections 4.5 and 4.6, deserves to be taken seriously.[417] The following is a brief summary.

There is very little reason to doubt that it is in principle possible to design AI (artificial intelligence) with human or superhuman levels of general intelligence. A more open question is whether we can achieve this within, say, the next twenty to one hundred years, but many experts think of this as highly plausible; see, e.g., Baum, Goertzel and Goertzel (2011), Müller and Bostrom (2014), and Bostrom (2014). Turing (1951) predicted that such a breakthrough might soon thereafter lead to a scenario not unlike what Good (1965) later coined an intelligence explosion, where the AI's intelligence level spirals quickly to levels far beyond human capabilities. Section 4.5 reviewed some of the more recent thinking that suggests such an intelligence explosion to be not unlikely. And once AIs far more intelligent than ourselves exist, we cannot realistically count on remaining

---

[415] Hoffman (1998).

[416] In contrast, on the billion-year time scale discussed in (ER3) of the previous section, the Venus syndrome is considered to be inevitable (at least in the absence of interference from humans, posthumans, or whatever intelligent agents might be around by then).

[417] Most commentators do *not* take it seriously, but they almost always fail to back up their attitudes with strong arguments; examples of this are given in Sections 4.6 and 4.7.

in control,[418] so from that point our fate will be in their hands, and depend on their goals and values. These goals and values were the topic of Section 4.6, where, falling back on the works of Yudkowsky, Bostrom and Omohundro, I emphasized how crucial the task is of instilling values in the AI that align with our own and that place high priority on human welfare. Moreover, we need to do this *before* the AI attains superhuman intelligence (because once it reaches those heights, it will surely not allow us to tamper with its goal system). This task is what Yudkowsky (2008b) calls the Friendly AI project. It appears to be extremely difficult, because human values seem to (and now I allow myself to slip into a language that is more scientifically fancy-schmancy sounding than our understanding of the topic warrants) occupy a very thin and irregularly shaped subset of the space of all possible values. The rest of the value space is mostly very dangerous territory (recall Yudkowsky's default scenario that "the AI does not hate you, nor does it love you, but you are made out of atoms which it can use for something else"),[419] whence failure to solve the Friendly AI problem may spell our doom.

(ER10) Grey goo

In Section 5.4 the grey goo scenario was outlined, where a self-replicating nanobot goes out of control and eats the entire biosphere, perhaps as fast as in a matter of weeks. Expert opinions differ as to whether such a scenario is sufficiently realistic to be a real concern, but perhaps the most common position is that while accidental grey goo is unlikely, there is more reason to worry about whether the Lex Luthor character introduced in Section 8.1 might launch it on purpose.

(ER11) Other nanotechnology hazards

The APM (atomically precise manufacturing) technology discussed in Section 5.2 can, if we ever get there, become an extraordinarily powerful general-purpose technology. An obvious application is the development and production of weapons, something that may lead to an acceleration of arms races by orders of magnitude compared to, e.g., what we saw during the cold war era. This would likely have a destabilizing effect that we might not survive; see Section 5.4 for more detail.

(ER12) Pandemics from engineered germs

As part of his argument for why the grey goo scenario (ER10) is not among our most pressing concerns, Sandberg (2014c) explains, in a passage quoted in

---

[418] To me this seems like one of the more obvious steps of the argument, yet a common reaction is to doubt precisely this point. I attribute this to a failure of imagination regarding what someone incomparably smarter than ourselves might say or do. *Of course* anyone with merely human-level intelligence who tries to imagine such a thing will fail; let me nevertheless ask the doubter to turn back to Footnote 261 for a human (and therefore feeble) attempt to imagine a key step of the AI takeover.

[419] Yudkowsky (2008b).

Section 5.4, that "it is tough to make a machine replicate: biology is much better at it, by default" and that "maybe some maniac would eventually succeed, but there are plenty of more low-hanging fruits on the destructive technology tree." This may be at least mildly reassuring as far as the existential risk arising specifically from grey goo is concerned, but not at all when it comes to the total risk level we are exposed to. Synthetic biology poses grave dangers.

In a 2002 report from the US National Academy of Sciences, the following warning was issued.[420]

Just a few individuals with specialized skills and access to a laboratory could inexpensively and easily produce a panoply of lethal biological weapons that might seriously threaten the US population. Moreover, they could manufacture such biological agents with commercially available equipment . . . and therefore remain inconspicuous. The deciphering of the human genome sequence and the complete elucidation of numerous pathogen genomes . . . allow science to be misused to create new agents of mass destruction.

In the same year, a group of researchers announced that they had synthesized the polio virus from readily purchased chemicals and a genetic blueprint available on the Internet.[421] Two later incidents were mentioned in Section 1.3: the reconstruction in 2005 of the Spanish flu virus,[422] and the 2012 laboratory construction of a variant of the bird flu virus that, unlike previously existing variants, is transmissible between mammals.[423] What is clear from these incidents and others, as well as from the debates that have followed, is that there are plenty of scientists who are eager to create terrible pathogens and to invent reasons why it is actually a good thing to do so and to make the results public (such as how the acquired knowledge will help us construct vaccines and in other ways prepare for handling outbreaks of other viruses). Faced with this, various concerns are striking, such as how certain we can be that these laboratories do not accidentally release their viruses into the environment – an issue that may become especially worrying when we consider applying an Ord–Hillerbrand–Sandberg-style analysis (see Section 8.1) to their security protocols.

Yet, it is not the spread of *pathogens* from these research groups that worries me most, but the spread of *knowledge*. The National Academy of Sciences quote above suggests why, and here is Martin Rees on the same topic:[424]

Biotechnology is plainly advancing rapidly, and by 2020 there will be thousands – even millions – of people with the capability to cause a catastrophic biological disaster. My concern is not only organized terrorist groups, but individual weirdos with the mindset of the people who now design computer viruses.

[420] Committee on Science and Technology for Countering Terrorism (2002), quoted in Rees (2003), p 54.
[421] Cello, Paul and Wimmer (2002).
[422] von Bubnoff (2005), van Aken (2007).
[423] Yong (2012), Ritter (2012).
[424] Rees (2002).

We may imagine an arms race between the attackers (terrorists and the individual weirdos that Rees speaks about) and the defenders (national or perhaps international security agencies), where the attackers design pathogens and the defenders devise vaccines and other countermeasures. If the attackers are able to work in secrecy, this particular arms race seems tilted in their favor, because they can carefully work out a single one out of an astronomical number of possible infectious agents, while the defenders need to be prepared to handle any one of them on short notice.

The great techno-optimist Freeman Dyson, in a 2007 essay entitled "Our biotech future," describes enthusiastically the following vision for where we might be around mid-century:[425]

I see a bright future for the biotechnology industry when it follows the path of the computer industry, . . . becoming small and domesticated rather than big and centralized. . . .

Domesticated biotechnology, once it gets into the hands of housewives and children, will give us an explosion of diversity of new living creatures, rather than the monoculture crops that the big corporations prefer. New lineages will proliferate to replace those that monoculture farming and deforestation have destroyed. Designing genomes will be a personal thing, a new art form as creative as painting or sculpture.

Few of the new creations will be masterpieces, but a great many will bring joy to their creators and variety to our fauna and flora. The final step . . . will be biotech games, designed like computer games for children down to kindergarten age but played with real eggs and seeds rather than with images on a screen. . . . The winner could be the kid whose seed grows the prickliest cactus, or the kid whose egg hatches the cutest dinosaur.

While cacti and dinosaurs are not the same thing as viruses, the question still arises: can we survive such a proliferation of genetic engineering technology? To be fair to Dyson, it should be mentioned that he does bring up the issue of risk:

The dangers of biotechnology are real and serious. . . . Five important questions need to be answered. First, can [the domestication of biotechnology] be stopped? Second, ought it to be stopped? Third, if stopping it is either impossible or undesirable, what are the appropriate limits that our society must impose on it? Fourth, how should the limits be decided? Fifth, how should the limits be enforced, nationally and internationally?

Dyson does not offer answers to these questions. Neither do I. But we need them, urgently.

(ER13) Scientific experiments

The first experimental nuclear detonation, codenamed Trinity, took place in the Jornada del Muerto desert in New Mexico on July 16, 1945, just weeks before the Hiroshima and Nagasaki bombings. Prior to the test Enrico Fermi, who was one of

---

[425] Dyson (2007).

the leading physicists working on the Manhattan project, suggested the possibility that the test would put an end to our existence by igniting the atmosphere, but concluded that it would not. Three of his colleagues looked into the matter and produced a report on it.[426] Here is Baum (2014), citing the report:

> They believed the chance of this happening to be exceptionally small, due to their understanding of the relevant physics. Still, they closed their report on the topic with the line "However, the complexity of the argument and the absence of satisfactory experimental foundations makes further work on the subject highly desirable". Thus the risk did give them some pause. Sure enough, they took the risk. As is now known, the Trinity test succeeded: the bomb worked, and the atmosphere did not ignite. The rest is history.

Are we in a similar situation today? Could experiments carried out at the RHIC particle accelerator at Brookhaven and the Large Hadron Collider at CERN inadvertently lead to the destruction of our planet? There has been plenty of speculation to this extent in popular media, but physicists have reassured us that the probability of such a catastrophe is microscopic; see, e.g., Wilczek (2008) for a survey and relevant references.

Before taking the physicists' risk estimates as a guarantee that the collider experiments are safe, it may be a good idea to expose their arguments to the Ord–Hillerbrand–Sandberg-type analysis discussed in Section 8.1. In other words, can we ignore the probability that the physicists are just mistaken and that catastrophe follows? This is the main concrete example in the Ord et al. (2010) paper, and they find it useful to divide up the possible physical mechanisms for catastrophe broadly in two categories, namely (a) the creation of a black hole that grows to eventually eat our planet, and (b) even more exotic phenomena, such as the creation of "strange matter," or the breakdown of vacuum to a even lower energy state.[427] Concerning (a), Ord et al. express satisfaction with the fact that Giddings and Mangano (2008) offer three separate and relatively independent

---

[426] Konopinski, Marvin and Teller (1946).

[427] Ord et al. offer the following executive summary of the (hypothetical) strange matter phenomenon, and refer to Witten (1984) and Jaffe et al. (2000) for more detail:

> Our ordinary matter is composed of electrons and two types of quarks: up quarks and down quarks. Strange matter also contains a third type of quark: the "strange" quark. It has been hypothesized that strange matter might be more stable than normal matter, and able to convert atomic nuclei into more strange matter (Witten 1984). It has also been hypothesized that particle accelerators could produce small negatively charged clumps of strange matter, known as strangelets. If both these hypotheses were correct and the strangelet also had a high enough chance of interacting with normal matter, it would grow inside the Earth, attracting nuclei at an ever higher rate until the entire planet was converted to strange matter – destroying all life in the process.

And here are their corresponding words on vacuum breakdown, with a reference to Turner and Wilczek (1982).

> The type of vacuum that exists in our universe might not be the lowest possible vacuum energy state. In this case, the vacuum could decay to the lowest energy state, either spontaneously, or if triggered by a sufficient disturbance. This would produce a bubble of "true vacuum" expanding outwards at the speed of light, converting the universe into a different state apparently inhospitable for any kind of life.

arguments for why a black hole disaster at the Large Hadron Collider is highly improbable; the probability that *all three* of those arguments are flawed can be judged to be much smaller than the error probability in each of them. They are less happy with the situation concerning (b). Recall the formula (19) for the probability $P(B)$ of a catastrophe, which we take here to be the probability of catastrophe during one year of collider experiments. That probability is bounded below by the term $P(\neg A)P(B|\neg A)$, where $\neg A$ is the event that the physics meant to assure us of the safety of the experiment is flawed, and B is the event of a catastrophe. What is the greatest acceptable value of $P(B)$? Well, we can safely assume that no ethics committee would allow a physics experiment that went on for years and that would predictably cause the death of 1000 innocent civilians per year (obviously, this puts the case mildly). It makes sense, then, that the *expected* number of deaths from one year of experimentation should not be allowed to exceed 1000. With a world population of about $7 \cdot 10^9$, keeping the expected death toll below 1000 means making $P(B)$ so small that $7 \cdot 10^9 P(B) < 1000$, so $P(B)$ must be kept below $1000/(7 \cdot 10^9)$, which is less than $1.5 \cdot 10^{-7}$.[428] Thus, we need to have

$$P(\neg A)P(B|\neg A) < 1.5 \cdot 10^{-7}. \tag{20}$$

Now, assigning reasonable values to the factor $P(\neg A)$ requires a good deal of guessing and handwaving, and the same goes for $P(B|\neg A)$, but some reflection upon the problem does seem to suggest that settling for values small enough to satisfy (20) would require the handwaving to involve a fair amount of epistemic recklessness (the Ord et al. discussion of this exercise supports that conclusion).

Employing the Ord–Hillerbrand–Sandberg formalism consistently seems to lead to higher levels of precaution than we are used to. It might even lead to paralysis. Should I use my kitchen knife to sharpen my pencil? The laws of physics as we know them suggest that such an action involves no risk of destroying our planet, but what if they are badly flawed? Applying Ord–Hillerbrand–Sandberg in such a case is clearly absurd, so when should we do it? Perhaps a good rule of thumb is to employ it only in situations extending beyond the commonplace and familiar. In the words of Sandberg (2014d), "if you do something that is within the envelope of what happens in the universe normally and there are no observed super-dangerous processes linked to it, then this activity is likely fine." Supercollider experiments seem like the kind of thing that warrant an Ord–Hillerbrand–Sandberg-style analysis and the degree of precaution that comes out

---

[428] Note that this takes into account only the population of humans alive today. But a catastrophe of the kind discussed here would also wipe out all future generations, and if we take those into account, we need to press $P(B)$ down to even smaller numbers; see Section 10.3, where I will discuss the ethical issue of whether we *should* take them into account.

of it. Another one – the recent successful operation by a research consortium named CUORE to cool a cubic meter of copper down to 0.006 Kelvin, with claims that this was the coldest cubic meter in the universe for over 15 days – is mentioned by Sandberg (2014d), whose musings about the kind of risks such an experiment might involve are mostly tongue-in-cheek, but not entirely:

How risky is it to generate such an outside of the envelope phenomenon? There is no evidence from the past. There is no cause for alarm given the known laws of physics. Yet this lack of evidence does not argue against risk either. Maybe there is an ice-9 like phase transition of matter below a certain temperature. Maybe it implodes into a black hole because of some macroscale quantum(gravity) effect. Maybe the alien spacegods get angry.

8.4 How badly in trouble are we?

The last two sections collected a wide variety of potential threats to the future existence of humanity. But how big is the risk? Does it all add up to a non-negligible probability that humanity is wiped out by the year 2100? The results of a survey among experts on global catastrophic risk, reported by Sandberg and Bostrom (2008a), suggest that the answer is yes. The median response, when these experts were asked to estimate the probability of such an event, was 19%. As I've emphasized elsewhere,[429] the exact figure should of course be taken with a grain of salt,[430] not only because the polled experts give their (informed but) subjective probability estimates rather than figures derived from empirical data by rigorous statistical methods, but also because the rather small set of respondents cannot in any straightforward way be seen as representing a larger population of experts.[431] Yet, it does indicate a cause for concern about the future of our species.

If we accept that we are at substantial risk of extinction, the obvious next question is which of the risks listed in Sections 8.2 and 8.3 (or whether there are others) contribute most to the overall risk. In Section 5.4 I quoted Sandberg's (2014c) "five biggest threats to human existence," and perhaps at this point the reader expects me to deliver my own list under the same heading. However, the reluctance to quantify probabilities and other numbers on shaky grounds that I've learned from

---

[429] Häggström (2014g).

[430] Most people I've discussed this with tend to take for granted that the "grain of salt" implies adjusting the figure *downwards*, but this is not at all clear to me, unless we assume that these experts consciously or unconsciously inflate their estimates as a means of promoting their own research area.

[431] The questionnaire was given to the participants of the Global Catastrophic Risk conference in Oxford in July 2008, which served as a kind of launch event for the collection by Bostrom and Ćirković (2008) in which many of the papers I've relied upon when preparing this chapter can be found. The list of participants at the conference and the list of contributors to the book are not identical, but the overlap is substantial, suggesting a bit of caution in interpreting the survey result as further evidence *on top of* what I've discussed earlier in the chapter.

my schooling as a statistician prevents me from sticking out my neck quite like that.[432] I'll risk saying two things about this, however:

(1) For most of the natural hazards discussed in Section 8.2, such as asteroids and supervolcanoes, we have access to (imperfect but far from useless) data over very long time spans, allowing us to estimate the "baseline risk level" stemming from these. Although the probability of human extinction arising from such causes during the coming century or so is non-negligible, it seems not to come anywhere near the probability that we self-destruct by means of nuclear war or some of the other more exotic emerging technologies discussed in Section 8.3. In other words, it seems to me that, in such a time frame, the total extinction risk from human-induced activities is far bigger than the total coming from natural hazards.

(2) Sandberg's top three are nuclear war (item (ER7)), engineered pandemic (item (ER12)) and superintelligence (corresponding roughly to item (ER9)), and these three probably do deserve to rank near the top of such a list.

Both of these conclusions harmonize well with the survey results of Sandberg and Bostrom (2008a) concerning more specific risk scenarios than just overall risk. Number five on Sandberg's (2014c) list is also worth highlighting: unknown unknowns.

---

[432] And the reader should avoid the temptation to interpret Sections 8.2 and 8.3 as the amount of ink spent on each specific scenario being proportional to the probability I assign to it. That is *not* what I meant.

CHAPTER 9

Space colonization and the Fermi Paradox

9.1 The Fermi Paradox

On a summer's day in 1950, physicist Enrico Fermi was having lunch at a restaurant in Los Alamos, New Mexico, with his colleagues Emil Konopinski, Edward Teller and Herbert York.[433] Conversation had touched on the possible existence of extraterrestrial intelligent life, on whether superluminal travel might turn out to be possible, and (jokingly) on whether aliens might be guilty of the mysterious disappearance of public trash cans that New York City experienced that summer. It had then drifted to more down-to-earth and mundane topics when Fermi suddenly exclaimed "where is everybody?" His colleagues around the table immediately understood that he was referring to the extraterrestrials, and he quickly followed up his questions with a rapid series of order-of-magnitude calculations to support the view that we ought to have been visited by aliens long ago and many times over. The exact nature of these calculations have been forgotten, but from Jones (1985) we learn that York seems to have recalled, more than three decades later, that they shared at least some elements with what later became known as the Drake equation; see Equation (21) later in this section.

Where is everybody? This is the **Fermi Paradox**, which should perhaps better have been called the Fermi *problem* if that term hadn't already been reserved for something else,[434] because there is not really any contradiction or apparent

---

[433] My account of their lunch conversation is based on Jones (1985) and Webb (2002).

[434] A **Fermi problem** (the concept is named after the same Fermi) is a pedagogical device designed to train students in identifying quantities that suffice for solving a given physics problem and whose order of magnitude can reasonably be guessed, and in making the necessary back-of-the-envelope calculations. Two classical examples are "How many piano tuners are there in Chicago?" and "Will your next breath contain at least one molecule from Ceasar's last breath?" A third one is "How many extraterrestrial civilizations currently interested in radio astronomy are there in the Milky Way?", but this one is arguably much harder than the first two, because answering it requires estimating quantities whose orders of magnitude are pretty much unknown to the current astrobiological state of the art; see Equation (21), which is set up to solve precisely this Fermi problem.

Here Be Dragons. First Edition. Olle Häggström.
© Olle Häggström 2016. Published in 2016 by Oxford University Press.

contradiction involved to warrant the term *paradox*. A more precise formulation of the problem posed by the Fermi Paradox is this: Why haven't we seen any evidence of an extraterrestrial civilization?[435] Another term for the Fermi Paradox is the **great silence.**

This question is one of the truly great riddles for science and philosophy, perhaps not quite on a par with "How can consciousness exist in a material universe?" and "Why does there exist something rather than nothing?", but in any case not far behind. We do not know its true answer, but there are many more or less plausible suggestions, 50 of which are treated in a very inviting manner in the popular book by Webb (2002).[436] Perhaps the emergence of life is such a rare event that our planet is the only one (or one among very few) where this has happened; perhaps it is normal for a planet to go through a very rough ride from asteroid impacts, supervolcano outbreaks and other climate change triggers,[437] to the extent that it takes a huge amount of luck to survive long enough to develop any sort of advanced life; perhaps the development of *intelligent* life is highly unlikely; perhaps human-level intelligence pops up every now and then but tends to self-destruct (possibly by one variant or another of the scenarios discussed in Section 8.3)[438] before it has time to make a noticeable impact on the universe; perhaps superhuman civilizations exist aplenty but have no interest in interstellar travel or in interstellar communication; perhaps they exist but choose to stay quiet out of fear of other such civilizations; perhaps they exist but have all chosen to transform themselves into pure energy (or dark matter) in forms that we are unable to recognize; perhaps they have *already* colonized our planet and we are their descendants; and so on and so forth.

The multitude of suggestions may seem overwhelming, but there are ways to think (more or less) systematically about them. The topic of the next section will be my favorite such framework, known as the Great Filter (Hanson, 1998), into which most of the suggested answers to the Fermi Paradox fit neatly. Before that, however, let us discuss the famous precursor of the Great Filter known as the **Drake equation** (21), first formulated by astronomer Frank Drake in preparation for a meeting in Green Bank, West Virginia, in 1961, on the (then-emerging)

---

[435] It will be assumed throughout this chapter that the question is correctly posed, i.e., that we indeed *haven't* seen any such evidence. I am, of course, aware that there are plenty of people out there who report having seen flying saucers manned by little green men from some faraway planet, and that some of these witnesses even report having been *kidnapped* by these little green men, but there is very little to suggest that these reports constitute real evidence of aliens visiting our planet. See, e.g., Sagan (1995).

[436] See also Ćirković (2009) for an equally rich but academically more rigorous review of the same topic.

[437] Recall from Section 2.2 that our own planet has been through a lot of such fuzz, possibly including one or more Snowball Earth phase.

[438] Most of those scenarios – various human-induced apocalypses – may potentially serve to explain the great silence, provided we can plausibly argue that all sufficiently advanced extraterrestrial civilizations will inevitably succumb to the same thing. One of them does not, however, namely the unfriendly AI scenario (**ER9**), because it merely replaces the question "Where are all those extraterrestrial biological creatures?" by (the seemingly equally difficult) "Where are all those extraterrestrial robots?"

undertaking known as SETI.[439] The main SETI methodology was (as it remains today) to use radio telescopes to look for signs of intelligent messages in electro-magnetic radiation reaching us from outer space. A key quantity for evaluating the prospects of that endeavor is the (approximate) number M of planets in our Milky Way galaxy that are home to civilizations engaging in radio astronomy. What the Drake equation does is to state M as a product of seven quantities that we need to get a handle on: if we can obtain a good estimate of each factor, then we have a reasonable estimate of M as well. The equation reads

$$M = Rf_p n_e f_l f_i f_c L, \tag{21}$$

where R is the average number of stars formed per year in our galaxy, f_p is the fraction of those stars having planets, n_e is the average number of potentially life-supporting planets of such stars, f_l is the fraction of such planets that actually do develop life at some point, f_i is the fraction of those life-bearing planets where life goes on to develop into *intelligent* life, f_c is the fraction of *those* planets where the intelligent life goes on to engage in radio astronomy, and L is the average number of years that such a civilization remains active in that area.[440] The trouble, how-ever, is that while we have a reasonable grip of some of the factors in the product $Rf_p n_e f_l f_i f_c L$, we are more or less in the dark about some of the others. The factors we understand better tend to be the ones showing up early in the expression – we pretty much know that R is around 7,[441] and the magnificent developments in re-cent years as regards discoveries of exoplanets make us increasingly confident that f_p is not very small[442] – whereas we have very little clue about the orders of mag-nitude of the quantities f_l, f_i, f_c and L showing up in the expression's second half. Conventional estimates (or, rather, guesses) tend to land M in the thousands or even millions,[443] which would mean that we have plenty of neighbors in the Milky Way that are active in radio astronomy. These estimates, however, involve fairly high values of, e.g., f_l and f_i that seem to be overly influenced by the observation that the required events (the emergence of life, and the emergence of intelligent life) did take place here on Earth,[444] and in fact I am not aware of any convincing argument to rule out the possibility that M is microscopic, which would mean that we are most likely alone in the Milky Way.

---

[439] Recall that, in Section 6.3, I briefly discussed SETI's status as a scientific undertaking.

[440] I think probabilistically about all these quantities. If we took M to mean, literally, the number of planets in our galaxy hosting a civilization doing radio astronomy, then we would know that $M \geq 1$, because the Earth is such a planet. What I mean, however, is for M to be the *average number of such planets to be expected* in a galaxy like the Milky Way. This opens the possibility that M might be orders of magnitude smaller than 1, meaning that the existence of a civilization like ours in the galaxy requires a great deal of luck. Similarly, by f_p I mean not the exact fraction of stars in the galaxy having planets, but the expected fraction. And so on.

[441] Wanjek (2006).

[442] See, e.g., Woo (2013).

[443] See, e.g., Webb (2002).

[444] Rather than discussing this in detail here, I'll refer to Section 9.2, where I treat the closely analogous problem of estimating the quantity p in the Great Filter.

9.2 The Great Filter

Robin Hanson begins his seminal paper "The Great Filter – are we almost past it?" by noting life's tendency to spread and adapt to fill every ecological niche it comes across.[445] This is both an empirical observation about the history of our planet, and a consequence of basic Darwinian principles. Humans have continued this tendency, filling new geographic and economic niches as they become available. Hanson mentions the case of how, "when imperial China closed itself to exploration for a time, other competing peoples, such as in Europe, eventually filled the gap." The purported Darwinian mechanisms here are not merely that the gap is filled, but that those who fill it carry with them a propensity for further expansiveness, either in their genes or (in the more recent human case more likely) culturally, in their memes.[446] These mechanisms should, according to Hanson, lead us to expect that[447]

when such space travel is possible, some of our descendants will try to colonize first the planets, then the stars, and then other galaxies. And we should expect such expansion even when most our descendants are content to navel-gaze, fear competition from colonists . . . , fear contact with aliens, or want to preserve the universe in its natural state. At least we should expect this as long as a society is internally-competitive enough to allow many members to have and act on alternative views. After all, even navel-gazing virtual reality addicts will likely want more and more mass and energy (really negentropy) to build and run better computers, and should want to spread out to mitigate local disasters.

Hanson then goes on to argue (i) that such interstellar and intergalactic colonization is in principle possible, and (ii) that the effects of the colonization would likely become noticeable to alien astronomers all over the visible universe, either by our direct presence on their home planets, or by galactic engineering projects resulting in structures that these astronomers will see are artifacts rather than naturally formed. In case readers find (i) and/or (ii) hard to believe, I ask for a bit of patience: reasons will be given in Sections 9.3 and 9.4 for why (i) and (ii) are reasonable conjectures, and in the meantime, let us accept them to see where they lead.

The visibility cuts both ways: if a human intergalactic civilization would be visible to alien astronomers, then an alien such civilization would be visible to us. But we do not observe any such civilization: all we see and hear is the great silence of the Fermi Paradox.

Consider now, along with Hanson, the trajectory from a potentially life-supporting planet in its youth, all the way up to the creation of an intergalactic

---

[445] Hanson (1998).

[446] For an introduction to memetics, see Blackmore (1999).

[447] In this quote, we may notice, in the formulation about "as long as a society is internally-competitive enough," the same anti-totalitarian sentiment as in his attack on Yudkowsky's Friendly AI idea: see Section 4.6, Footnote 273.

civilization originating from that planet. A lot can go wrong on that path, and it seems that it usually does, because starting from a huge number N of potentially life-supporting planets – perhaps on the order of $N = 10^{22}$ – not a single one has lead to a universally visible intergalactic civilization.[448] This is what Hanson calls **the Great Filter**: on this trajectory, there is so much that can go wrong that a planet that starts the journey has at most a tiny probability of making it all the way to the stage of intergalactic civilization, and the vast majority of planets that start the journey are eventually filtered out.

Apparently the journey involves one or more bottlenecks that are extremely unlikely for a planet to make it through. Besides the obvious candidate of the initial emergence of something self-replicating (maybe an RNA molecule) on which the Darwinian process of mutation, selection and reproduction can begin to act, Hanson mentions (with the caveat that the list is not meant to be complete) the emergence of prokaryotic single-cell life, of more complex (archaean and eukaryotic) single-cell life, of sexual reproduction, of multi-cell life, and of tool-using animals with big brains, plus the path to where we are now, and the possibly very tricky further journey past the existential threats listed in Section 8.3 on to the stage of large-scale space colonization.

To put Hanson's discussion in a precise mathematical framework, write r for the probability that a randomly chosen planet with potential for life actually makes it all the way through to the stage of an intergalactic civilization. The expected number of planets that make it all the way is then Nr. Our observations tell us that the actual number of planets that have made it is 0. Now, if Nr were substantially bigger than 1, a relation denoted $Nr \gg 1$, then with overwhelming probability at least one planet would have made it all the way. Hence, we have good reason to conclude that

$$Nr \not\gg 1, \tag{22}$$

where $\not\gg$ denotes "is not substantially bigger than."[449] Note that the factor r plays a similar role as several of the factors (namely f_l, f_i and f_c) do in the Drake equation (21), but clumped together into a single factor. From the specific point of view of humanity, it will, however, be of interest to factorize r as $r = pq$, where p

---

[448] Here's the back-of-the-envelope calculation for N: the Milky Way has some $3 \cdot 10^{11}$ stars, and with one potentially life-bearing planet for every three stars we get 10^{11} such planets in our galaxy, and multiplying with the number of galaxies in the visible universe, which is at least 10^{11}, we get at least 10^{22} such planets. The estimate $N = 10^{22}$ could easily be off by an order of magnitude one way or the other, but probably not by much more than that.

[449] I am being intentionally vague about the meaning of "substantially bigger then 1," but here is a quantitative consideration:

If the journey towards life and technological civilization is approximately independent on different planets – a plausible assumption – then the actual number of planets that make it all the way is approximately Poisson distributed (see, e.g., Olofsson and Andersson, 2012, for a definition and some properties) with mean Nr, meaning in particular that the probability of it taking value 0 is approximately e^{-Nr}. Taking $Nr = 5$ would yield $e^{-Nr} \approx 0.007$, which may be an acceptable coincidence, but taking $Nr = 20$ would yield $e^{-Nr} \approx 2 \cdot 10^{-9}$, which is harder to accept.

is the probability that a potentially life-supporting planet gives rise to life that develops all the way up to the level of technological civilization corresponding to humanity in the first few decades of the 21st century (i.e., now),[450] and q is the conditional probability that the civilization, having come that far, goes on to develop into an intergalactic civilization. Equation (22) thus becomes

$$Npq \ggg 1, \tag{23}$$

which can be considered the canonical way to express the Great Filter as a mathematical formula.[451] It is to some extent similar in spirit to the Drake equation (21), but it is worth pausing to reflect over the differences:

(i) The formula (23) for the Great Filter has fewer factors than the Drake equation (21). That makes it simpler (which is a virtue), but also less detailed in distinguishing different stages in the development of life and civilization. On the other hand, the factorization $r = pq$ that it does offer is ideally chosen so as to maximize what we might learn about the future prospects for humanity, more on which later in this section.

(ii) While the Drake equation modestly restricts attention to the Milky Way, the Great Filter deals with the entire visible universe. One possible reason for the discrepancy is that in the early days of the Fermi Paradox and the Drake equation, interstellar communication and travel seemed like radical enough ideas that the leading thinkers did not dare take the next step in the discussion, to the intergalactic counterparts. Only more recently has intergalactic colonization and visibility of artifacts come to be regarded as (probably) possible. More on this in Section 9.3.

Although overly narrow, the Milky Way perspective does have the advantage that certain considerations become simpler. Within the Milky Way, speed-of-light communication takes at most 100,000 years, and classical conservative estimates of the time it takes a civilization capable of interstellar colonization to take over the whole galaxy are a

---

[450] A natural reaction here may be to call for an exact definition. An alien civilization would presumably not develop on the exact same lines as humanity. At some point it may be ahead of us in some technologies, but lag behind in others, so that without an exact definition it remains unclear whether they are on our level, or ahead of us, or behind. I choose, however, to adopt the same relaxed attitude towards this definition as I did towards the definition of intelligence in Section 4.5. Ćirković (2009) takes a similar stance:

Although it is clear that philosophical issues are unavoidable in discussing the question of life and intelligence elsewhere in the universe, there is a well-delineated part which we shall leave at the entrance. Part of it is the misleading insistence on the definitional issues. The precise definition of both life and intelligence in general is impossible at present, as accepted by almost all biologists and cognitive scientists. This, however, hardly prevents any of them from their daily research activities.

[451] It is not written out explicitly in Hanson (1998), but it is obvious that Hanson has it in mind throughout the discussion. The formula appears explicitly in Aldous (2012).

couple of million years. These are negligible times on a cosmological time scale, and in relation to the estimate that the median age of Earth-like planets in the Milky Way is one to two billion years more than the age of the Earth, making it less plausible to think that super-advanced technological civilizations are out there but haven't had time to become visible to us – see Ćirković (2009). In contrast, when looking further afield, at galaxies sufficiently many billions of light years away, we are also looking substantially backwards in time, to times when Earth-like biology and human-like civilization had not had time to develop. I will ignore these considerations, however, as they will at most cut off an order of magnitude or so from the enormous number N – not enough to significantly affect the very crude quantitative considerations at play here.

(iii) Whereas the Drake equation, via the factors R (measured in $(\text{years})^{-1}$) and L (measured in years, so that the whole product becomes dimensionless, just as for the Great Filter), accounts for the longevity of a technological civilization, the Great Filter does not, because it is assumed that once the requisite level of civilization is attained, it will be sufficiently robust to survive for periods that for our purposes count as "forever" – billions of years.

(iv) Unlike the Drake equation, the Great Filter takes explicit account of the Fermi paradox by insisting that Npq cannot be a large number.

So what do we do with (23)? Well, noting that N is very large, and that Npq isn't very large, entails the conclusion that pq is very small, which can happen only if either p or q (or both of them) is very small. In either case, there is no supertechnological civilization out there. Now what? Aldous (2012) calls the last conclusion "prima facie (treating absence of evidence as evidence of absence)," and summarizes three typical reactions, admitting that he is "slightly caricaturing" them. He calls them "the science fiction view," "the hard science view," and simply "a third view." The first one he describes as follows:

The science fiction view. The bar scene in Star Wars is cool; it's nice to imagine there are aliens we could communicate with; so let's think of reasons why the prima facie conclusion might be wrong, in other words how there might be extraterrestrial intelligence that we don't observe.

In my opinion, the science fiction view comes close to contradicting the evolutionary arguments of Hanson (1998), and is probably wrong – but since this is such an extraordinarily difficult area of inquiry, I am very open to being wrong about the wrongness, and I therefore look favorably upon those scientists and other thinkers who decide to pursue that line of thought. See Webb (2002) and Ćirković (2009) for a multitude of ideas in this direction. Here's a fairly popular idea, expressed by Sagan and Newman (1983):

We think it possible that the Milky Way Galaxy is teeming with civilizations as far beyond our level of advance as we are beyond the ants, and paying about as much attention to us as we pay to the ants. Some subset of moderately advanced civilizations may be engaged in the exploration and colonization of other planetary systems; however, their mere existence makes it highly likely that their intentions are benign and their sensitivities about societies at our level of technological adolescence delicate.

The "mere existence" argument here seems to be that any species that has aggressive tendencies is bound to destroy itself in a nuclear Holocaust or some such thing (see Section 8.3), so those remaining will be nice and non-expansive. Regardless of the merit of this argument, there is so much to say against this scenario that I very much doubt it. In addition to the Darwinian arguments referred to in the first paragraph of this section, there is the obvious incentive for an advanced civilization that has the technological capability for colonization of space to do so, namely that resources such as land, minerals, energy and so on (things they are likely to need whatever their desires are) are limited on their home planet, but exist in abundance out there. Note also

(a) that even if a civilization decides initially to refrain from colonization of space, it might change its mind a hundred years later, or a thousand, or a million – time spans too short to explain the great silence, and

(b) that even if the vast majority of civilizations act in the way that Sagan and Newman suggest, it takes just one exception to fill every inhabitable corner of the galaxy.

See also Hanson (1998) and Bostrom (2008c). Hart (1975) speaks of "The Contemplation Hypothesis", defining it as the idea that "most advanced civilizations are primarily concerned with spiritual contemplation and have no interest in space explorations," but rejects it as an explanation of the great silence, elaborating eloquently on observations (a) and (b):

The Contemplation Hypothesis . . . might be a perfectly adequate explanation of why, in the year 600 000 BC, the inhabitants of Vega III chose not to visit the Earth. However, as we well know, civilizations and cultures change. The Vegans of 599 000 BC could well be less interested in spiritual matters than their ancestors, and more interested in space travel. A similar possibility would exist in 598 000 BC and so forth. Even if we assume that the Vegans' social and political structure is so rigid that no changes occur even over hundreds of thousands years, or that their basic psychological makeup is such that they always remain uninterested in space travel, there is still a problem, [as] it still would not explain why the civilizations which developed on Procyon VI, Sirius II, and Altair IV have also failed to come here. The Contemplation Hypothesis is not sufficient to explain [the great silence] unless we assume that it will hold for every race of extraterrestrials – regardless of its biological, psychological, social or political structure – and at every stage in their history after they achieve the ability to engage in space travel. That assumption is not plausible.

Next, here is Aldous' (2012) summary of the second item among the three views of the Great Filter he discusses:

The hard science view. One can hope to estimate N fairly accurately, but there is no conceivable way that either p or q could be estimated scientifically. So further theoretical speculation is futile – get a real job! – though devoting a very small proportion of the science budget to SETI to seek actual data may be reasonable.

In contrast to the science fiction view, which I am skeptical about but nevertheless think deserves further thought, I have no sympathy at all for the hard science view, which I find narrow-minded and unduly pessimistic. While we may not be able to give good and well-founded estimates of p or q *now*, this is not a reason to conclude that we will *never* be able to do so – unless we allow the hard science view to produce a self-fulfilling prophecy.

But it is Aldous' **third view** that I am most inclined to endorse, that Hanson (1998) argues for, and that earns the Great Filter (and, along with it, the Fermi Paradox) its relevance in this book. The conclusion that pq is very small implies that unless p is very small, q must be very small, which would be bad news for humanity, because it suggests that we will most likely *not* go on to colonize the universe, i.e., we seem to be at a dead end. Finding the right orders of magnitude of p and q is not only extremely interesting from a theoretical point of view, but also has practical importance for the future of humanity.[452] So we should go on and try to work out, as impartially and level-headedly as we can, good estimates of p and q. At the same time we have, for humanity's sake, strong reason to want q *not* to be small, and hence also to want p to be very small. This last conclusion is reflected in Hanson's statement that "the easier it was for life to evolve to our stage, the bleaker our future chances probably are."[453]

***

[452] Here is Hanson (1998) on what might be a sensible reaction to the (currently hypothetical) finding that q is small:

With such a warning in hand, we might, for example, take extra care to protect our ecosystems, perhaps even at substantial expense to our economic growth rate. We might be even especially cautious regarding the possibility of world-destroying physics experiments. And we might place a much higher priority on projects like Biosphere 2, which may allow some part of humanity to survive a great disaster.

This may not sound like a very strong statement, but given how Hanson otherwise typically comes across as an extreme techno-optimist with very little sympathy for environmentalism, it is still remarkable.

See also Shulman (2012) for an argument suggesting (but by no means logically implying) the futility of the kind of reactions suggested in the Hanson quote.

[453] Hanson (1998). Bostrom (2008c) expresses the same sentiment even more colorfully:

I hope that our Mars probes will discover nothing. It would be good news if we find Mars to be completely sterile. Dead rocks and lifeless sands would lift my spirit. Conversely, if we discovered traces of some simple extinct life form – some bacteria, some algae – it would be bad news. If we found fossils of something more advanced, perhaps something looking like the remnants of a trilobite or even the skeleton of a small mammal, it would be very bad news. The more complex the life we found, the more depressing the news of its existence would be. Scientifically interesting, certainly, but a bad omen for the future of the human race.

A little bit of moderation may be in order here. The last paragraph seems to be saying that a very small value of q spells doom for humanity. But we shouldn't be too certain about this. First, q is not meant to represent the probability that *humanity* goes on to form an intergalactic supercivilization, but the probability that a *typical civilization on the level of present-day humanity* does so. It could (at least in principle) turn out that humanity is sufficiently exceptional in some sufficiently relevant respect, that even a microscopic value of q is no longer much cause for alarm.[454] Second, there still remains the possibility that we (humanity) have a long and flourishing future ahead of us, without the creation of the kind of intergalactic civilization that the Great Filter refers to. It could be that there is more to physics than we are currently aware of, and that the typical long-term fate of civilizations is to migrate into black holes, or hidden dimensions, or anywhere else that is hidden from our sight; perhaps these other places are so rich in resources that those parts of the universe visible to us are by comparison an infertile and uninteresting scrapyard, best left behind. If this is the case, then those civilizations do not contribute to a large q, and an equally bright future might await (post-)humanity even if q should be very small.

Another, less dramatic and in a sense diametrically opposite, scenario in which humanity might prosper despite a small value of q is what we may call **the Bullerby Scenario** (after Astrid Lindgren's children's stories about the idyllic life in rural Sweden in the late 1940s). Here, humanity settles down into a peaceful and quiet steady state based on green energy, sustainable agriculture, and so on, and refrains from colonization of space and other radical technologies that might lead in that direction. I mention this possibility because it seems to be an implicit and unreflected assumption underlying much of current sustainability discourse, not because I consider it particularly plausible. In fact, given the Darwinian-style arguments discussed above, plus the paradigm of neverending growth that has come to reign both in the economy and in knowledge production (the scientific community), it seems very hard to imagine how such a steady state might come about, except possibly through the strict rule of a totalitarian global government (which I tend to consider incompatible with human flourishing).[455]

Be this as it may, let me go on to the difficult problem of estimating p and q in the Great Filter. Apart from the observation that at least one of them is very small, we know very little. The case of q will be discussed in Section 9.3. Estimating p seems to be, at least in large part, a matter of judging whether any of the breakthroughs in the development of life – emergence of the first self-reproducing RNA molecule (or other entity) on which Darwinian selection can begin to act, prokaryotic single-cell life, eukaryotic single-cell life – and so on,

---

[454] But see Shulman (2012).
[455] See also the argument by Hart (1975), quoted above.

constitutes a serious enough bottleneck (i.e., an event unlikely enough) to account for much of the Great Filter, in which case we can conclude that p is very small.[456] Hanson (1998) and Aldous (2012) have suggested clever ways in which we might draw conclusions about whether or not the breakthroughs are exceedingly unlikely by looking at their relative timing, but the sad truth is that the timings of these breakthroughs constitute such a small data set that any statistical conclusions drawn from them will be very weak.[457] Perhaps, in order to estimate the probability of the emergence of that first self-reproducing RNA molecule, we have to figure out the hairy details of how it must have looked and how it must have been assembled. And similarly for the emergence of prokaryotic single-cell life and for the other breakthroughs.

A tempting suggestion here is the following. Human-level intelligent life, including present-day technological civilization, *did* in fact arise on Earth – so isn't that strong evidence that p is not very small? Isn't observing that we did emerge on Earth and then speculating that p might be microscopic, say on the order of 10^{-22} for concreteness, about as crazy as tossing a coin, noticing that it came up heads, and then suggesting that maybe the heads-probability is 10^{-22}? Well, this last suggestion would really be crazy, but the situation with intelligent life on Earth is different. Bostrom (2008c) explains:

Whether intelligent life is common or rare, every observer is guaranteed to find themselves originating from a place where intelligent life did, indeed, arise. Since only the successes give rise to observers who can wonder about their existence, it would be a mistake to regard our planet as a randomly selected sample from all planets.

No matter whether p is small or large, every observer will find that they themselves and their technological civilization did come about, and since this holds independently of p, it seems that the observation has nothing to say about p.

Great Filter theoreticians (Hanson, 1998; Bostrom, 2008c; Aldous, 2012) seem to be in agreement about this, but one might nevertheless suggest a

---

[456] There is at least one other way that p might be very small, namely that, as suggested in Section 9.1, the climate of a life-bearing planet turns out to be in general such a horrific roller coaster ride, with asteroid impacts and various geological events leading to one catastrophe after another – with the Earth's turbulent past discussed in Section 2.2 being in fact an unusually calm planetary history – so that even though none of life's breakthroughs requires much of a miracle in itself, sustained conditions admitting the continued survival and evolution of complex life all the way to our level might be sufficiently unlikely to make p very small.

Or catastrophes might originate from life itself. Recall the discussion in Section 5.4 about the likelihood of a nanotechnological grey goo disaster, where I asked why biological evolution hadn't at some point created some "green goo" catastrophe creating a superefficiently reproducing microorganism eating up the rest of the biosphere and then starving to death. The answer could be that that is the usual fate of life on a planet, and that we have simply been extremely lucky on Earth to have avoided such an event (so far), again resulting in a very small p.

[457] This is similar to the situation discussed in Footnote 343.

counterargument, based on an analogy with the Kamprad thought experiment in Section 7.4. Recall how it went:

Kamprad tosses a fair coin. If heads, then he selects one world citizen at random (i.e., everyone has the same probability of being chosen) to donate $1M to, whereas if tails, he selects 1000 world citizens at random as recipients of $1M each.

The problem was to work out what conditional probability to attach to the event that the toss came up heads, if all you know is that you are a recipient of $1M. That probability turned out, via an application of Bayes' Theorem,[458] to be a mere 1/1001 – in contrast to the prior probability of heads, which was 1/2. This discrepancy is due to the fact that you have 1000 times better chances of getting your $1M in the case of tails than in the case of heads. But isn't this heads versus tails situation analogous to the small-p universe versus large-p universe dichotomy of the Great Filter? If p is large, then we have a much greater probability of existing, and thus of being able to observe our own existence, than if p is small. Assume for simplicity that there are only two possible values of p, namely $p = 10^{-21}$, meaning that there are only us and up to maybe a dozen other civilizations sparsely dispersed over the observable universe, and $p = 10^{-3}$, meaning that the universe is teeming with civilizations. These two cases give rise to very different total numbers of observers able to reflect on the Great Filter and related matters: let us say 10^{15} in the former case, and 10^{33} in the latter case. And assume that these two possibilities are a priori equally likely:

God Almighty (so to speak) tosses a fair coin. If heads, then he creates a sparsely populated universe with $p = 10^{-21}$ and a grand total of 10^{15} observers, whereas if tails, he creates a universe teeming with civilizations, with $p = 10^{-3}$ and 10^{33} observers.

You find yourself to exist: what, then, is your conditional probability that God Almighty's coin toss came up heads? Here it is very tempting to conclude that, analogously to the calculation (18) for the Kamprad case in Section 7.4, you have $10^{33}/10^{15} = 10^{18}$ times greater probability of existing given tails compared to given heads, so that analogously to (18) we would get a posterior probability close to 10^{-18} that the coin came up heads. In other words, your existence makes it overwhelmingly likely that the coin came up tails and that p has the comparatively large value of 10^{-3}.

This argument is worth meditating over, but is it convincing? Not really. It is far from clear how to make sense of your "probability of existing" – if it can be made sense of at all. As I emphasized in the discussion of Olum's counterargument to the Bayesian Doomsday Argument (Section 7.4), the analogy between the entirely straightforward Kamprad example and the much less clear-cut scenarios arising in Doomsday Argument and Great Filter discussions is problematic. A crucial difference is that while Kamprad has a well-defined population of $7 \cdot 10^9$ people

---

[458] See Equation (18) in Section 7.4.

from which to draw his winners, there is no obviously analogous population in the other cases. Assuming the existence of such a population seems tantamount to saying that God Almighty has a reservoir of perhaps 10^{100} souls from which he randomly draws one soul to insert in each body qualifying as an observer. This seems a rather shaky assumption, and it is unclear whether it is possible, via some Kamprad-like analogy or by any other means, to convincingly defend the idea that our existence here on Earth in itself tilts the balance of evidence in favor of a not-too-small value of p.

So a very small p remains a viable possibility. Ćirković (2009) has an interesting concern, focusing on biogenesis (the emergence of life), although the reasoning applies to any bottleneck that produces a very small p in the Great Filter:

If one concludes that the probability of biogenesis – even under favorable physical and chemical preconditions – [is] astronomically small, say 10^{-100}, but one still professes that it was [a] completely natural event, then a curious situation arises in which an opponent can argue that supernatural origin of life is clearly a more plausible hypothesis. Namely, even a fervent atheist and naturalist could not rationally claim that her probability of being wrong on this metaphysical issue is indeed smaller than 10^{-100}, knowing what we know about the fallibility of human cognition.

Ćirković's second sentence here makes good sense, and if we accept it, then a Bayesian approach (see Section 6.8) supports the view of the hypothetical opponent in the first sentence, but it is still not clear what to think of an extremely small value of p. Before turning to a tough case like $p = 10^{-100}$, we may note that a value of p that is of the order of $1/N$, i.e., $p \approx 10^{-22}$, is not a problem at all, because that would mean that there would typically be a planet or two somewhere in the visible universe on which a civilization on our level arises; the fact that we find ourselves on precisely such a planet is no more mysterious than the fact that we happen to be on one of these N potentially life-supporting planets despite these planets occupying a volume fraction of far less than 10^{-25} of the observable universe. As to the case $p = 10^{-100}$, this may still be consistent with naturalism by a similar reasoning, provided that we accept the view that the visible universe is embedded in a vastly bigger multiverse. This view comes in several variations, but some of them are fairly mainstream in the cosmological community; see Tegmark (2014a).

9.3 Colonizing the universe

Much of the last section was spent discussing what the order of magnitude of the probability p in the Great Filter formula (23) might be, without even coming close to pinpointing the order of magnitude. Let us now have a go at the other probability q and see if we can do better. Recall that q is the probability that a civilization that has reached the technological level of early 21st century humanity goes on to develop into an intergalactic civilization.

Until recently, most discussion of colonization of outer space was restricted to the Milky Way, but Armstrong and Sandberg (2013) have suggested that colonizing other galaxies and most of the visible universe is not much harder.[459] We'll get to their work in a minute, but let us first consider the more modest ambition of spreading our civilization throughout the Milky Way.

The classical approach is to conquer the galaxy via a kind of branching process. We send spaceships to a few nearby stars and establish colonies on suitably chosen planets orbiting these. Each such colony, once it has had the time to build the requisite infrastructure, then goes on to send spaceships to some further stars, and so on, branching out until we have filled the galaxy.[460] Reasonable estimates suggest that this process can be brought to a completion within at most a few million years, which on a cosmological time scale is a very short time.[461] But is such a project at all feasible?

Beckstead (2014) surveys this question. More precisely, he asks: if we are able to avoid the various threats to human existence outlined in Chapter 8 (plus those that we are still unaware of), and if our technological development is allowed to continue unhampered, will we eventually be able to do it? Opinions vary, and there are, as he points out, many potential obstacles, including . . .

very large energy requirements, health and reproductive challenges from microgravity and cosmic radiation, short human lifespans in comparison with great distances for interstellar travel, maintaining a minimal level of genetic diversity, finding a hospitable target, substantial scale requirements for building another civilization, economic challenges due to large costs and delayed returns, and potential political resistance. Each of these obstacles has various proposed solutions and/or arguments that the problem is not insurmountable.

Beckstead's overall finding is that, despite widespread claims about the impossibility of interstellar travel,[462] and despite his efforts to trace these claims back to rigorously argued sources, there seems to be very little of this:

I found no books or scientific papers arguing for in-principle infeasibility, and believe I would have found important ones if they existed. The blog posts and journalistic pieces arguing for the infeasibility of space colonization are largely unconvincing due to lack of depth and failure to engage with relevant counterarguments.

Another interesting observation Beckstead makes is that most of the arguments for why interstellar travel and colonizations would be infeasible go away if we assume a breakthrough in robotics and computer science – either artificial

---

[459] An early precursor on the study of intergalactic colonization is Fogg (1988).

[460] See, e.g., Webb (2002), Ćirković (2009) and Wright et al. (2014).

[461] These estimates were pioneered by Hart (1975). Wright et al. (2014) offer, based on deliberately very conservative assumptions, the more conservative upper bound of 10^8 to 10^9 years. Even these estimates do not come near the age of the Milky Way, which is almost as old as the universe, whose age is about $13.8 \cdot 10^9$ years.

[462] Brin (2006) speaks of "the Drake Doctrine" (named after the astronomer behind Equation (21)), which holds that interstellar travel is inherently impossible.

intelligence (as discussed in Section 4.5) or uploading (Sections 3.8 and 3.9) – that allows us to undertake the project without having to involve our own fragile biological bodies. He also offers a somewhat more detailed discussion of the various steps that we need to master. Roughly, these steps are:

(a) getting everything needed to build a colony (possibly including humans) into a spaceship,
(b) getting the spaceship to go fast in the right direction,
(c) maintaining the spaceship and the cargo sufficiently intact during the long voyage,
(d) slowing the spaceship down near the target, and
(e) building a colony at the target.

Of these, (e) stands out as the step that has received the least attention from space colonization optimists in terms of how to implement the details, so this might possibly be a weak link.

Let us move on to Armstrong and Sandberg (2013), who suggest in some detail how humanity might be ready, in perhaps as little as a century or two, to initiate a colonization of not only the Milky Way but of much of the visible universe, with the colonization frontier expanding at a substantial fraction of the speed of light.[463] Sandberg seems fairly convinced that something like this is doable; in an interview with Beckstead he expresses the view that it would probably require the discovery of some currently unknown physics for us to learn that space colonization is infeasible.[464] The Armstrong–Sandberg scenario should not be read as a prediction of what will actually happen, but merely as an indication that space colonization on cosmological scales is probably in principle possible. A reasonable corollary, if we trust that conclusion, is that q in the Great Filter is probably not very small, because if humanity is not so far from being able to do it, then it seems reasonable to assume that other civilizations at our level of technological maturity would also acquire the ability, and then the Darwinian-style arguments in Section 9.2 strongly suggest that some of them would probably go ahead with the expansion.

A central concept in the Armstrong–Sandberg scenario is that of a **von Neumann machine**, also called a self-replicating machine: a machine capable of making copies of itself. The word "machine" here does not need to imply that they are artifacts, and with this in mind, it becomes clear that von Neumann machines are all around us. A bacterium, for instance, is a von Neumann machine. Thus, John von Neumann's pioneering abstract work in the late 1940s (which became widely known through the classic *Scientific American* paper by Kemeny, 1955) on

---

[463] The latter (expanding at a substantial fraction of the speed of light) is not so important in the case of colonizing the galaxy, but for the grander task it is crucial, because the lower the speed is as a fraction of the speed of light, the larger the fraction of galaxies that will never be reached, due to the expansion of the universe.

[464] Beckstead (2014).

the concept was, in a sense, not necessary to establish that such machines were possible, but it served the purpose of making the biological phenomenon of reproduction seem less mysterious. It is important to note that a machine has the property of being a von Neumann machine *only relative to its environment*, i.e., a given machine has the requisite self-replication ability only in some range of environments. A minimum requirement on the environment is of course that it contains the energy and raw materials needed for the self-replication.

Artificial non-biological von Neumann machines currently exist only as software entities in computer simulations, but there is no doubt they can in principle be built. They can come in many sizes, from the self-replicating nanobots discussed in Chapter 5, up to (and beyond) the lunar factories of Freitas and Zachary (1981). They can, e.g., be spaceships; such von Neumann machines are also called **von Neumann probes**. Armstrong and Sandberg point out that "a spaceship with a variety of human couples, life support systems, large databases and an onboard factory would count as a von Neumann probe capable of building more copies of itself through manufacturing and reproduction." The spaceships used for colonizing the Milky Way in the classical scenario would count as von Neumann probes, with the self-replicating step taking place on the planets that are colonized and used as stepping stones for further colonization.

The Armstrong–Sandberg scheme builds on the same concept, but involves a braver-than-usual idea of the attainable scale of the launching endeavor: they consider sending one probe (or a thousand, or whatever number we prefer for redundancy) towards *each of the 10^{11} galaxies in the visible universe*.[465] Among the list of steps (a)–(e) discussed above in connection with Beckstead's paper, Armstrong and Sandberg devote much effort to handling (c), in particular the probes' very long exposure to risks of collisions with interstellar and intergalactic dust, and to the mechanics and energy requirements for the acceleration and deceleration phases (b) and (d). Here, I'll skip the discussion on (c) and focus on their solutions for (b) and (d). Due to the enormous number of probes and the near-luminal speeds they need to attain, the energy requirements are stupendous, and partly for this reason, it is important to keep the mass of the probes down. How small they can be made and still have the desired functionality is a wide-open question, and Armstrong and Sandberg consider a range of probe sizes, from 30 grams to 500 tons.[466] Sizes at the lower end require advances in nanotechnology and artificial intelligence[467] that are radical (but perhaps not implausible; see Chapters 4 and 5), while the upper end involves more conservative assumptions.

---

[465] This is an idealization and exaggeration on my part; Armstrong and Sandberg point out that even at 99% of the speed of light, the number of galaxies within our reach drops by more than one order of magnitude.

[466] These figures exclude the rocket and the fuel, which will make the entire spacecraft several orders of magnitude larger.

[467] Plus uploading (Sections 3.8 and 3.9) in case we want the space colonies to be populated by actual human beings.

But even the smaller size leads to energy demands corresponding to billions of years of humanity's current energy consumption.

Does this make the project infeasible? Not at all, if we build a **Dyson sphere!** A Dyson sphere is a gigantic system of solar power plants, arranged in orbits around the Sun in such a way as to intercept and collect all or nearly all of the energy released by the Sun.[468,469] The reader may be forgiven for thinking that this sounds like merely replacing a hopelessly large energy demand by a hopelessly large engineering project. But building a Dyson sphere is only infeasible if we cannot employ von Neumann machines to do it. With von Neumann machines, even stupendously big things can be built within a reasonable time, because the population of machines can maintain a constant doubling time, thus increasing exponentially, meaning that the time taken to reach a given population size k grows only logarithmically in k.[470]

Concretely, Armstrong and Sandberg propose disassembling the planet Mercury, and using the material (mainly) for building the Dyson spheres. This involves building von Neumann machines (not to be confused with those to be used at a later stage, when the Dyson sphere is completed, to travel to distant galaxies) capable of

(i) mining on Mercury,

(ii) transporting the material from Mercury to interplanetary space, and

(iii) building solar power plants out of that material.

One might construct a single species of von Neumann machine capable of doing all three things, but the more practical way would be to have three different kinds, one for each task. When the mining machines on Mercury have self-replicated for a number of generations, they have become sufficiently numerous for their energy requirements to be vast by our ordinary earthly standards. A key point in the Armstrong–Sandberg scheme to address this is that the machines for doing (ii) and (iii) will not wait until those for (i) have completed the disassembly of Mercury

---

[468] Dyson (1960), Sandberg (1996), Armstrong and Sandberg (2013).

[469] The most immediately appealing image of a Dyson sphere is that of a rigid spherical shell centered at a star (which of course does not necessarily need to be our Sun). Unfortunately such a sphere would be gravitationally unstable – even the slightest disturbance would cause it to fall into the star. That might be solved by an appropriately placed array of rockets to respond to each such disturbance by adjusting the sphere back into position, but there is another problem: the mechanical forces in the shell will be huge. Therefore, it is probably more realistic to suppose that the Dyson sphere will not literally be a sphere, but some other arrangement of solar power plants in orbits around the star. See, e.g., Sandberg (1996).

[470] This exponential increase can only be maintained for a limited amount of time – a well-known phenomenon, which is often attributed to the fact that we live on a finite planet. However, even if, hypothetically, we had von Neumann probes that could reproduce in vacuum without the need for raw materials, the fact that they are confined to three spatial dimensions and prohibited by the laws of physics to travel faster than the speed of light is enough to ensure that the population size as a function of time t eventually cannot grow faster than a constant times t^3, which is slower than any exponential growth. But the point here is that, even though a Dyson sphere is a very big thing, it takes *much* bigger scales for these fundamental physical constraints to significantly hamper the exponential growth.

before commencing their work, but instead get to work as soon as material from (i) becomes available. In this way, a growing Dyson sphere embryo can feed back energy to the machines down on Mercury. Thanks to the exponential growth in production rate, Armstrong and Sandberg estimate that the Dyson sphere can be completed in a matter of decades.

The next stage of their scenario is to use the energy from the Dyson sphere to launch the probes. (There is also a need for material to build the probes, but this is a negligible part compared to the material needed for the Dyson sphere.) Not only different probe sizes are considered, but also different speeds ($0.5c$, $0.8c$ and $0.99c$, where c is the speed of light) and different mechanisms for the eventual deceleration[471] of the probes: fission, fusion and matter–antimatter annihilation, ranging from the most straightforward to the most efficient but also the most speculative. A quantity of interest is how long it would take for the Dyson sphere to collect the required energy amount (with one probe per galaxy, i.e., no redundancy). For the case of fusion drive, $0.8c$ speed and a probe at the lower end of the size scale (30 grams), six hours of energy collection is enough – hence the phrase "Eternity in six hours" in the title of Armstrong and Sandberg (2013). If the probe size is instead at the high end (500 tons), then the energy requirement goes up proportionally, and the Dyson sphere will need to work for some 10,000 years. That is a long time from the human perspective we are used to, but compared to the cosmological time scales relevant to Fermi's paradox, and to the time needed for the probes to reach their destinations (which will typically be billions of years), it is merely the blink of an eye.

Once a probe finds a base planet in a distant galaxy, the next stage can begin: the creation of a colony, and preparations for similarly launching probes towards every star of the galaxy. If this is successful, then the time taken to colonize the entire galaxy will be negligible compared to the travel time from us to the galaxy.

Of course, as with all studies of not-yet-existent technologies, there are many details missing and many gaps to be filled in Armstrong's and Sandberg's description of their scenario. Their paper can by no means be seen as a rigorous and conclusive demonstration that full-scale intergalactic colonization is within our relatively near reach. Still, they do make a sufficiently plausible case that this it is doable to warrant taking the possibility very seriously. It is also a serious indication that the q of the Great Filter (23) is not very small.

Towards the end of their paper, Armstrong and Sandberg discuss the issue of whether we, or an extraterrestrial civilization with similar technological capabilities, will actually *want to* and *choose to* go ahead with such a large-scale intergalactic colonization. They are favorable to the Darwinian-style arguments discussed in Section 9.2, at least to the extent that if there are many human-level

---

[471] The deceleration is the hard part because the probes need to carry their own fuel. The acceleration part is much easier, and can be done without the probes carrying the fuel; Armstrong and Sandberg mention a variety of methods, including coilguns and quench guns.

technological civilizations out there, it is unlikely that *none of them* will go forth with such colonization. To these arguments, they add an interesting strategic perspective. Imagine a civilization on Planet X that has reached the technological maturity where they can choose to go ahead with some Armstrong–Sandberg-like plan for conquering the universe. They might[472] be perfectly happy to stay on Planet X, provided they could trust that nobody else is out there conquering the universe. They perceive the possibility that someone out there might be doing so as a threat, and this might lead them to initiate a colonization process purely for preemptive purposes.[473]

9.4 Dysonian SETI

Given the importance I attach to solving the problem posed by the Fermi paradox – why have we not seen any evidence of extraterrestrial civilizations? – and in particular to the issue of whether there are any such civilizations out there, the reader should not be surprised to learn that I am sympathetic to the SETI endeavor of searching for them. The classical and still dominant approach to SETI is to look for radio messages that these civilizations might be directing at us. But nothing has yet been found.[474] There are two alternative approaches, one known as **Dysonian SETI**,[475] and the other one as **METI** (Messaging to Extraterrestrial Intelligence) or sometimes also as active SETI. I strongly welcome the former, which will be treated briefly in the present section, and am deeply worried about the latter, which will be discussed in Section 9.5.

The idea of Dysonian SETI is that, rather than hoping that the aliens will be interested in sending us messages (it is certainly not obvious that the civilizations out there, even if they exist and have the requisite technology for sending us radio signals, will show any such interest), we could look for other signs of their existence.[476] In the words of Ćirković (2006), who advocates

---

[472] Although see the quote by Hart (1975) in Section 9.2.

[473] On the other hand, it is conceivable that such a colonization process could be judged to be dangerous to the population of Planet X, who might therefore decide not to go ahead with it. Von Neumann probes are susceptible to mutation and the resulting Darwinian evolutionary mechanisms. The probes might evolve into an aggressive species of machines – so called berserkers, after Saberhagen's (1967) science fiction stories involving such machines – that could turn against their creators.

[474] See, e.g., Wright et al. (2014) for a brief survey of the (null) results obtained so far. Brin (2006) offers a more informal summary:

> In a sense, SETI has only just begun. There is a *lot* of territory out there! Only a few of the blithely optimistic models that were offered early on (e.g. blatant and pervasive tutorial beacons) have been disproved so far. There's still plenty of room for interstellar cultures that are transmitting more quietly, by orders of magnitude. Quieter, but still possibly detectable, if we keep searching with better instruments. And patience.

[475] The term was proposed by Ćirković (2006).

[476] I am overstating the case against classical SETI a little bit here, because in addition to messages meant for us, messages not meant for us but leaking out into space may also be detected.

this approach, "the very existence of what we can term advanced technological civilizations . . . should provide us with the means for detecting them." The idea goes back to Dyson (1960), who proposed what may well be the most likely sign of such a civilization, namely Dyson spheres. The Armstrong–Sandberg scheme for large-scale space colonization discussed in Section 9.3 provides one possible incentive for the aliens to build a Dyson sphere. Pretty much anything they're up to will require energy, and if they refrain from interstellar colonization, then their star is their most obvious main source of energy. If they want to do as much as they can of whatever it is they're doing, they may not want to let much of their star's energy output go to waste, and the straightforward method for avoiding that is to build a Dyson sphere.

A Dyson sphere will hide the star from our astronomers' view (or else it would be failing in its purpose of avoiding energy waste), but basic thermodynamics dictates that the Dyson sphere will heat up to a temperature where it will emit blackbody radiation in the infrared spectrum.[477] So electromagnetic radiation with the right sort of infrared spectrum suggesting a Dyson sphere is what we should be looking for.[478]

There are other things to look for, beyond Dyson spheres and other mega-sized artificial objects emitting infrared blackbody radiation, that fit under the heading Dysonian SETI. For instance, Whitmire and Wright (1980) point out that it only takes a relatively modest amount of nuclear waste dumped into a star to change its electromagnetic spectrum noticeably away from what can be expected from a star that has not been used as such a dump; it therefore makes sense to look for such anomalies in stellar spectra.[479] Yet another idea might be to look for anomalous arrangements of stars: a sufficiently advanced technological civilization may attain the ability to move stars,[480] and it is hardly inconceivable that they, for some reason or other, will find the natural locations of stars in their neighborhood suboptimal.

To summarize my impression of Dysonian SETI, I think it is more interesting and probably also more promising than classical SETI, mainly because it does not stand and fall with the assumption that the aliens engage in interstellar signaling. But it seems worthwhile to continue pursuing both. The third option, however . . .

---

[477] The Dyson sphere will emit the same amount of energy that the star would have done, so my talk of "energy waste" is strictly speaking inaccurate: what the aliens really do by building the Dyson sphere is to make sure they can exploit most of the *negentropy* contained in their star's electromagnetic radiation, before letting the energy slip out of their system at longer wavelengths and lower negentropy.

[478] Wright et al. (2014) advocate this approach, and also summarize the results from searches that have been done so far. These are rather few, and with no clear cases suggesting any artificial objects out there.

[479] See, however, Freitas (1985) for a criticism of this approach, and alternatives based on looking for traces of artificial fusion (as opposed to fission as in the Whitmire–Wright study) energy.

[480] One possible technology for doing so, seemingly on about the same level of engineering difficulty as a Dyson sphere, is a kind of solar sail known as the Shkadov thruster; see, e.g., Badescu and Cathcart (2006).

9.5 Shouting at the cosmos

Like classical SETI, METI is based on the idea that extraterrestrial civilizations will want to communicate with us. The difference is that METI proponents are not content with just listening. If we wish to engage in dialogue with the aliens, why just sit and wait? Wouldn't it be better to initiate the dialogue? This is what they suggest.[481]

And action has been taken. Probably the most famous cases are the Pioneer plaques and the Voyager recording. The former are a pair of gold-anodized aluminum plaques on board the Pioneer 10 and the Pioneer 11 space probes (launched in 1972 and 1973, respectively, and on their way out of our solar system), with a pictorial message designed to give information about the origin of the probe and about humans. The latter are a pair of phonograph recordings on board Voyager 1 and Voyager 2 (both launched in 1977) with a similar purpose. These cripplingly slow space probes, traveling at a fraction less than 10^{-4} of the speed of light (relative to us), and not even aimed at any particular star, seem less likely to ever be found by extraterrestrials than the various radio messages that have been sent over the years. In 1974, the Arecibo radio telescope in Puerto Rico broadcast a message with rudimentary descriptions of our biochemistry, our physiology and our solar system, towards the globular star cluster M13 about 25,000 light years away. A number of similar METI actions have been taken since then.[482]

I consider these initiatives inexcusably reckless. It may well be that their most likely outcome is that the messages never reach any extraterrestrials – either because the extraterrestrials do not exist, or because our messages remain needles in the cosmic haystack – but *if* they do, it could well be terribly dangerous. Both in classical SETI and in METI, there is a more or less explicit assumption that extraterrestrials are friendly.[483] But how do we know that? What if there are civilizations (one or more) out there with the capacity for interstellar travel and the policy of preemptively wiping out newcomers? The Great Filter discussions in Section 9.2 hint that that is probably *not* the case, because if those civilizations had that capacity, they would probably already be here. But we do not *know* that. Our reasoning could be *wrong*. Extraterrestrial life and civilizations is a topic we know so extremely little about, that all the reasoning we do about them needs to come with a high level of epistemic humility.[484]

I'm not saying METI will always be the wrong way to go, just that we currently know too little to go ahead with it. David Brin, in his seminal 2006 essay

---

[481] Well-known proponents of METI include Zaitsev (2011) and Vakoch (2014).

[482] See Zaitsev (2011) for an overview.

[483] The quote from Sagan and Newman (1983) in Section 9.2 is fairly typical of this attitude.

[484] Recall from Section 8.1 the corrective equation (19) of Ord, Hillerbrand and Sandberg (2010). What probability should we assign to the event $\neg A$ that the reasoning leading to the Great Filter equation (22) is misleading in a way that gives us entirely the wrong ideas about extraterrestrial life? I think this is an area we understand so poorly that anything less than $P(\neg A) = 0.1$ would be injudicious.

"Shouting at the cosmos," is even more moderate, and asks merely that "all of those controlling radio telescopes forebear from significantly increasing Earth's visibility with deliberate skyward emanations, until their plans were first discussed before open and widely accepted international fora."[485] I suspect, however, that this moderate request is mostly just a rhetorical gambit, because it is hard to see how, with the current state of knowledge, rational discussions of the kind he suggests could lead to any other decision than "let's not do it."

A common response to fears about METI signals informing hostile aliens about our existence is that they already know about us, because of detectable leakage of our more everyday radio and radar activities, artificial night-time lighting of large urban areas, or the anomalously rapid changes in the composition of our atmosphere.[486] Settling this issue is far from straightforward, because detectability of Earth-originating signals depends on signal strength, on distance and on the size of the aliens' telescopes.[487] But suppose, for the sake of argument, that METI signaling does not impose any added risk of aliens detecting our existence, above and beyond what our other activities reveal. What, then, is the point of METI? Or, as Brin (2014) asks, "If the horses are gone, why are they so eager to open the barn door?", to which he adds that this exposes the hypocrisy of METI advocates. To which I agree.

Vakoch (2014) claims to have an answer, however. He suggests that those friendly aliens out there might have "research protocols that require silence if only undirected leakage radiation were detected, but that would authorize a reply to intentional efforts to make contact." This convoluted argument is debunked by Brin (2014) several times over. Just in case the reader wants to figure out for herself how to answer Vakoch (it is not difficult!) I'll leave out Brin's counterargument, apart from his sarcastic "trust me, I know exactly what is out there."

In the same discussion on METI as Brin (2014) and Vakoch (2014), in the journal *Cato Unbound*, we also have Hanson (2014d), who acknowledges the possibility that METI activities could lead to a space invasion apocalypse, but also that it could lead to enormous benefits. These possibilities, he says, need to have probabilities attached to them, and then be subjected to the kind of decision-theoretic analysis outlined in Section 6.7. The stupendous difficulty of finding well-grounded numbers to plug into the equations suggests that it would be good to tread lightly. Let us postpone any METI action until we understand the prospects and the risks much better, so that we are in a better position to make a wise decision. There's an obvious asymmetry here: METI activities cannot be

---

[485] Brin (2006).

[486] See Haqq-Misra et al. (2013) for a survey of such considerations.

[487] A civilization capable of building a Dyson sphere would surely also be able to build a similarly sized telescope (if they are motivated to do so).

undone,[488] whereas if we refrain from METI activities now we can make up for that by being more active in the future.

In another contribution to the same exchange, Hanson (2014c) asks whether we should also refrain from non-METI radio activities that nevertheless leak signals into outer space. Closing the barn door entirely seems like a highly unrealistic undertaking, and probably useless since the horses are mostly long gone. What we might do, however, is to refrain from those exceptionally strong signals we leak when doing radar astronomy (which involves directing microwaves towards other celestial bodies in the solar system and studying how these are reflected by those bodies). Here Hanson offers a rudimentary decision-theoretic analysis strongly suggesting such a moratorium.

We do not know whether there are technologically advanced extraterrestrial civilizations out there, and whether their intentions are benign. It is hard to see how we could (at our current state of knowledge) dismiss as implausible item (ER6) – genocide by extraterrestrials – from the Chapter 8 list of existential risks. My hunch would be that *if* a space invasion is to be expected, then the most likely case is that it is already too late to do anything about it, either because they are already on their way, or because they will be triggered by Earth-originating signals which have yet to reach them but which we have no way of intercepting. That is still no excuse for exposing us to additional risk through reckless METI initiatives.

One final caveat. The way I've discussed the various SETI approaches so far may give the impression that I'm contrasting the dangerous METI approach to the safe and harmless classical SETI approach. It is, however, not clear that words like "safe and harmless" are fully warranted. It is hardly inconceivable that extraterrestrial civilizations with ill intentions might send us information that could destroy us, such as easy recipes for doomsday weapons.[489] There have been extensive discussions in the SETI community on how to proceed upon detecting a signal from extraterrestrials, and there is even a widely (but not universally, and not adopted by governments) accepted protocol known as the *Declaration of Principles Concerning Activities Following the Detection of Extraterrestrial Intelligence*, although this particular concern does not usually play a central role.[490]

---

[488] A counterexample might be the Pioneer and Voyager probes, as it is conceivable that we would decide at some point in the future to launch a kind of rescue mission, to catch up with the probes and bring them back before they escape further into interstellar space (I'm not overly happy, however, with the ethics of actions that require such clean-up actions by future generations). Similarly catching up with our radio waves can't be done – assuming the impossibility of traveling faster than the speed of light.

[489] The reader may recall that that was the kind of information covered up by scientists in Yudkowsky's story *Three Worlds Collide* discussed in Section 1.1. The situation is also related to the so-called AI box problem – the problem of how to keep a superintelligent AI boxed in for safety purposes while still maintaining some level of communication with it (without which it would be hard to see the point of having such an AI) – illustrated in Footnote 261 of Section 4.6.

[490] See, e.g., SETI Permanent Committee (1989) and Michaud (2003).

CHAPTER 10

What do we want and what should we do?

10.1 Facts and values

The reader has probably noticed that all of the preceding chapters offer, on one hand, facts (and speculations about facts), and, on the other hand, value statements, i.e., statements about what is good, what is bad, and what ought to be done.[491] Going back to the last section (Section 9.5), we have in the former category the observation that the Pioneer and Voyager space probes move at very slow speeds compared to the speed of light, and that "radio and radar activities, artificial night-time lighting of large urban areas, [and] the anomalously rapid changes in the composition of our atmosphere" are all potentially detectable over interstellar distances. In the latter category is my statement that METI activities taking place today and in recent decades are "inexcusably reckless." Although I cannot guarantee that I never unintentionally slip between the two categories, I believe I have been mostly clear about what is what.

I am a firm believer in the distinction between *is* and *ought* – and in **Hume's law**, which states that the latter cannot be derived from the former.[492] But please note

---

[491] The distinction is not entirely straightforward. The statement "We ought to help the poor" may sound like a fact – namely a fact about what we ought to do. I will not go into the subtleties of this distinction (but see, e.g., Long, 2006), except to say that for the present discussion such (non-empirical) facts do not count as facts.

There is a further distinction that moral philosophers like to make, namely between on one hand value statements (what is valuable?) and on the other hand normative statements (what ought we to do?). I will ignore this distinction, thus in effect taking for granted the normative statement "we ought to promote what is valuable."

[492] Here is the famous passage from Hume (1739), where he emphasizes how common and easy it is to slip between the two categories:

In every system of morality, which I have hitherto met with, I have always remarked, that the author proceeds for some time in the ordinary ways of reasoning, and establishes the being of a God, or makes observations concerning human affairs; when all of a sudden I am surprised to find, that instead of the usual copulations of propositions, is, and is not, I meet with no proposition that is not connected with an ought,

Here Be Dragons. First Edition. Olle Häggström.
© Olle Häggström 2016. Published in 2016 by Oxford University Press.

that Hume's law does not say that facts are *irrelevant* for deriving statements about values and morality, only that facts alone are *insufficient*.[493] Consider the fact

```
(P1)    Arsenic is lethally poisonous
```

and the value statement

```
(C) It is wrong to put arsenic in one's neighbor's dinner.
```

Perhaps the reader feels tempted to say that (P1) implies (C), thus contradicting Hume's law, but that is not correct: (P1) does not imply (C). The tempting mistake is due to certain value statements that we take for granted, such as

```
(P2)    Thou shalt not kill.
```

Together, (P1) and (P2) imply (C). What Hume's law says in the arsenic case is that (C) is not implied either by (P1) alone or by any other collection of facts. It does not rule out that, provided we have accepted the truth of some other value statement, such as (P2), the fact (P1) can be of help for the derivation of further value statements such as (C).

Related to Hume's law is the idea of **metaethical moral relativism**, which is the idea that when people disagree about what is morally right, it is not the case that someone is objectively right and someone else is objectively wrong – there is no objective truth out there about what we ought to do, and moral disagreements are just matters of opinion.[494] I write *metaethical* moral relativism, to distinguish it from *descriptive* moral relativism and *normative* moral relativism; see, e.g., Gownsa (2012). Descriptive moral relativism is just the observation that people disagree on moral issues. Normative moral relativism is the extension of metaethical moral relativism that holds also that the non-existence of objective moral truths implies that we ought to tolerate the behavior of others even when it clashes with our own moral standards (cultural or otherwise). The distinction is important: I do tend to accept metaethical moral

---

or an *ought not*. This change is imperceptible; but is however, of the last consequence. For as this *ought*, or *ought not*, expresses some new relation or affirmation, 'tis necessary that it should be observed and explained; and at the same time that a reason should be given, for what seems altogether inconceivable, how this new relation can be a deduction from others, which are entirely different from it. But as authors do not commonly use this precaution, I shall presume to recommend it to the readers; and am persuaded, that this small attention would subvert all the vulgar systems of morality, and let us see, that the distinction of vice and virtue is not founded merely on the relations of objects, nor is perceived by reason.

[493] Hume's law is sometimes, still today, considered controversial. Purported counterexamples are sometimes presented, but all the examples I have seen are more or less obviously flawed; see, e.g., Häggström (2012c), criticizing Searle (1964).

[494] Alternatively, this position may be called **moral nihilism**. Metaethical moral relativism and moral nihilism are sometimes distinguished by saying that according to the former, there are many competing moral truths, whereas according to the latter, there is no moral truth. But the way I see it, the two positions collapse if we focus on *objective* truths, because if there are several competing truths, none of which has any privileged status over the others, then there is no objective truth.

relativism,[495] while strongly rejecting normative moral relativism. My skepticism about the existence of objective moral truths does not prevent me from vehemently opposing various cultural and other practices such as widow burning, female genital mutilation, and (to cite Section 9.5 again) METI.

My reason for being skeptical about the existence of objectively true moral statements is (and now I apologize for mostly repeating myself from Footnote 275 in Section 4.6) mainly an application of Occam's razor: the only empirical evidence that has been put forth in favor of the existence of an objective morality is the existence of our own moral intuitions, but since these can be accounted for by other factors[496] it seems superfluous to postulate an objective morality.[497,498]

The point of all this is that no facts in the world – be they about the potential consequences of emerging technologies, or about anything else – suffice for telling us what to do. Our actions are up to us, and our decisions need to be based as much on our values as on facts (or, rather, on our beliefs about facts). What kind of world do we want to live in? Surely that must influence our choices.

In fact, we don't even need to accept metaethical moral relativism to see that what we do is up to us.[499] For suppose that metaethical moral relativism is wrong and that there does exist an objectively right morality, and suppose furthermore that we somehow figure out how to access this objectively right morality, to find out in every situation what is the morally right way to proceed. Not even that would relieve us of our choices. To see this, suppose I am facing the choice of either spending a certain amount of money on a monster truck for myself, or

---

[495] I do so, however, in a soft tentative way, rather than maintaining it as a hard-line dogmatic position. My agnosticism on this issue is clear from my discussion in Section 4.6 of the Friendly AI concept known as moral rightness.

[496] See, e.g., Shermer (2004) and de Waal (2006) for popular introductions to evolutionary explanations of the propensity for moral behavior and moral reasoning exhibited by humans and other animals.

[497] Here's a more primitive argument which, to me at least, still has some force: Suppose you defend some moral position – let's say the position that the morally right thing to do is that which maximizes the total amount of pleasure minus suffering in the world – as being the objective truth about what we ought to do. Then I can play the devil's advocate by choosing the diametrically opposite position from yours, and say something along these lines:

> That's just your opinion. You favor pleasure and dislike suffering, but my preferences are the other way around. Therefore I think the right thing to do is that which *minimizes* the total amount of pleasure minus suffering. You see the symmetry of our positions? There's no way to break the symmetry without resorting to circular arguments (similarly to the case of anti-induction in Section 6.2). Which one is preferable is clearly just a matter of taste.

You will not believe I truly mean this, but I will insist, and in the end you will realize you have nothing to say to break the symmetry, other than "You moron!" But "You moron!" is not really much of an argument, so I win.

[498] Nevertheless, the issue of the existence or non-existence of an objectively true morality remains alive in academic philosophy, and the existence has been defended by first class contemporary thinkers such as Nagel (1997) and Parfit (2011).

[499] I am intentionally ignoring the annoying philosophical issue of whether we have free will, where even thinkers whom I hold in generally high regard seem stuck on the level of definitional quarrels; see, e.g., Dennett (2014).

donating the same amount to the humanitarian-aid non-governmental organization Médecins Sans Frontières. Even if I know that objective morality dictates that I give my money to Médecins Sans Frontières, I am still faced with a choice: to act morally by making the donation, or to act immorally by buying the monster truck. To know what is morally right is not the same thing as acting upon that knowledge.

Facts and values will influence our choices. The following are a few questions about value that have a potentially huge influence on decisions that may be crucial to the future of humanity.

(i) **How important is our own well-being compared to that of future generations?** Moral philosophers tend to think that a person's well-being is equally important regardless of when he lives: now or in the future. But there is a minority view known as axiological actualism, holding that ethical theory should assign things like levels of well-being or preference orderings only to *existing* persons, and not to *hypothetical future* persons.[500] Sections 10.2 and 10.3 will treat two radically different perspectives on the weighing of our own interests against those of future generations.

(ii) Provided that well-being of humans is what we value: **What is more important, the average level of well-being, or the total level (summed over all humans)?** Depending on the answer to this question, we get different answers to the question of which future is preferable, one containing relatively few people (say, a million) who all live superb lives, or one containing a much larger population (say, ten billion) living lives good enough to be worth living but not much more. This and similar questions belong to the difficult field of population ethics, and attempts to answer them in consistent fashion tend to have counterintuitive or repugnant consequences; see, e.g., Parfit (1984) and Arrhenius (2011).

(iii) **Does our planet's biosphere and its ecosystems have intrinsic value, or are they only valuable to the extent that they contribute to the well-being of humans?** This dichotomy represents two different strands of environmental ethics; see, e.g., Brennan and Lo (2011).

(iv) **Is experiencing physical reality the only existence that really counts, or would we be just as well off if we had the same experiences in a virtual reality?** Part of why I feel the environment is worth protecting is that I love hiking and cross-country skiing in the wilderness of northern Scandinavia. But what if I can get the exact same experience from some advanced virtual reality simulation? Would that make it less important to preserve the wilderness? And perhaps we should all simply migrate into virtual reality, voluntarily creating a *Matrix*-like situation?

---

[500] See, e.g., Miller (2003).

(v) **Should we go ahead and enhance cognitive and other human abilities without limits, or is there some Fukuyamaian Factor X that is crucial to the value of humans and humanity, and that risks being lost in such a process?** This and related issues were treated at some length in Sections 3.1–3.3.

(vi) **Comparing a future where humans prosper to one in which we are replaced by robots that prosper similarly, is the former automatically preferable to the latter?** This question is related to (v), as well as to the moral status of machines. Does a robot with human-level cognitive capacities have the same value as a human? Here, the issue of machine consciousness (discussed in Section 3.8) seems relevant. A future in which humans are replaced by unconscious machines that go on to colonize the rest of the universe seems pointless and absurd. But if the machines are conscious, perhaps there is reason to look more favorably on the scenario.

I do not have clear answers to these questions;[501] I'm merely mentioning them as questions we should ask ourselves when thinking about how to move forward.

10.2 Discounting

The question of how to weigh our own well-being against that of future generations (item (i) in the last section) has gained publicity in the last decade or so, thanks to the rise of the climate change issue in mainstream public discourse. The relevance of (i) to climate policy comes from the phenomenon that many of the proposed actions against climate change – say, the replacement of coal-fired power stations by solar energy for the production of electricity – are costly *now*, whereas most of the benefits of such actions come many decades hence, through the reduction of the risk of catastrophic climate change. To decide whether such actions are worth taking, we need to have a way of making relevant comparisons between present costs and future benefits; what we consider "relevant" here of course depends on how we answer (i).[502]

The standard way of doing that involves weighing the costs and the benefits using the so-called **social discount rate**, which will be explained below. While such

---

[501] This is not to say that I am clueless about value. There are plenty of other questions on value, such as. . .

(vii) **Which is better, a world where all people live in joy and happiness, or one in which the same people suffer unbearable pain?**

. . . which (pace Footnote 497) I *can* answer. The questions (i)–(vi) are chosen specifically for being nontrivial.

[502] A different approach here is to abandon cost–benefit analyses altogether, in favor of the ethical position that it is impermissible for us to act is such a way as to deteriorate the environment for future generations. That position may have something going for it, but in practice it is difficult or impossible to apply consistently.

calculations feature regularly in discussions on the economics and politics of climate change, they are much more rare in considerations of the various emerging technologies discussed in this book and the effect these may have on the future of humanity.[503] This a bit odd, because cost–benefit analyses with costs and benefits distributed over time – precisely the setting where discount rates are relevant – are just as important here as in considerations of climate change. For instance, if we take the risk of a catastrophic AI breakthrough (see Sections 4.6 and 8.3) seriously we might consider moving ahead more slowly and cautiously with AI research and development today, focusing less on making AI capable and more on making it safe.[504] Such a reorientation might, on one hand, imply costs to the economy today, lowering GDP (gross domestic product) growth that might otherwise have resulted from a less constrained approach to AI development, while, on the other hand, it might lower the risk for an AI catastrophe several decades into the future.

Economic calculations stretching more than a year or so into the future typically involve a discount rate r, that describes how much less value we attach to an asset if we have to wait a year before receiving it, compared to if we received it immediately.[505] Setting $r = 3\%$, for instance, means that if something today is worth \$1 to us, then we attach value \$$(1 - 0.03) = \$0.97$ to the prospect of receiving it a year from now.[506] Discounting, like bank interest rates, compounds multiplicatively, so that the prospect of receiving it two years from now is worth \$$(0.97 \cdot 0.97) \approx \$0.94$. More generally, the prospect of receiving an asset t years from now reduces its value by a factor $(1 - r)^t$ compared to if we had it now.

Various choices of r are possible. Often, when the time frame is just a few years, the choice between, say, $r = 3\%$ or $r = 6\%$ is relatively inconsequential, but on longer time scales the effect is dramatic. A positive r means that we attach more importance to the present than to the future, and the larger r is, the stronger this preference. Table 10.1 gives the remaining fraction of the value of an asset, given discount rate r, if we are forced to wait for it for 10 or 100 years, respectively.

In June 2007 I attended a very illuminating panel discussion in Stockholm, where, among others, prominent economists Nicholas Stern and Martin

---

[503] There are exceptions, however. For an interesting discussion exposing fundamentally different views on social discounting by two of the most frequently cited authors in this book, see the twin blog posts "Against discount rates" (Yudkowsky, 2008g) and "For discount rates" (Hanson, 2008f).

[504] Precisely this was recently suggested in a letter signed by well-known thinkers like Nick Bostrom, Erik Brynjolfsson, Stephen Hawking, Elon Musk, Peter Norvig, Steve Omohundro, Martin Rees, Anders Sandberg, Max Tegmark and Eliezer Yudkowsky, plus yours truly and thousands of others; see, e.g., Häggström (2015).

[505] The rest of this section is mostly based on Häggström (2007b).

[506] It is not a coincidence that I use money as the quantity to discount. Other quantities of value are possible, but those fond of discounting are typically economists, and they tend to think all value can be translated into money. For some valuable things, such as diamonds and automobiles, this is fairly unproblematic, whereas for others, such as human lives, human health, and ecosystems, it is rather more problematic.

Table 10.1 *The relative effects of different discount rates applied over different periods.*

r	10 years	100 years
0.1%	0.99	0.90
1%	0.90	0.37
1.4%	0.87	0.24
3%	0.74	0.048
6%	0.54	0.0021

Weitzman debated climate change economics in general, and discount rates in particular. Their disagreement about the urgency of cutting down on greenhouse gas emissions can be pinned down to their different choices of discount rates. Stern favors a discount rate of $r = 1.4\%$, as explained in his famous *Stern Review*, while Weitzman favors $r = 6\%$, a choice he motivates in his review of that very report.[507] A glance at Table 10.1 reveals that Stern, through his choice of discount rate, attaches a factor of about $\frac{0.24}{0.0021} \approx 120$ greater importance than Weitzman to the economy of our great grandchildren 100 years hence.[508] Given this, it is hardly surprising that they disagree about climate policy.[509]

During the panel discussion, Weitzman emphasized repeatedly that his own choice is mainstream in economics, roughly representative of some 95% of economists,[510] whereas Stern's position is on the extreme fringe. Weitzman judged Stern's advocacy of the lower discount rate to be comparable to a CEO of an oil company insisting on disregarding the positions of all climate scientists other than the 5% most optimistic ones concerning how sensitive climate is to our greenhouse gas emissions. This analogy is in my opinion a very poor one, because it ignores the fact–value distinction. The climate sensitivity issue is about *facts*, albeit uncertain, about the climate system, and in such issues non-specialists do not really have any better option than to trust mainstream science. The discounting issue, in contrast, is about how we *value* future generations' well-being and

---

[507] Stern (2007), Weitzman (2007).

[508] In Weitzman's own (slightly sarcastic) words:

> The disagreement over what interest rate to use for discounting is equivalent here in its impact to a disagreement about the estimated damage costs of global warming a hundred years hence of *two orders of magnitude*. Bingo!

[509] Later, Weitzman seems to have moved to a position closer to Stern's as regards the need for urgent action on climate. He motivates this as an insurance policy against the possibility that climate sensitivity (see Equation (4) in Section 2.3) may be at the high end of or even above the usual uncertainty ranges; see Weitzman (2011).

[510] This number agrees rather poorly with my own experience of academic economists, but I accept that Weitzman is in a much better position than myself for judging the distribution of opinions in that profession.

economic prosperity.[511] Compared to issues about facts, expert consensus on issues of value has much less of a privileged status in defining what opinion it is rational for non-experts to hold.

So where do these discount rates come from, anyway? Are they just pulled out of thin air? To some extent, yes,[512] but far from entirely. A framework that both Weitzman (2007) and Stern (2007) agree upon for selecting a discount rate r is the use of **Ramsey's formula**:

$$r = \eta g + \delta. \tag{24}$$

Here g denotes the average annual GDP per capita growth rate, $\eta \geq 0$ is the so-called risk aversion coefficient, and $\delta \geq 0$ represents our "pure" preference for our present well-being compared to that of future generations. This, of course, requires further explanation. Let us assume for the moment that we are impartial between our own well-being and that of future generations, therefore setting $\delta = 0$. This can still give rise to a positive discount rate r, provided that η and g are both positive. The idea is that if the average annual GDP increase g is positive, then future generations will be richer than we are, and therefore perhaps in less need of additional economic resources. It is plausible to think that an extra dollar in the wallet means more to someone who is poor than to someone who is rich, and the risk aversion coefficient η is meant to represent this. The marginal utility $U'(x)$ of an extra dollar as a function of how much money x one already has should then be a decreasing function of x, and if it is proportional to $1/x^b$ for some fixed b, then we set $\eta = b$.[513] The extreme case $\eta = 0$ means that a beggar benefits no more (but also no less) from an extra dollar than does a billionaire. Instead, taking $\eta > 0$ means that the extra dollar *is* worth more to the beggar than to the billionaire. The larger η, the more marked is this effect. An intuitively understandable special case is $\eta = 1$, which means that an increase of one's financial assets by a given percentage means as much regardless of how much one had to begin with, so that an extra \$1 is worth as much to someone who has \$100 as an extra \$10 is to someone who has \$1000.[514] In any case, if g correctly represents future annual

---

[511] Weitzman commits further blurring of the fact–value distinction when he says that "a major problem with Stern's numbers is that people are not observed to behave as if they are operating with [those numbers]" (Weitzman, 2007). Observations of people's behavior do fall in the domain of facts, but such facts do not imply anything about how we should value the future. The empirical observation that people exhibit impatience in their daily dealings does not imply that it is *morally right* to do so. Nordhaus (2007) makes the same mistake. See Caney (2008) for further critique of the Weitzman–Nordhaus approach.

[512] Recall the "why game" that every child eventually discovers, and that I discussed in Section 3.2. Tell me which discount rate you favor, and I'll ask "Why?" Whatever you answer, my reply will be "Why?" and eventually you will have to terminate the discussion by answering "Just because it seems obvious to me," thereby implicitly admitting that your choice of discount rates is based on premises pulled out of thin air.

[513] The marginal utility $U'(x)$ is the derivative of the utility function $U(x)$ discussed in Section 6.7.

[514] In other words, while $\eta = 0$ holds the utility of an *absolute* increase in financial assets constant, $\eta = 1$ holds the utility of a *relative* increase in financial assets constant.

GDP increase, and η correctly represents how our utility depends on how much money we have, then discounting at rate $r = \eta g$ is necessary for optimizing the total amount of utility, summed over all present and future people.[515] If, on top of this, we have a preference for increasing our own well-being or utility rather than that of future generations, we can account for that by choosing a positive value of δ.

Stern derives his $r = 1.4\%$ by setting $\eta = 1$, $g = 1.3\%$ and $\delta = 0.1\%$,[516] and plugging these values into Ramsey's formula (24). Weitzman instead prefers what he calls a "trio of twos": $\eta = 2$, $g = 2\%$ and $\delta = 2\%$, so that (24) yields $r = 6\%$. He then criticizes Stern for artificially producing a small value of r through his unconventionally small values of η, g and δ. At the Stockholm meeting, he was especially scornful about Stern's choice of $\delta = 0.1\%$, but Stern struck back, pointing out that Weitzman's $\delta = 2\%$ amounts to a half-life of 35 years, and the unreasonableness of valuing the lifetime amount of well-being of someone born in 1970 twice as high as that of someone born in 2005.

On that particular issue (how to choose δ) I am much more sympathetic to Stern's point of view than to Weitzman's, feeling a bit dismayed by the idea of valuing people's lives differently depending merely on when in history they happen to be born. It is harder to take a stand on the value of η. But rather than calling for a small η, one might voice doubts concerning the use of a positive value of the future annual GDP increase g. It is true that we've experienced mostly economic growth for a very long time, and it is true that further technological progress has the potential to give us further economic growth, perhaps even on levels unimagined by mainstream economists, given, e.g., a Hansonian transition to an economy of uploads (Section 3.9), a benign intelligence explosion (Sections 4.5 and 4.6), a Drexlerian breakthrough in atomically precise manufacturing (Section 5.2), or some combination of these. But it is also true that we are depleting several of our

---

[515] Note that the risk aversion coefficient η is defined in terms of an individual's utility as a function of her economic resources, but we apply it to society as a whole. There's a gap here. The gap can be filled by assuming that the future distribution of economic resources in society will just be a scaled version of today's (for instance, at some future time when GDP per capita has doubled, the income distribution would be that which we would get today if everyone doubled their income). This assumption is likely to be false, as there are strong indications that economic inequality is increasing and will continue to do so; see the discussion at the end of Section 4.4, as well as Piketty (2014). The assumption still makes sense as a simplification and a way to separate out the issue of economic inequality within a generation (or, more precisely, within society at a given time t) as something for that generation to sort out on its own.

[516] He objects to assigning unequal weight to the well-being of present or future people. This suggests setting $\delta = 0$. His choice to nevertheless set $\delta = 0.1\%$ is based on his estimate of an annual probability of 0.1% that humanity goes extinct through one disaster or another (see Chapter 8 for a list of plausible candidates); the idea is to avoid taking into account the well-being of generations that will not even exist. What worries me about this kind of reasoning is that it can become self-fulfilling: if we have a high estimate of the probability that humanity goes extinct in the relatively near future, and this estimate leads us to a larger value of the discount rate r, then this r can lead to a more short-sighted policy that increases the probability of human extinction (thus in a way affirming that r should be large, but in a way that I for obvious reasons am not comfortable with).

planet's natural resources in a way that can hurt us badly (Section 2.3) and that we are facing a variety of global catastrophic risks (Chapter 8). It is highly uncertain which of these tendencies will dominate. To postulate a positive g is to simply take for granted that things will go well for us, and that strikes me as reckless.

So to summarize my view of the quantities going into the right-hand side of Ramsey's formula (24), I am highly suspicious about positive values of δ and of g. That leaves η, which I do think should be positive, but note that if neither δ nor g is positive, then a positive η, no matter how large, will fail to result in a positive discount rate r. The bottom line here is that I seem to be committed to the view that it is wrong to postulate a positive discount rate r. I am, however, not entirely comfortable with that conclusion, due to the paradoxical and perhaps unwanted effects (such as ruling out consumption entirely in favor of investing for the future) that setting r to 0 or something very small tends to have on economic decisions; see Weitzman (2007).

A further difficulty here is that if we decide to compromise by setting r very small but not quite at 0, with the idea of giving reasonable weight to the economy of future generations, then we in fact only have to switch to longer time scales to notice that the future is discounted away almost entirely. If, for instance, we take the (from the point of view of mainstream economics) extremely small discount rate $r = 0.1\%$, then we see from Table 10.1 that this corresponds to retaining 90% of value a hundred years from now, which may sound relatively reasonable. But look what happens 10,000 years from now: the fraction of value retained after such a time period is $(1 - 0.001)^{10,000} \approx 0.000045$, meaning, in frank terms, that we do not care about the economy and welfare of our great-great-...-great-grandchildren 10,000 years hence.

The level of discount rate that looks reasonable tends to depend on the time frame. On short time scales, high values of r tend to seem plausible,[517] whereas longer time scales seem to require lower discount rates. This observation has led to suggestions to instead use so-called **hyperbolic discounting** (briefly and more informally discussed in Section 2.10), which is a deviation from the usual constant-rate discounting where the rate is allowed to decrease over time; see, e.g., Weitzman (2001) and Karp (2005). With a constant discount rate r, the value (to us, now) $V(t)$ of something we must wait for for t years decays exponentially:

$$V(t) = V(0)e^{-\rho t},$$

---

[517] This becomes shockingly obvious when we look at very short time scales, such as days or even hours. I suspect many readers will find the following experience of mine familiar. It has sometimes happened that, late on Friday evening, I decide to have a fifth glass of wine, knowing full well that the increase in immediate well-being that extra glass gives me is more than compensated for by the decrease in well-being the next morning – that is, unless I apply the appropriate discount rate. If we assume, not unreasonably, that my decrease in well-being the next morning is twice the extra well-being I feel immediately, then (as a quick back-of-the-envelope calculation shows) a discount rate that, if we applied it over a whole year, would reduce value by a stupendous factor like 10^{200} is needed to motivate my behavior.

where $\rho = -\log(1-r)$. This standard case is therefore called **exponential discounting**. In hyperbolic discounting, $V(t)$ still shrinks over time, but the curve has a different shape:

$$V(t) = \frac{V(0)}{1+ct},$$

where $c > 0$ is a constant that we are free to choose as we like. This corresponds to a time-variable discount rate $r(t)$ that starts out with $r(0) = 1 - e^{-c}$, and then gradually decreases, approaching 0 as $t \to \infty$. This decrease is tantamount to saying that we are asking future people to think and act more long-term than we do.

For all its faults, exponential discounting is superior to hyperbolic discounting in the following respect, called **time consistency**. If we decide to apply exponential discounting at a given rate r, and if things go as we've planned until time t, then we can expect people at time t to agree with us about what to do in the further future, provided they apply the same principle of exponential discounting, with the same r. If instead we decide that the right principle for discounting the future is hyperbolic discounting with parameter c, and if the future people at time t agree that that is the right principle, but insist (reasonably enough) on applying the principle from their own vantage point on the time axis, then they will revise our plans for the post-t future in favor of something more short-sighted. For them to stick to our plans, we either have to have a way of *forcing* them to do so,[518] or hope that they will switch to a discounting principle that emphasizes long-term objectives better than we do. The point here is that with exponential discounting (unlike the hyperbolic case) it doesn't matter that people at time t will privilege their own "now" in the same way that we do ours – they will still have the same preferences about the post-t future as we do.

My view of the use of hyperbolic discounting in cost–benefit analyses affecting the future of humanity is this. As Ainslie (2001) explains beautifully, hyperbolic discounting is a psychological phenomenon that profoundly affects many parts of our lives.[519] But from the point of view of rationality, it is a defect or a miscalibration of our brains, causing predictable inconsistencies over time of our preferences. The idea of elevating such a defect to being normative for how we should plan the long-term future of humanity strikes me as grotesque.

To summarize, both exponential and hyperbolic discounting are highly problematic.[520] Yet, it is doubtful whether we can do without discounting in serious

---

[518] As with the time lock in Footnote 519 below.

[519] A wide range of empirical findings suggest that we are psychologically predisposed to spontaneously value the future in a way that fits hyperbolic discounting better than exponential. Phenomena like putting time locks on one's own fridge, or hiding away money in pension funds inaccessible for the next few decades, are hard to understand without reference to the internal conflict that hyperbolic discounting models. See also the brief discussion of hyperbolic discounting in Section 2.10.

[520] One could of course define other ways to let the discount rate depend on t, giving discounting schemes that are neither exponential nor hyperbolic, but the search for some third alternative that avoids all the downsides discussed here is not looking promising. As soon as we deviate from exponential discounting, time inconsistency follows.

decision-making affecting our long-term future. The topic of the next section is an approach that avoids discounting, but which is limited to a particular aspect of the future, albeit an important one, namely the avoidance of existential catastrophe.

10.3 Existential risk prevention as global priority?

Preventing an existential catastrophe that would wipe out humanity is important. But just how important? Any reader who has followed me all the way to this final chapter will probably agree with me that the issue deserves more attention than it currently gets from decision-makers, scientists and other thinkers. Recall, from Section 1.5, Nick Bostrom's remark that 20 times as many academic publications are published on the topic of snowboarding, compared to those on risks of human extinction.[521] The implicit message here is that although snowboarding may be an important topic, surely existential risk prevention is *at least* as important, and probably *even more* important.

This, however, understates the message in the Bostrom paper in which the remark is made. Its title is "Existential risk prevention as global priority." Bostrom argues that the prevention of an existential catastrophe is even more important than we typically believe, and deserves to be elevated to the top of the international agendas of scientific efforts and public discourse. The core of his argument was formulated three decades ago, by Derek Parfit in his wonderful book *Reasons and Persons*:[522]

I believe that if we destroy mankind, as we now can, this outcome will be *much* worse than most people think. Compare three outcomes:

(1) Peace.
(2) A nuclear war that kills 99% of the world's existing population.
(3) A nuclear war that kills 100%.

(2) would be worse than (1), and (3) would be worse than (2). Which is the greater of these two differences? Most people believe that the greater difference is between (1) and (2). I believe that the difference between (2) and (3) is *very much* greater. (p 453)

This is a thought experiment where the exact details, such as the catastrophe being nuclear war (the obvious choice in 1984 when Parfit wrote his book) rather than some other event from the list in Section 8.3, do not matter.[523] It is equally inconsequential to the force of the argument if we pick a nice round number like

---

[521] Bostrom (2013).

[522] Parfit (1984). The book offers an amazing variety of brilliant and, as it would turn out, influential philosophical ideas. The influence on the present book goes far beyond the quote given here; see, e.g., the discussion on teleportation in Section 3.8, that builds heavily on Parfit's ideas of personal identity.

[523] Someone might doubt this in view of the statement in (ER7) of that same section that a global nuclear war is unlikely to directly kill 100% or even 99% of humanity. If we include indirect effects during the next few years after the war, such as famine resulting from nuclear winter, and the kind of social collapse discussed in Section 8.1, then Parfit's scenarios (2) and (3) do not strike me as implausible.

7 billion for the world's current population, so let's do that. That would mean that Parfit's (2) represents the killing of 6.93 billion people, which, in relative terms, is nearly as bad as the killing in (3) of 7 billion people. This straightforward comparison is what prompts Parfit's remark about what "most people believe."

Parfit's point, however, is that, unlike (2), scenario (3) does not merely wipe out all people living today, but also the potential existence of *all future generations*.[524] For someone who, like Parfit, thinks a prosperous future for humanity would be a very good thing, the wiping out of that potential is a very bad thing. To give a hint towards quantifying its badness, Parfit mentions that "the Earth will remain inhabitable for at least another billion years" (p 453). Bostrom (2013) goes on with the obvious back-of-the-envelope calculation: if humanity enters a mode of sustainable existence on our planet with a population of one billion people and an average life span of 100 years, then the total number of human lives during the next billion years is 10^{16}. That is *a lot of lives*, and with this way of thinking, the difference in value between Parfit's scenarios (2) and (3) is more than a million times the corresponding difference between (1) and (2).[525,526] Bostrom then goes on to suggest that this 10^{16} number has huge consequences.

> This implies that the expected value of reducing existential risk by a mere *one millionth of one percentage point* is at least a hundred times the value of a million human lives. (25)

This is a remarkable and counterintuitive conclusion.[527] It serves as a superb pedagogical device for driving home the point that avoiding existential

---

[524] This is the point where an axiological actualist (see item (i) in Section 8.1) would interject: "Who cares?"

[525] Failure to grasp this point probably goes a long way towards explaining why people in general do not take the issue of preventing existential catastrophe as seriously as it deserves. Yudkowsky (2008a) makes a psychological observation that may serve as a further explanation of this:

In addition to standard biases, I have personally observed what look like harmful modes of thinking specific to existential risks. The Spanish flu of 1918 killed 25–50 million people. World War II killed 60 million people. 10^7 is the order of the largest catastrophes in humanity's written history. Substantially larger numbers, such as 500 million deaths, and *especially* qualitatively different scenarios such as the extinction of the entire human species, seem to trigger a *different mode of thinking* – enter into a "separate magisterium". People who would never dream of hurting a child hear of an existential risk, and say, "Well, maybe the human species doesn't really deserve to survive."

See also his discussion in the same paper on the more general phenomenon of scope neglect, quoted in Section 8.1, Footnote 393.

[526] If future generations insist (despite the possibilities, outlined in Chapter 3, of developing into a posthuman existence) on having biological bodies similar to our own, then, to keep estimates conservative, we may have to cut down a bit on Parfit's "another billion years" of human existence, due to the gradual increase in the Sun's luminosity; see (ER3) in Section 8.2. Not more than one order of magnitude, however, leaving us with at least 10^{15} future human lives, which is still enough to make the point that human extinction is many orders of magnitude worse than a naive calculation involving only people alive today would suggest.

[527] Nick Beckstead, in his remarkable PhD thesis *On the Overwhelming Importance of Shaping the Far Future*, arrives at similar conclusions, via arguments similar to Bostrom's but carried out at greater length and greater depth (Beckstead, 2013).

catastrophe is extremely important and a problem that deserves more attention than it has hitherto received. However, the idea of elevating it to a policy recommendation to be taken literally makes me very uneasy, for reasons that I will return to in Section 10.4.

Next, note that the 10^{16} figure used to derive Bostrom's observation (25) is highly conservative, in that it assumes that humanity will be stuck on Earth. What if (as discussed in Section 9.3) we go on to colonize the rest of the universe? This easily leads to conclusions even more extreme than (25). Bostrom suggests that if we do take space colonization into account, then 10^{34} is a more reasonable estimate of the number of future man-years, provided that we stick to biological bodies similar to those we have now, whereas if (Section 3.9) we transfer our minds to computer hardware, then 10^{54} is more like it.[528,529] These numbers are of course very uncertain and could be off by many orders of magnitude, but all we need to understand here is that they are stupendously large, and lead, if we reason similarly as in Bostrom's statement (25), to conclusions far far more extreme.

Note that the last paragraph implicitly assumes that the colonization of the universe by humans (or posthumans) is a good thing – that a universe imbued by human civilization is better than one that is not. That is a very common assumption in futuristic discussions, but not one that is self-evident. What if such a colonization process would actually constitute a catastrophe of cosmic proportions? Computer scientist and ethicist Brian Tomasik has, in a couple of recent essays "The importance of wild-animal suffering" and "Applied welfare biology and why wild-animal advocates should focus on not spreading nature," highlighted a way in which it could turn into such a catastrophe.[530] In the first of these, he begins with quoting Dawkins (1995):

---

[528] Bostrom (2013) is rather concise in motivating these numbers. Here's the footnote used to back up the 10^{34} figure:

> This is based on an accelerating universe with a maximal reachable co-moving distance of 4.74 Gpc, a baryonic matter density of $4.55 \cdot 10^{28}$ kg/m^3, a luminosity ratio of stars $\approx$100, and 1 planet per 1,000 stars being habitable by 1 billion humans for 1 billion years (Gott et al., 2005; Heyl, 2005). Obviously the values of the last three parameters are debatable, but the astronomical size of the conclusion is little affected by a few orders-of-magnitude change.

And here's the one for 10^{54}:

> This uses an estimate by the late futurist Robert Bradbury that a star can power 10^{42} operations per second using efficient computers built with advanced nanotechnology. Further, it assumes (along with the cosmological estimates mentioned in the previous footnote) that the human brain has a processing power of 10^{17} operations per second and that stars on average last 5 billion years. It does not assume any new star formation. See also Ćirković (2004).

[529] To translate these numbers (10^{34} and 10^{54}) to human lives of centennial duration, we just have to trim them by two orders of magnitude, giving 10^{32} and 10^{52}. It is unclear, however, how relevant such a unit of measurement (number of 100-year human lives) is in this context, because in a technological civilization that can take us to the stars and the galaxies, it is implausible to think that we would not have the ability to prolong our lives considerably (Section 3.7), and it is just barely conceivable that we would choose deliberately not to do so.

[530] Tomasik (2009, 2013).

The total amount of suffering per year in the natural world is beyond all decent contemplation. During the minute it takes me to compose this sentence, thousands of animals are being eaten alive; others are running for their lives, whimpering with fear; others are being slowly devoured from within by rasping parasites; thousands of all kinds are dying of starvation, thirst and disease. (p 131)

Tomasik then goes on to argue that "animal advocates should consider focusing their efforts to raise concern about the suffering that occurs in the natural environment."[531] In the second essay, he points out that if, when we colonize of the universe, we decide to bring wildlife along, and to terraform planets where this wildlife can then spread, then we risk multiplying the already enormous amount of suffering described by Dawkins immensely. Humans (at least many of them, including myself) tend to value nature and wildlife highly; hence there may be a temptation to bring those to the planets we settle down on. Tomasik's suggestion is that we ought to resist that temptation due to the vast amount of animal suffering we would otherwise create.

10.4 I am not advocating Pascal's Wager

Recall from the previous section Bostrom's conclusion (25) about how reducing the probability of existential catastrophe by even a minuscule amount can be more important than saving the lives of a million people. While it is hard to find any flaw in his reasoning leading up to the conclusion, and while if the discussion remains sufficiently abstract I am inclined to accept it as correct, I feel extremely uneasy about the prospect that it might become recognized among politicians and decision-makers as a guide to policy worth taking literally. It is simply too reminiscent of the old saying "If you want to make an omelet, you must be willing to break a few eggs," which has typically been used to explain that a bit of genocide or so might be a good thing, if it can contribute to the goal of creating a future utopia.[532] Imagine a situation where the head of the CIA explains to the US president that they have credible evidence that somewhere in Germany, there is a lunatic who is working on a doomsday weapon and intends to use it to wipe out humanity, and that this lunatic has a one-in-a-million chance of succeeding. They have no further information on the identity or whereabouts of this lunatic. If the president has taken Bostrom's argument to heart, and if he knows how to do the arithmetic, he may conclude that it is worthwhile conducting a full-scale nuclear assault on Germany to kill every single person within its borders.[533]

There are answers to this thought experiment. First, it seems highly unrealistic that the CIA would be able to produce precisely that information, including

---

[531] See also Footnote 189 in Section 3.9.

[532] This quote is often misattributed to Joseph Stalin, but in fact seems to go back to the 18th-century French politician François de Charette; see Vuolo (2013).

[533] In a private conversation in 2007, Martin Weitzman offered me a similar example to argue for the untenability of stipulating a discount rate of zero.

a credible estimate of the probability that the lunatic succeeds in building and launching his doomsday weapon. Second, surely the annihilation of Germany would be bad for international political stability and increase existential risk from global nuclear war by more than one in a million; if the president factors this into his calculation, then he will understand that nothing is gained from the nuclear assault.

To this, in turn, I have a couple of responses. One is to ask whether we can trust that our world leaders understand these points. Another is that even if, hypothetically, I found myself in the situation of the thought experiment being literally true and free from caveats about political instability resulting from the nuclear assault, and if I were the president's advisor, then I would probably tell him this: "To hell with the expected utility calculation showing that you should wipe out Germany – there are things *that you simply cannot do*, no matter how much future value is at stake!" And by reflecting over this example, I realize that although I strongly reject axiological actualism (see item (i) in Section 10.1), there lives a little bit of an axiological actualist inside of me.

Having said all this, let me repeat that I still accept Bostrom's conclusion (25) as an excellent pedagogical device for showing that existential risk prevention is even more important than we naively or spontaneously think, and that even a very small reduction in the probability of an existential catastrophe can be worth a considerable cost. There are other arguments pointing in the same direction. One such argument, emphasized by Taleb et al. (2014), is the following.

Write p for the annual probability of an existential catastrophe wiping out humanity. No $p > 0$ is acceptable (so the argument goes), because no matter how small p is, the risk accumulates over time and the probability of humanity surviving approaches 0 over sufficiently long time intervals, so we are doomed.[534] For instance, $p = 0.001$ may sound small, but if the annual extinction probability stays the same, then the probability of survival for the next t years will be 0.37 for $t = 1000$, 0.0067 for $t = 5000$, and 0.000045 for $t = 10,000$. Even smaller values of p lead to longer time scales over which we are doomed, but we are still doomed. Ergo, we need to reduce p to 0.

This Taleb et al. argument relies on the assumption that the annual extinction probability p is constant over time. Their conclusion that positive p implies doom is escapable if we allow for the possibility that p is time dependent and can be expected to go down in the future. If we somehow manage to convince ourselves that we live in a particularly dangerous but not-too-long transition period, then a positive value of p, say $p = 0.001$, might be considered tolerable if the transition period is estimated to last for at most a few decades. So are we living in such a period? Perhaps we are: as argued in Sections 8.3 and 8.4, existential risk today is abnormally high. But for the transition period reply to work we also need to be able to say with some confidence that the period will soon be over. Perhaps

---

[534] In mathematical terms, what I'm saying here is that if the annual probability of catastrophe is p, then the probability of humanity surviving for t years is $(1-p)^t$, and if $p > 0$ then $\lim_{t \to \infty}(1-p)^t = 0$.

there is such a light at the end of the tunnel,[535] but to take that for granted seems reckless. A techno-optimist might say that future generations will be in a much better position to reduce existential risk than we are, but it seems to me morally impermissible to expect them to do that while not doing our best ourselves to reduce existential risk. This gives the Taleb et al. argument some force.

What has been said so far in this section and in Section 10.3 – in particular with phrases such as "no matter how small the probability is" – is likely to remind readers versed in the history of philosophy (or simply remembering Footnote 193 in Section 3.10) of **Pascal's Wager**. The French 17th-century mathematician and philosopher Blaise Pascal basically invented the decision-theoretic framework discussed in Section 6.7 in order to analyze the question of whether or not he ought to believe in God (and along with that, attend church and whatever else a good Christian is supposed to do). He noted that the gain from believing in God, as compared to not doing so, is

$$\begin{cases} \text{positive and infinite (from getting to spend} \\ \text{eternity in heaven instead of in hell)} \qquad \text{if God exists} \\ \\ \text{negative but finite (from having to go to church} \\ \text{rather than enjoying various sinful activities)} \qquad \text{if God does not exist} \end{cases}$$

and furthermore that no matter how small a non-zero probability is attached to God's existence, the calculation analogous to what we did in Section 6.7 comes out infinitely in favor of believing in God.[536]

Very few thinkers take Pascal's Wager seriously today. I, for one, do not – and I'll come back to why in a minute. Let me first note that what makes Pascal's Wager relevant in a discussion of existential risk is mostly this: those who, for one reason or another, are critical of the study of existential risk – as presented in Chapter 8, Section 10.3, and several other parts of this book – tend to rejoice in comparing that study to Pascal's Wager.[537] A recent blog post by Swedish bioethicist Christian Munthe may serve as an example.[538] Munthe says that "what drives the argument [for why we should do something to prevent a particular existential catastrophe] is the (mere) possibility of a massively significant outcome, and the (mere) possibility of a way to prevent that particular outcome, thus doing masses of good"; he goes on to compare this to the (mere) possibility that Pascal's God exists, and he poses the rhetorical question that is also the title of his blog post: "Why aren't existential risk/ultimate harm argument advocates all attending mass?"

---

[535] It could for instance be that humanity enters a new era of lower existential risk as soon as we begin large-scale colonization of space, perhaps along the lines described in Section 9.3.

[536] See, e.g., Hájek (2012).

[537] See, however, Beckstead (2013) for a serious and in-depth treatment on how one can handle conclusions along the lines of Bostrom's (25) without succumbing to the madness of Pascal's Wager.

[538] Munthe (2015).

Crucial to Munthe's argument is the clumping together of a wide class of very dissimilar concepts as "the (mere) possibility of a massively significant outcome." Unfortunately, his blog post lacks specific pointers to the literature, but it is clear from the context that what is under attack here are publications like Bostrom and Ćirković (2008) and Bostrom (2013, 2014) – all of them heavily referenced (usually approvingly) in the present book. So, for concreteness, let us consider the main existential risk scenario discussed in Bostrom (2014), namely

(S1)   The emergence of a superintelligent AI that has
         goals and values in poor alignment with our own,
         and that wipes us out as a result.

This is, if I understand Munthe correctly, a "(mere) possibility of a massively significant outcome." This makes sense if we take it to mean a scenario that we still do not understand well enough to assign it a probability (not even an approximate one) to be plugged into a decision-theoretic analysis à la Section 6.7. But Munthe conflates this with another sense of "(mere) possibility," more on which shortly.

First, however, consider Munthe's parallel to Pascal's Wager, and note that Pascal didn't have in mind just any old god, but a very specific one, in a scenario that I take the liberty of summarizing as follows:

(S2)   The god Yahweh created man and woman with original
         sin. He later impregnated a woman with a child that
         was in some sense also himself. When the child had
         grown up, he sacrificed it (and thus, in some sense,
         himself) to save us from sin. But only those of us
         who worship him and go to church. The rest will
         be sent to hell.

Munthe and I appear to be in full agreement that (S2) is a far-fetched and implausible hypothesis, unsupported by either evidence or rational argument. And, as Munthe points out,

there are innumerable possible versions of the god that lures you with threats and promises of damnation and salvation, and what that particular god may demand in return, often implying a ban on meeting a competing deity's demands, so the wager doesn't seem to tell you to try to start believing in any particular of all these (merely) possible gods.

In particular, Pascal's Wager succumbs to the following scenario (S3), which appears at least as plausible as (S2) but under which Pascal's choice of going to church becomes catastrophically counterproductive:

(S3)   There exists an omnipotent deity, Baal, who likes
         atheists and lets them into heaven. The only people
         whom he sends to hell instead are those who make
         him jealous by worshiping some other deity.

Next, for Munthe's argument to work, he needs scenario (S1) to share (S2)'s far-fetchedness, implausibility and lack of rational support. He writes that

there seems to be an innumerable amount of thus (merely) possible existential risk scenarios, as well as innumerable (merely) possibly workable technologies that might help to prevent or mitigate each of these, and it is unlikely (to say the least) that we have resources to bet substantially on them all, unless we spread them so thin that this action becomes meaningless.

Is (S1) really just one of these "innumerable" risk scenarios? There are strong arguments that it is not, and that it does not have the properties of being far-fetched and implausible. It may perhaps, at first sight, *seem* far-fetched and implausible to someone who is as unfamiliar with the existential risk literature as Munthe appears to be. Bostrom (2014) and others offer a well-argued and rational defense of the need to consider (S1) as a serious possibility. If Munthe wishes to argue that (S1) does *not* merit such attention, he needs to concretely address those arguments (or at least the somewhat sketchy account of them that I offer in Sections 4.5 and 4.6 of the present book), rather than tossing around vague and unwarranted parallels to Pascal's Wager.

The crucial point is this. There are cases where a phenomenon seems to be possible, although we do not yet understand it well enough to be ready to meaningfully assign it a probability – not even an approximate probability. In such cases, we may speak of a "(mere) possibility." But we must not be misled, along with Munthe, by this expression into thinking that this implies that the phenomenon has extremely low probability.[539] In some cases, such as (S2), such a conclusion appears correct, whereas in others, such as (S1), it is manifestly unwarranted. Treating all "(mere) possibilities" as equally unlikely is a coarse-grained way of thinking that will lead to nothing but confusion.[540]

---

[539] This is analogous to the warning voiced by Sotala and Yampolskiy (2015) and repeated in Section 4.5 against the tempting mistake of treating the huge uncertainty about how near or how far an AI breakthrough is as indicating that it is very far.

[540] This is not the only example of deficient reasoning in Munthe (2015). That blog post also offers the statement that the existential risk literature under attack is "a case of the emperor's new clothes." In particular, we have the following, says Munthe:

The fact that there are possible threats to human civilizations, the existence of humanity, life on earth or, at least, extended human well-being, is not exactly news, is it? Neither is there any kind of new insight that some of these are created by humans themselves.

Well, of course. A journal like the *Bulletin of the Atomic Scientists* has pointed out humanity's capacity for self-destruction over and over since it stared in 1945. So let's assume, counterfactually but for the sake of argument, that the present-day existential risk literature in general, and a paper like Bostrom's "Existential risk prevention as global priority" (Bostrom, 2013), are devoid of original content. So what? Munthe's criticism here misunderstands the existential risk literature as primarily being a product of individual researchers' attempts to succeed in the academic publications game, for which things like formal novelty and fancy theory-building are crucial. That is (although I can really only speak for myself) not how I perceive most of the leading contributors to this field. Instead, they seem to be motivated primarily by an urgent conviction that the topic of existential risk from emerging technologies is *important*, and the desire to bring it to the rest of academia's and the world's attention. From

10.5 What to do?

What to do? It would be too much of an anticlimax if I merely gave the truthful answer "I don't know," so let me expand on it.

Our future is not certain, and we have nothing to gain from being fatalistic about it.[541] We need to realize that actions that we take now and in the next few decades may very well mean the difference between, on one hand, an outcome where we are the seed of a civilization that survives and thrives for millions or even billions of years, perhaps also expanding to the stars, and, on the other hand, one in which we already live in the end times. Any serious attempt to quantify how much is at stake is (as we saw in Section 10.3) bound to give astounding answers.

So what do we need to do in order to improve our chances of a happy outcome? Which technologies should we most eagerly pursue, which ones are less important, and with which ones should we tread lightly? Might there even be ones that need to be avoided entirely? We do not know, because we have a very limited understanding of the territories that lie ahead of us, their prospects and their dangers. So first of all we need to map those territories.

My hope is that, 10 or 15 years from now, either I or someone else will be able to write a book covering the same territories as the present book, but from the position of knowing much more about the possible and likely prospects and pitfalls of future technologies – so much more that it can serve as useful background material for some actual policy recommendations. For that to happen, much more effort has to be put into the serious futurology of emerging technologies. So there we have a first policy recommendation![542]

What else can we say about which research directions and technologies should move forward as fast as can be done, and which ones should be held back? Well, if two research areas seem to involve non-negligible existential risk, and one of

---

this perspective, whether or not a particular result or a particular argument has appeared before in the literature is largely irrelevant.

Munthe's blog post, in all its vagueness, offers no constructive ideas on how this field could improve to please him better. It is therefore hard to read him as suggesting anything other than a pushback of the field (currently occupying about one twentieth as big a part of the academic literature as the topic of snowboarding does; see Section 1.5) to a less prominent position, thus leaving more room on the academic marketplace for other areas, including Munthe's own. I couldn't disagree more.

[541] This paragraph and the next are adapted from Häggström (2014g).

[542] This, however, comes close to a type of policy advice that I tend to find rather lame and predictable: a researcher A working in area B points out the importance of directing more resources to area B. (In my case, while it is true that I have most of my research qualifications in other fields, it is also true that I am in the process of becoming increasingly absorbed by future studies.) Faced with such a statement, it may be a good idea to question A's motives. Is it the case that

(i) A considers area B important because she works in B, or that

(ii) A works in area B because she considers B important?

Of course, we typically find a bit of both, in a reinforcing feedback loop. But which of the two causal influences dominates? In case of (i), there may be reason to question A's motives. But then again, in the end a fair evaluation of A's proposal needs to be based more on her actual arguments than on speculations about her underlying motives.

them is a major driver of today's technological and economic progress, while the other is a marginal scientific area that has little or nothing to contribute to the economy or human well-being in a short-to-medium time perspective, then it is of course much easier to propose bans or restrictions on the latter than on the former. Very much for this reason, my views on what to do about METI and about AI, respectively, diverge significantly.

METI, as argued in Section 9.5, may have the potential to bring about our doom. Until we understand the situation better, I have no problem advocating a total ban on implementing METI procedures, i.e., on actively trying to send messages to extraterrestrial civilizations. By all means, go ahead and *theorize* all you want about METI; it's the *implementation* that I condemn.

AI is a whole different story. As argued in Sections 4.5, 4.6 and 8.3, the prospect of an AI breakthrough leading to the creation of a superintelligence that goes on to destroy humanity is sufficiently plausible and realistic to merit taking seriously. Recall also from Section 4.5 the warning from Sotala and Yampolskiy that the huge uncertainty among today's experts about how near or how far we are from an AI breakthrough "suggests that it is unjustified to be highly certain of [artificial general intelligence] being near, but also of it *not* being near. We thus consider it unreasonable to have a confident belief in the first proposition."[543] It is understandable if this leads some readers to think it might be a good idea to call for a moratorium on AI research and development. To ban such activities across the board would, however, be such a huge intrusion in one of the most dynamic and prosperous sectors of the economy that it seems like a politically completely unrealistic proposal, especially given how little we still understand – works like Bostrom (2014) and Sotala and Yampolskiy (2015) notwithstanding – about AI risk.[544] It is not even clear if such a ban would be desirable even if it were politically possible, because there may well be ways in which progress in such a multi-purpose field as AI helps us alleviate various existential risks, so we can't even be sure about whether such a moratorium increases or decreases existential risk. Much more realistic and sensible, at present, would be to work towards a reorientation of priorities *within* AI research and development – away

---

[543] Sotala and Yampolskiy (2015). Very recently there have been reports about the high-profile IT entrepreneur and billionaire Elon Musk voicing concerns that "the risk of something seriously dangerous happening is in the five year timeframe," with claims about having inside information; see, e.g., Cook (2014). It seems, however, that he has failed to back up his warning with anything substantial, whence it is unclear whether it deserves being taken seriously.

[544] More generally, the difficulty of turning insights about what would be good policy into real political action should not be underestimated. Here is Bostrom (2014):

> Any abstract point about "what should be done" must be embodied in the form of a concrete message, which is entered into the arena of rhetorical and political reality. There it will be ignored, misunderstood, distorted, or appropriated for various conflicting purposes; it will bounce around like a pinball, causing actions and reactions, ushering in a cascade of consequences, the upshot of which need bear no straightforward relationship to the intentions of the original sender. (p 238)

See also Häggström (2014e).

from a single-minded focus on making AI capable, and towards more emphasis on making it safe.[545]

Several factors contribute to the difficulty of deciding on a good AI policy. One is the huge uncertainty that faces us concerning what the technology may bring about. Another is the fact that possible consequences seem to range from extremely good to catastrophically bad: this is what geographer and futurologist Seth Baum refers to in the title of his recent paper "The great downside dilemma for risky emerging technologies."[546] Another technology with a similar dilemma is synthetic biology, with the risk of devastating pandemics playing the role of the great downside; see (ER12) in Section 8.3. Synthetic biology seems to be a case where we are likely to end up in deep trouble fairly soon unless we look seriously into the problem of how it can be sensibly regulated.

And what if regulation is not enough? What if mass surveillance turns out to be needed in order to prevent terrorists from causing global catastrophe through the synthesis and release of some new deadly virus? (Similar concerns arise for the distributed manufacturing that nanotechnology promises; see Chapter 5.) This is not a straightforward issue. Here is Sandberg (2013):

It is not obvious whether we should wish for better surveillance or less: it is not clear it is a force for good or evil in general. Surveillance can prevent or solve crimes, alert society to dangers, provide information for decision-making and so on. It can also distort our private lives, help crime, enable powerful authoritarian and totalitarian forces, and violate human rights. But there doesn't seem to exist any knock-down argument that dominates the balance: in many cases messy empirical issues and the current societal context likely determines what we should wish for, did we know the full picture.

Moving on to a more positive note, it is worth pointing out, along with Baum (2014), that there are a number of technologies that offer huge benefits without any great downside dilemma. One class of such technologies is, as emphasized already in Section 2.6, green energy: solar cells, wind power and so on, including energy storage technologies to handle the intermittency problem (we need energy not only when the sun is shining or the wind is blowing). It makes good sense, in my opinion, to use subsidies and other means to promote research and development in this area.

Then there is nuclear fusion, which Baum describes as "perhaps the Holy Grail of sustainable design," because "it promises a clean, safe, abundant energy source." Someone with a taste for nitpicking might be tempted to point out that nuclear fusion is strictly speaking not a sustainable energy source, because it does use up fuel,[547] but the point is that deuterium and other candidate fuels exist in sufficient

---

[545] Recall that this is what is proposed in the open letter mentioned in Section 10.2, Footnote 504, with many high-profile signatories.

[546] Baum (2014).

[547] But then again, not even solar energy is sustainable when considered over sufficiently long time scales (billions of years).

abundance that it dwarfs our supplies of uranium used for nuclear fission, as well as those of fossil fuels, by orders of magnitude. And unlike fission, it produces no significant radioactive waste. If we succeed in taming fusion so as to employ it as a large-scale energy source, the abundance of energy creates great opportunities. For instance, we could use it for seawater desalination, thereby ending all water shortage. And it would short-circuit one of the main obstacles to the air capture idea touched upon in Section 2.8, i.e., industrial processes for removing our excess CO_2 from the atmosphere.[548]

Unfortunately, we do not have any reliable timeline for when nuclear fusion will be a feasible energy production technology.[549] The situation is somewhat similar to that of AI, discussed briefly in Section 4.5. Both have existed as active research areas for more than half a century, both have had their ups and downs about how close to achieving their ultimate goal they are judged to be, and in neither case are we presently facing any unequivocal signs that the goal will be reached within the next decade or two.[550] My judgment, in the fusion case, is nevertheless that, in view of how much it could mean to society if the technology became feasible, it is well worth continuing or (preferably) increasing efforts on it.

Baum's final example of a potential future technology involving no obvious great downside dilemma is space colonization. But there is much to be gained. As discussed in Section 9.2, there is the simple observation that most material resources are to be found not on Earth, but elsewhere. In Baum's words: "The opportunities for civilization are, quite literally, astronomically greater beyond Earth than on it." There is also the issue of resilience: a planet-sized catastrophe of one sort or another (see Chapter 8) may wipe out humanity – unless some of us live in self-sustaining space colonies.[551] This all seems to suggest that space colonization should be a research priority, but that conclusion may be overly hasty, missing the crucial point that great downside dilemmas are not the only possible reason to abstain from a given technology. Another reason might be that the technology is so costly that, compared to pursuing other research directions, it would be money not well spent. The energy cost of lifting matter (such as ourselves, our machines, or raw materials) from the Earth or some other planet into space is large,[552]

---

[548] And who knows what other highly energy-consuming activities we will think of engaging in once we have access to the abundance of energy provided by nuclear fusion? Perhaps something really bad. If we wish to ascribe some great downside dilemma to nuclear fusion, this might be a good starting point. And while I do think this question merits further thought, I don't think it warrants, at present and in itself, any hesitation about supporting nuclear fusion research.

[549] Baum is even open to the possibility that it will never be doable.

[550] For more on the history and future prospects of fusion research, see, e.g., Hickman (2011) and Lehnert (2013).

[551] As a third benefit, Baum mentions "the marvelous inspiration that humanity can draw from marveling at its cosmic achievements." But that is not so obvious: what if we just turn blasé?

[552] Murphy (2011a,b) explains this very well.

possibly to the extent of making large-scale[553] space colonization prohibitively expensive until the day we master self-replicating probe technology as in the Armstrong–Sandberg scenario outlined in Section 9.3.

These are some very tentative recommendations concerning a few specific areas and technologies. But for the full spectrum of emerging technologies, including in particular the troublesome fields of biotechnology, nanotechnology and artificial intelligence, we simply need to understand more about the prospects and consequences of future technological progress before we are in a position to make wise decisions. We need to know more.

On the other hand, the exhortation "we need to know more" must not be taken as an excuse for procrastination. Our knowledge today is limited, imperfect and uncertain, but that will presumably always be the case, so accepting this as an argument for postponement and indecision has the unacceptable consequence that we will *never* need to take action. Not making a decision is also a decision, and that decision may have consequences. It may well be that we are standing at or very near a decisive turning point that can lead either to our prompt extinction, or to a future where we flourish beyond our wildest dreams, perhaps on cosmic scales. Let us not sit idly by as the future unfolds.

---

[553] By "large-scale," I mean here not the cosmic scale colonization discussed in Section 9.3, but something along the more modest lines of building space colonies elsewhere in our solar system to be inhabited by, say, a billion people.

REFERENCES

Ainslie, G. (2001) *Breakdown of Will*, Cambridge University Press, Cambridge, UK.

van Aken, J. (2007) Ethics of reconstructing Spanish Flu: Is it wise to resurrect a deadly virus? *Heredity* **98**, 1–2.

Aldous, D. (2012) The great filter, branching histories and unlikely events, *The Mathematical Scientist* **37**, 55–64.

Alpert, M. and Raiffa, H. (1982) A progress report on the training of probability assessors, in *Judgment under Uncertainty: Heuristics and Biases* (eds Kahneman, D., Slovic, P. and Tversky, A.), Cambridge University Press, Cambridge, UK, pp 294–305.

Angel, R. (2006) Feasibility of cooling the Earth with a cloud of small spacecraft near the inner Lagrange point (L1), *Proceedings of the National Academy of Sciences* **103**, 17184–17189.

Angelica, A. (2011) Alcor update from Max More, new CEO, *Kurzweil Accelerating Intelligence Blog*, January 13. http://www.kurzweilai.net/alcor-update-from-max-more-new-ceo.

Angell, M. (2011) The epidemic of mental illness: Why?, *The New York Review of Books*, June 23.

Annas, G., Andrews, L. and Isasit, R. (2002) Protecting the endangered human: Toward an international treaty prohibiting cloning and inheritable alterations, *American Journal of Law and Medicine* **28**, 151–178.

Archer, D. (2007) *Global Warming: Understanding the Forecast*, Blackwell, Oxford.

Armstrong, S. (2010) The AI in a box boxes you, *Less Wrong*, February 2, http://lesswrong.com/lw/1pz/the_ai_in_a_box_boxes_you/.

Armstrong, S. (2011) Anthropic decision theory for self-locating beliefs, arXiv 1110.6437, http://arxiv.org/abs/1110.6437.

Armstrong, S. and Sandberg, A. (2013) Eternity in six hours: Intergalactic spreading of intelligent life and sharpening the Fermi paradox, *Acta Astronautica* **89**, 1–13.

Armstrong, S. and Sotala, K. (2012) How we're predicting AI – or failing to, in Romportl et al. (2012), pp 52–75.

Arratia, R. and Goldstein, L. (2010) Size bias, sampling, the waiting time paradox, and infinite divisibility: When is the increment independent?, arXiv 1007.3910, http://arxiv.org/abs/1007.3910.

Arrhenius, G. (2008) Life extension versus replacement, *Journal of Applied Philosophy* **25**, 211–227.

Arrhenius, G. (2011) The impossibility of a satisfactory population ethics, in *Descriptive and Normative Approaches to Human Behavior* (eds Colonius, H. and Dzhafarov, E.), World Scientific, Singapore, pp 1–26.

Asimov, A. (1950) *I, Robot*, Gnome Press, New York.

Augustin, L. et al. (2004) Eight glacial cycles from an Antarctic ice core, *Nature* **429**, 623–628.

Bach, A. (1988) The concept of indistinguishable particles in classical and quantum physics, *Foundations of Physics* **18**, 639–649.

Bach, A. (1997) *Indistinguishable Classical Particles*, Springer, New York.

Bacon, F. (1620) *Novum Organum*.

Badescu, V. and Cathcart, R. (2006) Use of class A and class C stellar engines to control sun movement in the galaxy, *Acta Astronautica* **58**, 119–129.

Baker, A. (2010) Simplicity, *The Stanford Encyclopedia of Philosophy* (ed. Zalta, E.), http://plato.stanford.edu/entries/simplicity/.

Baker, K. (2002) Q & A/ Colin McGinn / Life of the mind, *SFGate*, http://www.sfgate.com/books/article/Q-A-Colin-McGinn-Life-of-the-mind-2837010.php.

Balmaseda, M., Trenberth, K. and Källén, E. (2013) Distinctive climate signals in reanalysis of global ocean heat content, *Geophysical Research Letters* **40**, 1754–1759.

Barrat, J. (2013) *Our Final Invention: Artificial Intelligence and the End of the Human Era*, Thomas Dunne Books, New York.

Barzun, J. (1964) *Science: The Glorious Entertainment*, Harper and Row, New York.

Baum, R. (2003) Nanotechnology: Drexler and Smalley make the case for and against "molecular assemblers," *Chemical & Engineering News* **81**(48), 37–42.

Baum, S. (2014) The great downside dilemma for risky emerging technologies, *Physica Scripta* **89**, 128004.

Baum, S., Goertzel, B. and Goertzel, T. (2011) How long until human-level AI? Results from an expert assessment, *Technological Forecasting & Social Change* **78**, 185–195.

Baum, S., Maher, T. and Haqq-Misra, J. (2013) Double catastrophe: Intermittent stratosphere geoengineering induced by societal collapse, *Environment, Systems and Decisions* **33**, 168–180.

Beckstead, N. (2013) On the Overwhelming Importance of Shaping the Far Future, PhD thesis, Rutgers University, https://rucore.libraries.rutgers.edu/rutgers-lib/40469/.

Beckstead, N. (2014) Will we eventually be able to colonize other stars? Notes from a preliminary review, Future of Humanity Institute, http://www.fhi.ox.ac.uk/will-we-eventually-be-able-to-colonize-other-stars-notes-from-a-preliminary-review/.

Benyamin, B. et al. (2014) Childhood intelligence is heritable, highly polygenic and associated with FNBP1L, *Molecular Psychiatry* **19**, 253–258.

Berger, T.W., Song, D., Chan, R.H., Marmarelis, V.Z., LaCoss, J., Wills, J., Hampson, R.E., Deadwyler, S.A. and Granacki, J.J. (2012) A hippocampal cognitive prosthesis: Multi-input, multi-output nonlinear modeling and VLSI implementation, *IEEE Transactions on Neural Systems and Rehabilitation Engineering* **20**, 198–211.

Bergström, L. (1994) Notes on the value of science, in *Logic, Methodology and Philosophy of Science IX: Proceedings of the Ninth International Congress of Logic, Methodology and Philosophy of Science* (eds Prawitz, D., Skyrms, B. and Westerståhl, D.), Elsevier, Amsterdam, pp 499–522.

Best, B. (2008) Scientific justification of cryonics practice, *Rejuvenation Research* **11**, 493–503.

Bethe, H. (1991) *The Road from Los Alamos*, American Institute of Physics, New York.

Bindeman, I. (2006) The secrets of supervolcanoes, *Scientific American* **294**(6), 36–43.

Binmore, K. (2009) *Rational Decisions*, Princeton University Press, Princeton, NJ.

Blackford, R. (2014) Introduction II: Bring on the machines, in Blackford and Broderick (2014), pp 11–25.

Blackford, R. and Broderick, D. (2014) *Intelligence Unbound: The Future of Uploads and Machine Minds*, Wiley Blackwell, Chichester.

Blackmore, S. (1999) *The Meme Machine*, Oxford University Press, Oxford.

Blackmore, S. (2012) She won't be me, *Journal of Consciousness Studies* **19**, 16–19.

Bognar, G. (2012) When philosophers shoot themselves in the leg, *Ethics, Policy & Environment* **15**, 222–224.

Bostrom, N. (1999) The Doomsday argument is alive and kicking, *Mind* **108**, 539–550.

Bostrom, N. (2002) *Anthropic Bias: Observation Selection Effects in Science and Philosophy*, Routledge, New York.

Bostrom, N. (2003a) Are you living in a computer simulation?, *Philosophical Quarterly* **53**, 243–255.

Bostrom, N. (2003b) *The Transhumanist FAQ: A General Introduction*, Version 2.1, World Transhumanist Organisation, http://www.transhumanism.org/resources/FAQv21.pdf.

Bostrom, N. (2003c) Ethical issues in advanced artificial intelligence, in *Cognitive, Emotive and Ethical Aspects of Decision Making in Humans and in Artificial Intelligence, Vol. 2* (eds Smit, I. et al.), International Institute of Advanced Studies in Systems Research and Cybernetics, pp. 12–17.

Bostrom, N. (2005a) A history of transhumanist thought, *Journal of Evolution and Technology* **14**(1), 1–25.

Bostrom, N. (2005b) In defense of posthuman dignity, *Bioethics* **19**, 202–214.

Bostrom, N. (2005c) The fable of the dragon-tyrant, *Journal of Medical Ethics* **31**, 273–277.

Bostrom, N. (2006) What is a singleton?, *Linguistic and Philosophical Investigations* **5**, 48–54.

Bostrom, N. (2008a) Why I want to be a posthuman when I grow up, in *Medical Enhancement and Posthumanity* (eds Gordijn, B. and Chadwick, R.), Springer, New York, pp 107–137.

Bostrom, N. (2008b) Dignity and enhancement, in Pellegrino et al. (2008), pp 173–206.

Bostrom, N. (2008c) Where are they? Why I hope the search for extraterrestrial life finds nothing, *MIT Technology Review* May/June issue, 72–77. http://www.nickbostrom.com/extraterrestrial.pdf.

Bostrom, N. (2009) Pascal's mugging, *Analysis* **69**, 443–445.

Bostrom, N. (2011) Information hazards: A typology of potential harms from knowledge, *Review of Contemporary Philosophy* **10**, 44–79.

Bostrom, N. (2012) The superintelligent will: Motivation and instrumental rationality in advanced artificial agents, *Minds and Machines* **22**, 71–85.

Bostrom, N. (2013) Existential risk prevention as global priority, *Global Policy* **4**, 15–31.

Bostrom, N. (2014) *Superintelligence: Paths, Dangers, Strategies*, Oxford University Press, Oxford.

Bostrom, N. and Ćirković, M. (2003) The Doomsday argument and the self-indication assumption: Reply to Olum, *Philosophical Quarterly* **53**, 83–91.

Bostrom, N. and Ćirković, M. (2008) *Global Catastrophic Risks*, Oxford University Press, Oxford.

Bostrom, N. and Savulescu, J. (2009) *Human Enhancement*, Oxford University Press, Oxford.

Brennan, A. and Lo, Y.-S. (2011) Environmental ethics, *The Stanford Encyclopedia of Philosophy* (ed. Zalta, E.), http://plato.stanford.edu/entries/ethics-environmental/.

Brian, K. (2013) The amazing story of IVF: 35 years and five million babies later, *The Guardian*, July 12.

Brin, D. (2002) *Kiln People*, Tor Books, New York.

Brin, D. (2006) Shouting at the cosmos: How SETI has taken a worrisome turn into dangerous territory, http://www.davidbrin.com/shouldsetitransmit.html.

Brin, D. (2014) "Trust me, I know exactly what's out there!", *Cato Unbound*, December 10. http://www.cato-unbound.org/2014/12/10/david-brin/trust-me-i-know-exactly-whats-out-there.

Bringsjord, S. (2012) Belief in the singularity is logically brittle, *Journal of Consciousness Studies* **19**, 14–20.

Bringsjord, S., Bringsjord, A. and Bello, P. (2012) Belief in the singularity is fideistic, in Eden et al. (2012), pp 395–408.

Brooks, R. (2014) Artificial intelligence is a tool, not a threat, *Rethink Robotics Blog*, November 10, http://www.rethinkrobotics.com/artificial-intelligence-tool-threat/.

Brynjolfsson, E. and McAfee, A. (2014) *The Second Machine Age: Work, Progress, and Prosperity in a Time of Brilliant Technologies*, W.W. Norton, New York.

Bubnoff, A. von (2005) The 1918 flu virus is resurrected, *Nature* **437**, 794–795.

Buck Louis, G. et al. (2008) Environmental factors and puberty timing: Expert panel research needs, *Pediatrics* **121**, S192–S207.

Butler, T. (2006) Galileo's paradox, *Suitcase of Dreams*, http://www.suitcaseofdreams.net/Paradox_Galileo.htm.

Caldeira, K. and Wood, L. (2008) Global and Arctic climate engineering: Numerical model studies, *Philosophical Transactions of the Royal Society A* **366**, 4039–4056.

Callaway, E. (2014) "Smart genes" prove elusive, *Nature*, September 8. http://www.nature.com/news/smart-genes-prove-elusive-1.15858.

Camus, A. (1942) *Le Mythe de Sisyphe*, Gallimard, Paris.

Caney, S. (2008) Human rights, climate change, and discounting, *Environmental Politics* **17**, 536–555.

Carter, B. (1983) The anthropic principle and its implications for biological evolution, *Philosophical Transactions of the Royal Society of London A, Mathematical and Physical Sciences* **310**, 347–363.

Carvalho, M., Carmo, H., Costa, V.M., Capela, J.P., Pontes, H., Remião, F., Carvalho, F. and de Lourdes Bastos, M. (2012) Toxicity of amphetamines: An update, *Archives of Toxicology* **86**, 1167–1231.

Cello, J., Paul, A.V. and Wimmer, E. (2002) Chemical synthesis of poliovirus cDNA: Generation of infectious virus in the absence of natural template, *Science* **297**, 1016–1018.

Center for Responsible Nanotechnology (2003) Grey goo is a small issue, briefing document, December 14, http://www.crnano.org/BD-Goo.htm.

Chaitin, G. (2006) The limits of reason, *Scientific American* **294**, 71–84.

Chalmers, D. (1996) *The Conscious Mind*, Oxford University Press, Oxford.

Chalmers, D. (2003) *The Matrix as Metaphysics*, for the philosophy section of the official Matrix website, http://consc.net/papers/matrix.pdf.

Chalmers, D. (2010) The singularity: A philosophical analysis, *Journal of Consciousness Studies* **17**, 7–65.

Chalmers, D. (2012) The singularity: A reply to commentators, *Journal of Consciousness Studies* **19**, 141–167.

Chalmers, D. (2014) Uploading: A philosophical analysis, in Blackford and Broderick (2014), pp 102–118.

Chivers, T. (2014) The Flynn effect: Are we really getting smarter?, *The Telegraph*, October 31.

Chorost, M. (2006) *Rebuilt: How Becoming Part Computer Made Me More Human*, Mariner Books, New York.

Chorost, M. (2011) *World Wide Mind: The Coming Integration of Humanity, Machines and the Internet*, Free Press, New York.

Christian, B. (2011) Mind vs. Machine, *The Atlantic*, March.

Ćirković, M. (2004) Forecast for the next eon: Applied cosmology and the long-term fate of intelligent beings, *Foundations of Physics* **34**, 239–261.

Ćirković, M. (2006) Macroengineering in the galactic context: A new agenda for astrobiology, in *Macro-Engineering. A Challenge for the Future* (eds Badescu, V., Cathcart, R.B. and Schuiling, R.D.), Springer, New York.

Ćirković, M. (2009) Fermi's paradox – the last challenge for Copernicanism?, *Serbian Astronomical Journal* **178**, 1–20.

Clark, A. (2008) *Supersizing the Mind: Embodiment, Action, and Cognitive Extension*, Oxford University Press, Oxford.

Clark, A. and Chalmers, D. (1998) The extended mind, *Analysis* **58**, 7–19.

Cohen, B. (1998) Howard Aiken on the number of computers needed for the nation, *IEEE Annals of the History of Computing* **20**, 27–32.

Cohon, R. (2010) Hume's moral philosophy, *The Stanford Encyclopedia of Philosophy* (ed. Zalta, E.), http://plato.stanford.edu/entries/hume-moral/.

Colleton, L. (2008) The elusive line between enhancement and therapy and its effect on health care in the US, *Journal of Evolution and Technology* **18**, 70–78.

Collini, S. (2012) *What are Universities For?*, Penguin, London.

Committee on Science and Technology for Countering Terrorism (2002) *Making the Nation Safer: The Role of Science and Technology in Countering Terrorism*, The National Academies Press, Washington, DC.

Committee to Review the National Nanotechnology Initiative (2006) *A Matter of Size: Triennial Review of the National Nanotechnology Initiative*, The National Academies Press, Washington, DC.

Condorcet, Marquis de (1822) *Esquisse d'un tableau historique des progrès de l'esprit humain*, Masson et Fils, Paris. The English translation from which the quote in Chapter 3 was taken is from the Internet Modern History Sourcebook: http://www.fordham.edu/halsall/mod/condorcet-progress.asp.

Cook, J. (2014) Elon Musk: Robots could start killing us all within 5 years, *Business Insider UK*, November 17.

Copeland, J. (2008) The Church–Turing thesis, *The Stanford Encyclopedia of Philosophy* (ed. Zalta, E.), http://plato.stanford.edu/entries/church-turing/.

Crews, F. (2007) Talking back to Prozac, *New York Review of Books*, December 6.

Croll, J. (1875) *Climate and Time, in Their Geological Relations*, Daldy, Isbister, & Co, London.

Crowther, M. (2013) Magnus Carlsen is the new world chess champion, *The Week in Chess*, November 22.

Crutzen, P. (2002) Geology of mankind, *Nature* **415**, 23.

Curtis, V., de Barra, M. and Aunger, R. (2011) Disgust as an adaptive system for disease avoidance behaviour, *Philosophical Transactions of the Royal Society B* **366**, 389–401.

Danaher, J. (2014a) Chalmers vs Pigliucci on the philosophy of mind-uploading (1): Chalmers's optimism, *Philosophical Disquisitions*, September 17, http://philosophicaldisquisitions. blogspot.se/2014/09/chalmers-vs-pigliucci-on-philosophy-of.html.

Danaher, J. (2014b) Chalmers vs Pigliucci on the philosophy of mind-uploading (2): Pigliucci's pessimism, *Philosophical Disquisitions*, September 17, http://philosophicaldisquisitions. blogspot.se/2014/09/chalmers-vs-pigliucci-on-philosophy-of_19.html.

Danaher, J. (2014c) Sex work, technological unemployment and the basic income guarantee, *Journal of Evolution and Technology* 24, 113–130.

Danaylov, N. and Hanson, R. (2013a) Details matter . . . and for that you need social science, *Singularity Weblog*, https://www.singularityweblog.com/robin-hanson-on-singularity-1-on-1/.

Danaylov, N. and Hanson, R. (2013b) Social science or extremist politics in disguise?!, *Singularity Weblog*, https://www.singularityweblog.com/robin-hanson-social-science-or-extremist-politics-in-disguise/.

Dar, A. (2008) Influence of supernovae, gamma-ray bursts, solar flares, and cosmic rays on the terrestrial environments, in Bostrom and Ćirković (2008), pp 238–261.

Dar, A., De Rujula, A. and Heinz, U. (1999) Will relativistic heavy ion colliders destroy our planet?, *Physics Letters B* 470, 142–148.

Davis, M. (2011) *The Universal Computer: The Road from Leibniz to Turing*, Turing Centenary Edition, CRC Press, Boca Raton, FL.

Dawkins, R. (1982) *The Extended Phenotype*, Oxford University Press, Oxford.

Dawkins, R. (1995) *River Out of Eden*, Basic Books, New York.

Degnan, G.G., Wind, T.C., Jones, E.V. and Edlich, R.F. (2002) Functional electrical stimulation in tetraplegic patients to restore hand function, *Journal of Long-Term Effects of Medical Implants* 12, 175–188.

Dennett, D. (1982) The myth of the computer: An exchange, *New York Review of Books*, June 24.

Dennett, D. (1989) Murmurs in the cathedral (Review of R. Penrose, *The Emperor's New Mind*), *Times Literary Supplement*, September 29, 55–57.

Dennett, D. (1991) *Consciousness Explained*, Little Brown, New York.

Dennett, D. (2005) *Sweet Dreams: Philosophical Obstacles to a Science of Consciousness*, MIT Press, Cambridge, MA.

Dennett, D. (2008) Commentary on Kraynak, in Pellegrino et al. (2008), pp 83–88.

Dennett, D. (2013) *Intuition Pumps and Other Tools for Thinking*, Allen Lane, London.

Dennett, D. (2014) Reflections on free will, *Naturalism.Org*, January 24, http://www.naturalism. org/Dennett_reflections_on_Harris's_Free_Will.pdf.

DeSantis, A., Webb, E. and Noar, S. (2008) Illicit use of prescription ADHD medications on a college campus: A multimethodological approach, *Journal of American College Health* 57, 315–324.

Deutsch, D. (2011) *The Beginning of Infinity: Explanations that Transform the World*, Allen Lane, London.

Devlin, P. (1965) *The Enforcement of Morals*, Oxford University Press, Oxford.

Devlin, B., Daniels, M. and Roeder, K. (1997) The heritability of IQ, *Nature* 388, 468–471.

Diaconis, P., Holmes, S. and Montgomery, R. (2007) Dynamical bias in the coin toss, *SIAM Review* 49, 211–235.

Dimberu, P. (2011) Immortal jellyfish provides clues for regenerative medicine, *Singularity Hub*, April 25, http://singularityhub.com/2011/04/25/immortal-jellyfish-provides-clues-for-regenerative-medicine/.

Djerassi, C. (2014) The divorce of coitus from reproduction, *New York Review of Books*, September 25.

Dobzhansky, T. (1973) Nothing in biology makes sense except in the light of evolution, *American Biology Teacher* 35, 125–129.

Drexler, E. (1986) *Engines of Creation: The Coming Era of Nanotechnology*, Doubleday, New York.

Drexler, E. (2003a) An open letter on assemblers, Foresight Institute, http://www.foresight.org/nano/Letter.html.

Drexler, E. (2003b) Toward closure, Foresight Institute, http://www.foresight.org/nano/Letter2.html.

Drexler, E. (2013) *Radical Abundance: How a Revolution in Nanotechnology Will Change Civilization*, PublicAffairs, New York.

Drexler, E., Forrest, D., Freitas, R., Hall, S., Jacobstein, N., McKendree, T., Merkle, R. and Peterson, C. (2001) On physics, fundamentals, and nanorobots: A rebuttal to Smalley's assertion that self-replicating mechanical nanorobots are simply not possible, Institute for Molecular Manufacturing, http://www.imm.org/publications/sciamdebate2/smalley/.

Dummett, M. (1981) Ought research to be unrestricted?, *Grazer Philosophische Studien* 12, 281–298.

Dyson, F. (1960) Search for artificial stellar sources of infrared radiation, *Science* 131, 1667–1668.

Dyson, F. (1979) *Disturbing the Universe*, Harper and Row, New York.

Dyson, F. (2007) Our biotech future, *The New York Review of Books*, July 19.

Easterbrook, S. (2004) *What is engineering?*, http://www.cs.toronto.edu/~sme/CSC340F/slides/03-engineering.pdf.

Eden, A., Moor, J., Soraker, J. and Stenhart, E. (2012) *Singularity Hypotheses: A Scientific and Philosophical Assessment*, Springer, New York.

Einstein, A. (1954) *Ideas and Opinions*, Crown Publishers, New York.

Emonson, D. and Vanderbeek, R. (1995) The use of amphetamines in US Air Force tactical operations during Desert Shield and Storm, *Aviation, Space, and Environmental Medicine* 66, 260–263.

Engelbart, D. (1962) Augmenting human intellect: A conceptual framework, Summary report AFOSR-3223, Stanford Research Institute, Menlo Park, CA. http://www.dougengelbart.org/pubs/augment-3906.html.

Fagerström, T. (2014) Kunskapsförakt och grön ideologi tar makten över jordbruket, *Sans* 1/2014.

Fenton, E. (2010) The perils of failing to enhance: A response to Persson and Savulescu, *Journal of Medical Ethics* 36, 148–151.

Feynman, R.P. (1959) There's plenty of room at the bottom, American Physical Society annual meeting, Pasadena, CA. Transcript at http://www.zyvex.com/nanotech/feynman.html.

Feynman, R.P. (1985) *Surely You're Joking, Mr. Feynman? Adventures of a Curious Character*, W.W. Norton, New York.

de Finetti, B. (1937) Foresight: Its logical laws, its subjective sources, in *Studies in Subjective Probability* (eds Kyburg, H. and Smokler, H.), Kreiger Publishing, Huntington, NY.

Flynn, J. (1999) Searching for justice: The discovery of IQ gains over time, *American Psychologist* 54, 5–20.

Fogg, M.J. (1988) The feasibility of intergalactic colonisation and its relevance to SETI, *Journal of the British Interplanetary Society* 41, 491–496.

Franck, S., Bounama, C. and von Bloh, W. (2005) Causes and timing of future biosphere extinction, *Biogeosciences Discussions* 2, 1665–1679.

Freitas, R. (1985) Observable characteristics of extraterrestrial technological civilizations, *Journal of the British Interplanetary Society* 38, 106–112.

Freitas, R. (2001) The gray goo problem, *Kurzweil Accelerating Intelligence*, March 20, http://www.kurzweilai.net/the-gray-goo-problem.

Freitas, R. (2002) Death is an outrage, *Fight aging!*, December 8. https://www.fightaging.org/archives/2002/12/death-is-an-outrage-1.php.

Freitas, R. (2007) The ideal gene delivery vector: Chromallocytes, cell repair nanorobots for chromosome replacement therapy, *Journal of Evolution and Technology* 16, 1–97.

Freitas, R. (2009) Welcome to the future of medicine, *Studies in Health Technology and Informatics*, 149. 251–256.

Freitas, R. and Zachary, W. (1981) A self-replicating, growing lunar factory, in *Space Manufacturing 4*, (eds Gray, J. and Hamdan, L.), American Institute of Aeronautics and Astronautics, New York, pp 109–119. http://www.rfreitas.com/Astro/GrowingLunarFactory1981.htm.

Frey, C.B. and Osborne, M. (2013) The future of employment: How susceptible are jobs to computerisation?, preprint, http://www.oxfordmartin.ox.ac.uk/downloads/academic/The_Future_of_Employment.pdf.

Fuegi, J. and Francis, J. (2003) Lovelace & Babbage and the creation of the 1843 "notes," *IEEE Annals of the History of Computing* 25(4), 16–26.

Fukuyama, F. (2002) *Our Posthuman Future – Consequences of the Biotechnology Revolution*, Farrar, Straus and Giroux, New York.

Gardiner, S. (2006) A perfect moral storm: Climate change, intergenerational ethics and the problem of moral corruption, *Environmental Values* 15, 397–413.

Gardner, M. (2001) A skeptical look at Karl Popper, *Skeptical Inquirer* 25, 13–14.

Gazzaniga, M.S. (1998) The split brain revisited, *Scientific American* 279, 34–39.

Gershenfeld, N. (2012) How to make almost anything: The digital fabrication revolution, *Foreign Affairs* 91(6), 43–57.

Giddings, S. and Mangano, M. (2008) Astrophysical implications of hypothetical stable TeV-scale black holes, *Physical Review D* 78, 035009.

Gilks, W., Richardson, S. and Spiegelhalter, D. (1996) *Markov Chain Monte Carlo in Practice*, Chapman & Hall, London.

Goertzel, B. (2010) *A Cosmist Manifesto*, Humanity+ Press, Los Angeles.

Goertzel, B. and Pitt, J. (2014) Nine ways to bias open-source artificial general intelligence toward friendliness, in Blackford and Borderick (2014), pp 61–89.

Goldacre, B. (2012) *Bad Pharma: How Drug Companies Mislead Doctors and Harm Patients*, 4th Estate, London.

Goldowsky, H. (2014) How to catch a chess cheater: Ken Regan finds moves out of mind, *Chess Life*, June issue. http://www.uschess.org/content/view/12677/763.

Good, I.J. (1965) Speculations concerning the first ultraintelligent machine, *Advances in Computers*, vol 6 (eds Alt, F. and Rubinoff, M.), Academic Press, New York.

Goodman, N. (1955) *Fact, Fiction, and Forecast*, Harvard University Press, Cambridge, MA.

Gordon, R. (2012) Is U.S. economic growth over? Faltering innovation confronts the six headwinds, Working paper, National Bureau of Economic Research, http://www.nber.org/papers/w18315.pdf.

Gore, A. (2006) *An Inconvenient Truth*, Rodale, Emmaus, PA.

Gore, A. (2013) *The Future*, Random House, New York.

Gott, R. (1993) Implications of the Copernican principle for our future prospects, *Nature* **363**, 315–319.

Gott, R., Juric, M., Schlegel, D., Hoyle, F., Vogeley, M., Tegmark, M., Bahcall, N. and Brinkmann, J. (2005) A map of the universe, *Astrophysical Journal* **624**, 463–483.

Gowans, C. (2012) Moral relativism, *The Stanford Encyclopedia of Philosophy* (ed. Zalta, E.), http://plato.stanford.edu/entries/moral-relativism/.

Grayson, M. (2012) Ageing, *Nature* **492**, S1.

Gross, D. (2013) Texas company makes metal gun with 3-D printer, *CNN*, November 9. http://edition.cnn.com/2013/11/08/tech/innovation/3d-printed-metal-gun/index.html.

Gunson, D. and McLachlan, H. (2013) Risk, Russian-roulette and lotteries: Persson and Savulescu on moral enhancement, *Medicine, Health Care and Philosophy*, **16**(4), 877–884.

Haack, S. (2003) *Defending Science – Within Reason: Between Scientism and Cynicism*, Prometheus, New York.

Habermas, J. (2003) *The Future of Human Nature*, Blackwell, Oxford.

Häggström, O. (2005) Till Harvardrektorns och det fria sanningssökandets försvar, *Qvartilen*, no. 4. http://www.math.chalmers.se/~olleh/skolans_sak/Summers.pdf.

Häggström, O. (2007a) Uniform distribution is a model assumption, http://www.math.chalmers.se/~olleh/reply_to_Dembski.pdf.

Häggström, O. (2007b) Ramseys ekvation och planetens framtid, *Nämnaren* **2007**(4), 43–46.

Häggström, O. (2008) *Riktig vetenskap och dåliga imitationer*, Fri Tanke, Stockholm.

Häggström, O. (2010a) Recension: Superfreakonomics, *Uppsalainitiativet*, January 10, http://uppsalainitiativet.blogspot.se/2010/01/recension-superfreakonomics.html.

Häggström, O. (2010b) Book review: The Cult of Statistical Significance, *Notices of the American Mathematical Society* **57**, 1129–1130.

Häggström, O. (2011a) Klimatvetenskap, klimatdebatt, klimathuliganism, in *Climategate och hotet mot isbjörnarna* (ed. Almqvist, K. and Gröning, L.), Axel och Margaret Ax:son Johnson, Stockholm. http://www.math.chalmers.se/~olleh/Klimathuliganismen_Engelsberg.pdf.

Häggström, O. (2011b) Book review: Nonsense on Stilts, *Notices of the American Mathematical Society* **58**, 582–584.

Häggström, O. (2011c) Ett smörgåsbord av dumheter, *Axess* 2/2011. http://www.math.chalmers.se/~olleh/Fara.pdf.

Häggström, O. (2012a) Dör man av att teleporteras?, *Häggström hävdar*, January 1, http://haggstrom.blogspot.se/2012/01/dor-man-av-att-teleporteras.html.

Häggström, O. (2012b) Tidskriften Arena till relativismens försvar, *Häggström hävdar*, May 28, http://haggstrom.blogspot.se/2012/05/tidskriften-arena-till-relativismens.html.

Häggström, O. (2012c) Om Humes lag och Searles motexempel, *Häggström hävdar*, August 16, http://haggstrom.blogspot.se/2012/08/om-humes-lag-och-searles-motexempel.html.

Häggström, O. (2013a) Why the empirical sciences need statistics so desperately, in *European Congress of Mathematics, Krakow, 2–7 July, 2012* (eds Latala, R. et al.), European Mathematical Society Publishing House, Zürich, pp 347–360.

Häggström, O. (2013b) No nonsense – my reply to David Sumpter, *Häggström hävdar*, October 23, http://haggstrom.blogspot.se/2013/10/no-nonsense-my-reply-to-david-sumpter.html.

Häggström, O. (2013c) Statistisk signifikans och Armageddon, *Nämnaren* **2013**(1), 37–47.

Häggström, O. (2013d) Statistical significance is not a worthless concept, *Häggström hävdar*, February 5, http://haggstrom.blogspot.se/2013/02/statistical-significance-is-not.html.

Häggström, O. (2013e) Artificial intelligence and Solomonoff induction: What to read?, *Häggström hävdar*, August 20. http://haggstrom.blogspot.se/2013/08/artificial-intelligence-and-solomonoff.html.

Häggström, O. (2013f) Silverprediktioner, *Häggström hävdar*, January 18, http://haggstrom.blogspot.se/2013/01/silverprediktioner.html.

Häggström, O. (2013g) Om kryonik, *Häggström hävdar*, June 24. http://haggstrom.blogspot.se/2013/06/om-kryonik.html.

Häggström, O. (2013h) Reading the Hanson–Yudkowsky debate, *Häggström hävdar*, October 15. http://haggstrom.blogspot.se/2013/10/reading-hanson-yudkowsky-debate.html.

Häggström, O. (2014a) David Keith har skrivit en läsvärd liten bok om geoengineering, *Häggström hävdar*, February 27, http://haggstrom.blogspot.se/2014/02/david-keith-har-skrivit-en-lasvard.html.

Häggström, O. (2014b) On the value of replications: Jason Mitchell is wrong, *Häggström hävdar*, July 9, http://haggstrom.blogspot.se/2014/07/on-value-of-replications-jason-mitchell.html.

Häggström, O. (2014c) Superintelligence odds and ends I: What if human values are fundamentally incoherent?, *Häggström hävdar*, September 12, http://haggstrom.blogspot.se/2014/09/superintelligence-odds-and-ends-i-what.html.

Häggström, O. (2014d) Superintelligence odds and ends II: The Milky Way preserve, *Häggström hävdar*, September 13, http://haggstrom.blogspot.se/2014/09/superintelligence-odds-and-ends-ii.html.

Häggström, O. (2014e) Superintelligence odds and ends III: Political reality and second-guessing, *Häggström hävdar*, September 15, http://haggstrom.blogspot.se/2014/09/superintelligence-odds-and-ends-iii.html.

Häggström, O. (2014f) Om Lennart Bengtssons beklämmande färd ner i klimatförnekarträsket, *Häggström hävdar*, May 11. English translation published at *Rabett Run*, May 16, 2014. http://rabett.blogspot.se/2014/05/laffaire-bengtsson.html.

Häggström, O. (2014g) Emerging technologies and the future of humanity, *Physica Scripta* **89**, 120201.

Häggström, O. (2014h) Om Singulariteten i DN, *Häggström hävdar*, June 22, http://haggstrom.blogspot.se/2014/06/om-singulariteten-i-dn.html.

Häggström, O. (2015) An open letter calling for research on robust and beneficial AI, *Häggström hävdar*, January 13, http://haggstrom.blogspot.se/2015/01/an-open-letter-calling-for-research-on.html.

Hájek, A. (2012) Pascal's Wager, *The Stanford Encyclopedia of Philosophy* (ed. Zalta, E.), http://plato.stanford.edu/entries/pascal-wager/.

Haldane, J.B.S. (1924) *Daedalus; or, Science and the Future*, E.P. Dutton, New York.

Hansen, J. (2009) *Storms of My Grandchildren: The Truth About the Coming Climate Catastrophe and Our Last Chance to Save Humanity*, Bloomsbury, London.

Hansen, J. (2013) Making things clearer: Exaggeration, jumping the gun, and the Venus syndrome, http://www.columbia.edu/~jeh1/mailings/2013/20130415_Exaggerations.pdf.

Hansen, J. et al. (2008) Target atmospheric CO_2: Where should humanity aim? *The Open Atmospheric Science Journal* **2**, 217–231.

Hansen, J. and Sato, M. (2012) Paleoclimate implications for human-made climate change, in *Climate Change*, Springer, Vienna, 21–47.

Hanson, R. (1994) If uploads come first: The crack of a future dawn, *Extropy* 6(2), 10–15.

Hanson, R. (1998) The Great Filter – are we almost past it?, http://hanson.gmu.edu/greatfilter.html.

Hanson, R. (2008a) When life is cheap, death is cheap, *Overcoming Bias*, November 24, http://www.overcomingbias.com/2008/11/when-life-is-ch.html.

Hanson, R. (2008b) Engelbart as UberTool?, *Overcoming Bias*, November 13, http://www.overcomingbias.com/2008/11/engelbarts-uber.html.

Hanson, R. (2008c) Economics of the singularity, *IEEE Spectrum*, June, 37–42.

Hanson, R. (2008d) Outside view of singularity, *Overcoming Bias*, June 20, http://www.overcomingbias.com/2008/06/singularity-out.html.

Hanson, R. (2008e) Catastrophe, social collapse and human extinction, in Bostrom and Ćirković (2008), pp 363–377.

Hanson, R. (2008f) For discount rates, *Overcoming Bias*, January 21, http://www.overcomingbias.com/2008/01/protecting-acro.html.

Hanson, R. (2012) Em need for speed, *Overcoming Bias*, April 16, http://www.overcomingbias.com/2012/04/fast-em-bosses.html.

Hanson, R. (2013) Is social science extremist?, *Overcoming Bias*, February 21, http://www.overcomingbias.com/2013/02/is-social-science-extremist.html.

Hanson, R. (2014a) What will it be like to be an emulation?, in Blackford and Broderick (2014), pp 298–309.

Hanson, R. (2014b) Regulating infinity, *Overcoming Bias*, August 17, http://www.overcomingbias.com/2014/08/regulating-infinity.html.

Hanson, R. (2014c) Should Earth shut the hell up?, *Cato Unbound*, December 3. http://www.cato-unbound.org/2014/12/03/robin-hanson/should-earth-shut-hell.

Hanson, R. (2014d) Pascal's alien wager, *Cato Unbound*, December 10. http://www.cato-unbound.org/2014/12/10/robin-hanson/pascals-alien-wager.

Hanson, R. and Yudkowsky, E. (2013) *The Hanson–Yudkowsky AI-Foom Debate*, Machine Intelligence Research Institute, Berkeley, CA. http://intelligence.org/ai-foom-debate/.

Haqq-Misra, J., Busch, M., Som, S. and Baum, S. (2013) The benefits and harm of transmitting into space, *Space Policy* 29, 40–48.

Hardin G. (1968) The tragedy of the commons, *Science* 162, 1243–1248.

Harford, T. (2006) *The Undercover Economist*, Little, Brown & Co., New York.

Harris, J. (2009) Enhancements are a moral obligation, in Bostrom and Savulescu (2009), pp 131–154.

Hart, M.H. (1975) An explanation for the absence of extraterrestrials on Earth, *Quarterly Journal of the Royal Astronomical Society* 16, 128–135.

Hatton, T.J. and Bray, B.E. (2010) Long run trends in the heights of European men, 19th–20th centuries, *Economics & Human Biology* 8, 405–413.

Hawking, S., Russell, S., Tegmark, M. and Wilczek, F. (2014) Transcendence looks at the implications of artificial intelligence – but are we taking AI seriously enough?, *The Independent*, May 1.

Hays, J.D., Imbrie, J. and Shackleton, N. (1976) Variations in the Earth's orbit: Pacemaker of the ice ages, *Science* 194, 1121–1132.

Hempel, C.G. (1945) Studies in the logic of confirmation I, *Mind* **54**, 1–26.

Heyl, J.S. (2005) The long-term future of space travel, *Physical Review D* **72**, 1–4.

Hickman, L. (2011) Fusion power: Is it getting any closer? *The Guardian*, August 23.

Hinson, J.M. and Staddon, J.E.R. (1983) Matching, maximizing and hill-climbing, *Journal of the Experimental Analysis of Behavior* **40**, 321–331.

Hodges, A. (1983) *Alan Turing: The Enigma*, Burnett Books, London.

Hoffman, D. (1998) Cold-war doctrines refuse to die, *Washington Post*, March 15. http://www.washingtonpost.com/wp-srv/inatl/longterm/coldwar/shatter031598a.htm.

Hoffman, D. (1999) "I Had A Funny Feeling in My Gut," *Washington Post*, February 19. http://www.washingtonpost.com/wp-srv/inatl/longterm/coldwar/shatter021099b.htm.

Hofstadter, D. (1979) *Gödel, Escher, Bach: An Eternal Golden Braid*, Basic Books, New York.

Hofstadter, D. (2007) *I am a Strange Loop*, Basic Books, New York.

Hofstadter, D. and Dennett, D. (1981) *The Mind's I*, Basic Books, New York.

Hume, D. (1739) *A Treatise of Human Nature*, John Noon, London.

Hume, D. (1748) *Philosophical Essays Concerning Human Understanding*, A. Millar, London.

Husfeldt, T. (2015) Universal computation told in quotes, http://thorehusfeldt.net/2015/01/25/universal-computation-told-in-quotes/.

Hutter, M. (2005) *Universal Artificial Intelligence: Sequential Decision Based on Algorithmic Probability*, Springer, New York.

Huxley, A. (1932) *Brave New World*, Chatto & Windus, London.

Ilieva, I., Hook, C. and Farah, M. (2015) Prescription stimulants' effects on healthy inhibitory control, working memory, and episodic memory: A meta-analysis, *Journal of Cognitive Neuroscience*, to appear.

Imbrie, J. and Imbrie, K.P. (1979) *Ice Ages: Solving the Mystery*, Harvard University Press, Cambridge, MA.

Indermühle, A. et al. (1999) Holocene carbon-cycle dynamics based on CO_2 trapped in ice at Taylor Dome, Antarctica, *Nature* **398**, 121–126.

IPCC (2005) *IPCC Special Report on Carbon Dioxide Capture and Storage* (eds Metz, B. et al.), Cambridge University Press, Cambridge, UK.

IPCC (2013) *Climate Change 2013: The Physical Science Basis. Contribution of Working Group I to the Fifth Assessment Report of the Intergovernmental Panel on Climate Change*, http://www.climatechange2013.org/.

Jaffe, R.L., Busza, W., Wilczek, F. and Sandweiss, J. (2000) Review of speculative "disaster scenarios" at RHIC, *Reviews in Modern Physics* **72**, 1125–1140.

Jaynes, E.T. (2003) *Probability Theory: The Logic of Science*, Cambridge University Press, Cambridge, UK.

Jebari, K. (2012) *Crucial Considerations: Essays on the Ethics of Emerging Technologies*, licentiate thesis, Royal Institute of Technology, Stockholm. http://kth.diva-portal.org/smash/get/diva2:573601/FULLTEXT02.

Jebari, K. (2013) Brain machine interface and human enhancement – an ethical review, *Neuroethics* **6**, 617–625.

Jebari, K. (2014a) *Human Enhancement and Technological Uncertainty: Essays on the Promise and Peril of Emerging Technology*, PhD thesis, Royal Institute of Technology, Stockholm.

Jebari, K. (2014b) Of Malthus and Methuselah: Does longevity treatment aggravate global catastrophic risks?, *Physica Scripta* **89**, 128005.

Jobs, S. (2005) Commencement address, *Stanford Report*, June 14. http://news.stanford.edu/news/2005/june15/jobs-061505.html.

Johansson, D., O'Neill, B., Tebaldi, C. and Häggström, O. (2015) Equilibrium climate sensitivity in light of observations over the warming hiatus, *Nature Climate Change*, to appear.

Jones, E. (1985) Where is everybody?, *Physics Today* **38**(8), 11–13. http://scitation.aip.org/content/aip/magazine/physicstoday/article/38/8/10.1063/1.2814654.

Joy, B. (2000) Why the future doesn't need us, *Wired*, April. http://archive.wired.com/wired/archive/8.04/joy.html.

Kahneman, D. (2011) *Thinking, Fast and Slow*, Farrar, Straus and Giroux, New York.

Karp, L. (2005) Global warming and hyperbolic discounting, *Journal of Public Economics* **89**, 261–282.

Kasparov, G. (2010) The chess master and the computer, *New York Review of Books*, February 11.

Kass, L. (1985) *Toward a More Natural Science: Biology and Human Affairs*, Free Press, New York.

Kass, L. (1994) *The Hungry Soul: Eating and the Perfecting of Our Nature*, Simon and Schuster, New York.

Kass, L. (1997) The wisdom of repugnance, *The New Republic*, June 2, 17–26.

Kass, L. (2002) *Life, Liberty, and Defense of Dignity: The Challenge for Bioethics*, Encounter Books, San Francisco.

Kates, R. (1962) Hazard and choice perception in flood plain management, Research paper No. 87, Dept. of Geography, University of Chicago, Chicago, IL. http://ipcc-wg2.gov/njlite_download.php?id=7421.

Keith, D. (2000) Geoengineering the climate: History and prospect, *Annual Review of Energy and the Environment* **25**, 245–284.

Keith, D. (2009) Why capture CO_2 from the atmosphere?, *Science* **325**, 1654–1655.

Keith, D. (2013) *A Case for Climate Engineering*, MIT Press, Cambridge, MA.

Kemeny, J. (1955) Man viewed as a machine, *Scientific American* **192**, 58–67.

Keynes, J.M. (1931) Economic possibilities for our grandchildren, in *Essays in Persuasion*, Macmillan, London.

Kilbourne, E.D. (2008) Plagues and pandemics: Past, present and future, in Bostrom and Ćirković (2008), pp 287–307.

Knutti, R. and Hegerl, G. (2008) The equilibrium sensitivity of the Earth's temperature to radiation changes, *Nature Geoscience* **1**, 735–743.

Koch, C. (2012) *Consciousness: Confessions of a Romantic Reductionist*, MIT Press, Cambridge, MA.

Komlos, J. and Baur, M. (2004) From the tallest to (one of) the fattest: The enigmatic fate of the American population in the 20th century, *Economics & Human Biology* **2**, 57–74.

Konopinski, E.J., Marvin C. and Teller, E. (1946) Ignition of the atmosphere with nuclear bombs, Report LA-602, Los Alamos Laboratory, http://library.lanl.gov/cgi-bin/getfile?00329010.pdf.

Kringelbach, M., Jenkinson, N., Owen, S. and Aziz, T. (2007) Translational principles of deep brain stimulation, *Nature Reviews Neuroscience* **8**, 623–635.

Krugman, P. (1996) Ricardo's difficult idea, http://web.mit.edu/krugman/www/ricardo.htm.

Krugman, P. (1997) The accidental theorist, *Slate*, January 24. http://www.slate.com/articles/business/the_dismal_science/1997/01/the_accidental_theorist.html.

Kurzweil, R. (1999) *The Age of Spiritual Machines: When Computers Exceed Human Intelligence*, Viking, New York.

Kurzweil, R. (2005) *The Singularity Is Near: When Humans Transcend Biology*, Viking, New York.

Kurzweil, R. (2006) Nanotechnology dangers and defenses, *Kurzweil Accelerating Intelligence*, March 27, http://www.kurzweilai.net/nanotechnology-dangers-and-defenses.

Lacis, A., Schmidt, G., Rind, D. and Ruedy, R. (2010) Atmospheric CO_2: Principal control knob governing Earth's temperature, *Science* 330, 356–359.

Latour, B. (1987) *Science in Action: How to Follow Scientists and Engineers through Society*, Harvard University Press, Cambridge, MA.

Lehmann, E. and Romano, J. (2008) *Testing Statistical Hypotheses* (third edition), Springer, New York.

Lehnert, B. (2013) Half a century of fusion research towards ITER, *Physica Scripta* 87, 018201.

Leibniz, G.W. (1714) *La Monadologie*. http://classiques.uqac.ca/classiques/Leibniz/La_Monadologie/leibniz_monadologie.pdf.

Leslie, J. (1998) *The End of the World: The Science and Ethics of Human Extinction*, Routledge, New York.

Levitt, S. and Dubner, S. (2005) *Freakonomics: A Rogue Economist Explores the Hidden Side of Everything*, William Morrow, New York.

Levitt, S. and Dubner, S. (2009) *SuperFreakonomics: Global Cooling, Patriotic Prostitutes, and Why Suicide Bombers Should Buy Life Insurance*, William Morrow, New York.

Levy, F. and Murnane, R.J. (2004) *The New Division of Labor: How Computers are Creating the Next Job Market*, Princeton University Press, Princeton, NJ.

Liao, M., Sandberg, A. and Roache, R. (2012) Human engineering and climate change, *Ethics, Policy & Environment* 15, 206–221.

Lindley, D. (1990) The 1988 Wald Memorial Lectures: the present position in Bayesian statistics (with comments and a rejoinder by the author), *Statistical Science* 5, 44–89.

Lloyd, S. (2000) Ultimate physical limits to computation, *Nature* 406, 1047–1054.

Long, R. (2006) Review of Hilary Putnam's The Collapse of the Fact/Value Dichotomy and Other Essays, *Reason Papers* 28, 125–131.

López-Otín, C., Blasco, M.A., Partridge, L., Serranon, M. and Kroemer, G. (2013) The hallmarks of aging, *Cell* 153, 1194–1217.

Lynas, M. (2011) *The God Species: Saving the Planet in the Age of Humans*, Fourth Estate, London.

Macklin, R. (2003) Dignity is a useless concept, *BMJ* 327, 1419–1420.

Markoff, J. (2011) Armies of expensive lawyers, replaced by cheaper software, *New York Times*, March 4.

Markoff, J. (2013) Obama seeking to boost study of human brain, *New York Times*, February 17.

Marks, D. (2010) IQ variations across time, race and nationality: An artifact of differences in literary skills, *Psychological Reports* 106, 643–664.

Martin, J. (2006) *The Meaning of the 21st Century: A Vital Blueprint for Ensuring our Future*, Transworld, London.

Matloff, G. (2012) Deflecting asteroids, *IEEE Spectrum*, March 28. http://spectrum.ieee.org/aerospace/space-flight/deflecting-asteroids.

McCarthy, J., Minsky, M., Rochester, N. and Shannon, C. (1955) A Proposal for the Dartmouth Summer Research Project on Artificial Intelligence, http://www-formal.stanford.edu/jmc/history/dartmouth/dartmouth.html.

McCue, T.J. (2014) 3D printed prosthetics, *Forbes*, August 31.

McGinn, C. (2004) *Consciousness and its Objects*, Clarendon, Oxford.

McKibben, B. (2003) *Enough: Genetic Engineering and the End of Human Nature*, Bloomsbury, London.

McKibben, B. (2009) Can 350.org save the world? *Los Angeles Times*, May 15.

McNeil, D. (2014) White House to cut funding for risky biological study, *New York Times*, October 17.

Merkle, R. (1994) The molecular repair of the brain, *Cryonics* 15, no. 1 and 2. http://www.merkle.com/cryo/techFeas.html.

Metzinger, T. (2000) *Neural Correlates of Consciousness: Empirical and Conceptual Questions*, MIT Press, Cambridge, MA.

Metzinger, T. (2003) *Being No One*, MIT Press, Cambridge, MA.

Michaud, M. (2003) Ten decisions that could shake the world, *Space Policy* 19, 131–136.

Midgley, M. (2012) Death and the human animal, *Philosophy Now*, March/April.

Miller, D. (2003) Axiological actualism and the converse intuition, *Australian Journal of Philosophy* 81, 123–125.

Miller, J. (2012) *Singularity Rising: Surviving and Thriving in a Smarter, Richer, and More Dangerous World*, Benbella, Dallas, TX.

Milman, O. (2013) Top 20 things politicians need to know about science, *The Guardian*, November 20.

Minsky, M. (1967) *Computation: Finite and Infinite Machines*, Prentice-Hall, Englewood Cliffs, NJ.

Mitchell, J. (2014) On the emptiness of failed replications, http://wjh.harvard.edu/~jmitchel/writing/failed_science.htm.

Monbiot, G. (2006) *Heat: How to Stop the Planet Burning*, Allen Lane, London.

Moody, T. (1994) Conversations with zombies, *Journal of Consciousness Studies* 1, 196–200.

Mooney, C. and Kirshenbaum, S. (2009) *Unscientific America: How Scientific Illiteracy Threatens our Future*, Basic Books, New York.

Moore, G. (1965) Cramming more components onto integrated circuits, *Electronics Magazine* 38, 114117.

More, M. (2005) The proactionary principle, Version 1.2, July 29, http://www.maxmore.com/proactionary.html.

Moreno-Cruz, J., Ricke, K. and Keith, D. (2012) A simple model to account for regional inequalities in the effectiveness of solar radiation management, *Climatic Change* 110, 649–668.

Muehlhauser, L. (2011) *Facing the Intelligence Explosion*, http://intelligenceexplosion.com/2011/preface/.

Muehlhauser, L. (2013) When will AI be created?, Machine Intelligence Research Institute, http://intelligence.org/2013/05/15/when-will-ai-be-created/.

Muehlhauser, L. (2014) 3 misconceptions in Edge.orgs conversation on "The Myth of AI," Machine Intelligence Research Institute, November 18, https://intelligence.org/2014/11/18/misconceptions-edge-orgs-conversation-myth-ai/.

Müller, V. and Bostrom, N. (2014) Future progress in artificial intelligence: A poll among experts [short version], *AI Matters* 1(1), 9–11.

Munthe, C. (2015) Why aren't existential risk/ultimate harm argument advocates all attending mass?, *Philosophical Comment*, February 1, http://philosophicalcomment.blogspot.se/2015/02/why-arent-existential-risk-ultimate.html.

Murphy, T. (2011a) Why not space?, *Do the Math*, October 12, http://physics.ucsd.edu/do-the-math/2011/10/why-not-space/.

Murphy, T. (2011b) Stranded resources, *Do the Math*, October 25, http://physics.ucsd.edu/do-the-math/2011/10/stranded-resources/.

Nagel, T. (1997) *The Last Word*, Oxford University Press, Oxford.

Nair, P. (2008) As IVF becomes more common, some concerns remain, *Nature Medicine* 14, 1171.

Napier, W. (2008) Hazards from comets and asteroids, in Bostrom and Ćirković (2008), pp 222–237.

Needleman, R. (2014) Such a doll: Get yourself scanned and printed in 3D, *Yahoo Tech*, April 23. https://www.yahoo.com/tech/such-a-doll-get-yourself-scanned-and-printed-in-3d-83592431558.html

Neisser, U. (1997) Rising scores on intelligence tests, *American Scientist* 85, 440–447.

von Neumann, J. and Morgenstern, O. (1944) *The Theory of Games and Economic Behavior*, Princeton University Press, Princeton.

Nordhaus, W. (2007) A review of the Stern Review on the Economics of Climate Change, *Journal of Economic Literature* 45, 686–702.

Nozick, R. (1974) *Anarchy, State, and Utopia*, Basic Books, New York.

Olofsson, P. and Andersson, M. (2012) *Probability, Statistics, and Stochastic Processes* (second edition), Wiley, New York.

Olum, K. (2002) The doomsday argument and the number of possible observers, *Philosophical Quarterly* 52, 164–184.

Omohundro, S. (2008) The basic AI drives, *Artificial General Intelligence 2008: Proceedings of the First AGI Conference* (eds Wang, P., Goertzel, B. and Franklin, S.), IOS, Amsterdam, pp 483–492.

Omohundro, S. (2012) Rational artificial intelligence for the greater good, in Eden et al. (2012), pp 161–175.

Ord, T., Hillerbrand, R. and Sandberg, A. (2010) Probing the improbable: Methodological challenges for risks with low probabilities and high stakes, *Journal of Risk Research* 13, 191–205.

Oreskes, N. and Conway, E. (2010) *Merchants of Doubt: How a Handful of Scientists Obscured the Truth on Issues from Tobacco Smoke to Global Warming*, Bloomsbury, New York.

Ottaviani J. and Myrick, L. (2011) *Feynman*, First and Second, New York.

Pappano, L. (2012) The year of the MOOC, *New York Times*, November 2.

Parfit, D. (1984) *Reasons and Persons*, Oxford University Press, Oxford.

Parfit, D. (2011) *On What Matters*, Oxford University Press, Oxford.

Parnas, D. (2007) Stop the numbers game, *Communications of the ACM* 50(11), 19–21.

PBL Netherlands Environmental Assessment Agency (2011) *Long-Term Trend in Global CO_2 Emission*, http://www.pbl.nl/en/publications/2011/long-term-trend-in-global-co2-emissions-2011-report.

Pellegrino, E. et al. (2008) *Human Dignity and Bioethics: Essays Commissioned by the President's Council on Bioethics*, President's Council on Bioethics, Washington, DC. https://bioethicsarchive.georgetown.edu/pcbe/reports/human_dignity/.

Penrose, R. (1989) *The Emperor's New Mind: Concerning Computers, Minds and the Laws of Physics*, Oxford University Press, Oxford.

Perlman, R.M. (1954) The aging syndrome, *Journal of the American Geriatrics Society* 2, 123–129.

Persson, I. and Savulescu, J. (2008) The peril of cognitive enhancement and the urgent imperative to enhance the moral character of humanity, *Journal of Applied Philosophy* 25, 162–177.

Persson, U. (2014) *Karl Popper, falsifieringens profet*, CKM Förlag, Stockholm.

Petragila, M. et al. (2007) Middle paleolithic assemblages from the Indian subcontinent before and after the Toba super-eruption, *Science* 317, 114–116.

Phoenix, C. and Drexler, E. (2004) Safe exponential manufacturing, *Nanotechnology* 15, 869–872.

Phoenix, C. and Treder, M. (2008) Nanotechnology as global catastrophic risk, in Bostrom and Ćirković (2008), pp 481–503.

Pierrehumbert, R. (2009) An open letter to Steve Levitt, *RealClimate*, October 29, http://www.realclimate.org/index.php/archives/2009/10/an-open-letter-to-steve-levitt/.

Pigliucci, M. (2010) *Nonsense on Stilts: How to Tell Science from Bunk*, University of Chicago Press, Chicago, IL.

Pigliucci, M. (2014a) If there's a movement that resembles . . ., *Twitter*, August 7. https://twitter.com/mpigliucci/status/497421238554603520.

Pigliucci, M. (2014b) The plausibility of cryonics doesn't quite . . ., *Twitter*, August 7. https://twitter.com/mpigliucci/status/497424869475483648.

Pigliucci, M. (2014c) Uploading: A philosophical counter-analysis, in Blackford and Broderick (2014), pp 119–130.

Piketty, T. (2014) *Capital in the Twenty-First Century*, Harvard University Press, Cambridge, MA.

Pinker, S. (2008) The stupidity of dignity, *The New Republic*, May 28. http://claradoc.gpa.free.fr/doc/73.pdf.

Pinker, S. (2011) *The Better Angels of Our Nature: Why Violence Has Declined*, Viking, New York.

Pinker, S. (2014) Response to Jaron Lanier's "The Myth of AI," *Edge*, November 19, http://edge.org/conversation/the-myth-of-ai.

Popper, K. (1934) *Logik der Forschung: Zur Erkenntnistheorie der modernen Naturwissenschaft*, Mohr Siebeck, Tübingen.

Popper, K. (1959) *The Logic of Scientific Discovery*, Hutchinson, London.

Popper, K. (1994) *The Myth of the Framework: In Defence of Science and Rationality*, Routledge, London.

Potsdam Institute for Climate Impact Research and Climate Analytics (2012) *Turn Down the Heat: Why a 4° Warmer Planet Must be Avoided*, The World Bank, Washington DC. http://www.worldbank.org/en/news/feature/2012/11/18/Climate-change-report-warns-dramatically-warmer-world-this-century.

Rampino, M. (2008) Super-volcanism and other geophysical processes of catastrophic import, in Bostrom and Ćirković (2008), pp 205–221.

Ramsey, F. (1931) Truth and probability, in *Foundations of Mathematics and Other Logical Essays* (ed. Frank Ramsey), Harcourt, New York.

Rathmanner, S. and Hutter, M. (2011) A philosophical treatise of universal induction, *Entropy* 13, 1076–1136.

Rees, M. (2002) By 2020, bioterror or bioerror will lead to one million casualties in a single event, *Long Bets*, http://longbets.org/9/.

Rees, M. (2003) *Our Final Century: Will the Human Race Survive the Twenty-first Century?*, William Heinemann, London.

Ricroch, A., Bergé, J. and Kuntz, M. (2011) Evaluation of genetically engineered crops using transcriptomic, proteomic, and metabolomic profiling techniques, *Plant Physiology* 155, 1752–1761.

Riha, D. (2014) Geek tech: Apollo guidance computer vs. iPhone 5s, *The Daily Crate*, February 1.

Ritter, M. (2012) Bird flu: Study published after terrorism debate, *Huffington Post*, June 21.

Roache, R. and Clarke, S. (2009) Bioconservatism, bioliberalism, and the wisdom of reflecting on repugnance, *Monash Bioethics Review* 26, paper 04.

Rockström, J. et al. (2009) A safe operating zone for humanity, *Nature* 461, 472–475.

Romportl, J., Ircing, P., Zackova, E., Polak, M. and Schuster, R. (2012) *Beyond AI: Artificial Dreams*, University of West Bohemia, Pilsen. http://www.kky.zcu.cz/en/publications/1/JanRomportl_2012_BeyondAIArtificial.pdf.

Rothstein, B. (2005) *Social Traps and the Problem of Trust*, Cambridge University Press, Cambridge, UK.

Royal Society (2005) *Ocean Acidification Due to Increasing Atmospheric Carbon Dioxide*, http://royalsociety.org/policy/publications/2005/ocean-acidification/.

Royal Society (2009) *Geoengineering the Climate*, http://royalsociety.org/policy/publications/2009/geoengineering-climate/.

Russell, B. (1945) *A History of Western Philosophy*, George Allen & Unwin, London.

Russell, P. (1926) *Benjamin Franklin, the First Civilized American*, Brentano's, New York.

Saberhagen, F. (1967) *Berserker*, Ballantine, New York.

Sagan, C. (1995) *The Demon-Haunted World: Science as a Candle in the Dark*, Random House, New York.

Sagan, C. and Newman, W. (1983) The solipsist approach to extraterrestrial intelligence, *Quarterly Journal of the Royal Astronomical Society* 24, 113–121.

Salsburg, D. (2001) *The Lady Tasting Tea: How Statistics Revolutionized Science in the Twentieth Century*, W.H. Freeman, New York.

Sample, I. (2013) Scientists reveal the full power of the Chelyabinsk meteor explosion, *The Guardian*, November 7.

Sandberg, A. (1996) Dyson sphere FAQ, http://www.aleph.se/Nada/dysonFAQ.html.

Sandberg, A. (2001) Morphological freedom – why we not just want it, but need it, http://www.aleph.se/Nada/Texts/MorphologicalFreedom.htm.

Sandberg, A. (2013) Secret snakes biting their own tails: Secrecy and surveillance, *Oxford Martin School*, June 12, http://www.oxfordmartin.ox.ac.uk/opinion/view/215.

Sandberg, A. (2014a) Ethics of brain emulations, *Journal of Experimental and Theoretical Artificial Intelligence* 26, 439–457.

Sandberg, A. (2014b) Being nice to software animals and babies, in Blackford and Broderick (2014), pp 279–297.

Sandberg, A. (2014c) The five biggest threats to human existence, *The Conversation*, May 29, http://theconversation.com/the-five-biggest-threats-to-human-existence-27053.

Sandberg, A. (2014d) Cool risks outside the envelope of nature, *Andart II*, October 22, http://aleph.se/andart2/risk/existential-risk-risk/cool-risks-outside-the-envelope-of-nature/.

Sandberg, A. and Bostrom, N. (2008a) Global catastrophic risks survey, Future of Humanity Institute technical report #2008-1.

Sandberg, A. and Bostrom, N. (2008b) Whole brain emulation: A roadmap, Future of Humanity Institute technical report #2008-3.

Sandel, M. (2004) The case against perfection: What's wrong with designer children, bionic athletes, and genetic engineering, *The Atlantic Monthly*, April.

Sankaran, A.V. (2003) Neoproterozoic "snowball earth" and the "cap" carbonate controversy, *Current Science* 84, 871–873.

Savage, L. (1951) *The Foundations of Statistics*, Wiley, New York.

Savulescu, J., Bostrom, N. and de Grey, A. (2009) Why we need a war on aging, *Practical Ethics*, January 31. http://blog.practicalethics.ox.ac.uk/2009/01/why-we-need-a-war-on-aging/.

Savulescu, J. and Persson, I. (2012) Moral enhancement, *Philosophy Now*, July/August.

Schmidt, G. (2006) Current volcanic activity and climate?, *Real Climate*, May 16, http://www.realclimate.org/index.php/archives/2006/05/current-volcanic-activity-and-climate/.

Schröder, K.-P. and Smith, R. (2008) Distant future of the Sun and Earth revisited, *Monthly Notices of the Royal Astronomical Society* 386, 155–163.

Schulte, P. et al. (2010) The Chicxulub asteroid impact and mass extinction at the Cretaceous–Paleogene boundary, *Science* 327, 1214–1218.

Scully, T. (2012) To the limit, *Nature* 492, S2–S3.

Searle, J. (1964) How to derive "ought" from "is," *Philosophical Review* 73, 43–58.

Searle, J. (1980) Minds, brains and programs, *Behavioral and Brain Sciences* 3, 417–457.

Searle, J. (1982a) The myth of the computer, *New York Review of Books*, April 29.

Searle, J. (1982b) The myth of the computer: An exchange (reply to Dennett), *New York Review of Books*, June 24.

Searle, J. (2004) *Mind: A Brief Introduction*, Oxford University Press, Oxford.

Searle, J. (2014) What your computer can't know, *New York Review of Books*, October 9.

SETI Permanent Committee (1989) Declaration of Principles Concerning Activities Following the Detection of Extraterrestrial Intelligence, International Academy of Astronautics. http://avsport.org/IAA/protdet.htm.

Sharvy, R. (1985) It ain't the meat it's the motion, *Inquiry* 26, 125–134.

Shaw, G.B. (1921) *Back to Mathuselah*, Constable, London.

Shearer, D., Mulvhill, B., Klerman, L., Wallander, J., Hovinga, M. and Redden, D. (2002) Association of early childbearing and low cognitive ability, *Perspectives on Sexual and Reproductive Health* 34, 236–243.

Shelley, M. (1818) *Frankenstein; or, The Modern Prometheus*, Lackington, Hughes, Harding, Mavor & Jones, London.

Shermer, M. (2004) *The Science of Good and Evil*, Henry Holt, New York.

Shulman, C. (2008) "Evicting" brain emulations, *Overcoming Bias*, November 23, http://www.overcomingbias.com/2008/11/suppose-that-ro.html.

Shulman, C. (2012) Future filter fatalism, *Overcoming Bias*, December 22. http://www.overcomingbias.com/2012/12/future-filter-fatalism.html.

Shulman, C. and Bostrom, N. (2012) How hard is artificial intelligence? Evolutionary arguments and selection effects, *Journal of Consciousness Studies* 19, 103–130.

Shulman, C. and Bostrom, N. (2014) Embryo selection for cognitive enhancement: Curiosity or game-changer?, *Global Policy* 5, 85–92.

Silver, N. (2012) *The Signal and the Noise: The Art and Science of Prediction*, Allen Lane, London.

Simon, H. (1965) *The Shape of Automation for Men and Management*, Harper & Row, New York.

Singer, P. (2005) Ethics and intuitions, *The Journal of Ethics* 9, 331–352.

Skyrms, B. (1984) *Pragmatics and Empiricism*, Yale University Press, New Haven.

Smalley, R. (2001) Of chemistry, love and nanobots, *Scientific American* 285(3), 76–77.

Smith, R.W. (1989) The Cambridge network in action: The discovery of Neptune, *Isis* 80, 395–422.

Smith, W.J. (2013) That new time transhumanism religion, *National Review Online*, June 28.

Snell, M. (2003) The spread of the Black Death through Europe, *About Education*, http://historymedren.about.com/od/theblackdeath/a/black_death_maps.htm.

Soifer, R. (2014) Humans can always pull the plug, *Financial Times*, July 16.

Sokal, A. (2008) *Beyond the Hoax: Science, Philosophy and Culture*, Oxford University Press, Oxford.

Solomonoff, R. (1985) The time scale of artificial intelligence: Reflections on social effects, *North-Holland Human Systems Management* 5, 149–153.

Sotala, K. (2013) A brief history of ethically concerned scientists, *Less Wrong*, February 9. http://lesswrong.com/lw/gln/a_brief_history_of_ethically_concerned_scientists/.

Sotala, K. and Yampolskiy, R. (2015) Responses to catastrophic AGI risk: A survey, *Physica Scripta* 90, 018001.

Sparrow, R. (2013) In vitro eugenics, *Journal of Medical Ethics*, doi:10.1136/medethics-2012-101200.

Stanford, K. (2013) Underdetermination of scientific theory, *The Stanford Encyclopedia of Philosophy* (ed. Zalta, E.), http://plato.stanford.edu/archives/win2013/entries/scientific-underdetermination/.

Steinfeld, H., Gerber, P., Wassenaar, T., Castel, V., Rosales, M. and de Haan, C. (2006) *Livestock's Long Shadow*, Food and Agriculture Organization of the United Nations, Rome.

Stern, N. (2007) *The Economics of Climate Change: The Stern Review*, Cambridge University Press, Cambridge, UK.

Strannegård, C., Amirghasemi, M. and Ulfsbäcker, S. (2013) An anthropomorphic method for number sequence problems, *Cognitive Systems Research* 22–23, 27–34.

Strannegård, C., Nizamani, A.R., Engström, F. and Häggström, O. (2014) Symbolic reasoning with bounded cognitive resources, *36th Annual Conference of the Cognitive Science Society*, to appear. http://gup.ub.gu.se/records/fulltext/199298/199298.pdf.

Strogatz, S. (2010) The Hilbert Hotel, *New York Times*, May 9.

Sumpter, D. (2013) Why "intelligence explosion" and many other futurist arguments are nonsense, *Häggström hävdar*, October 22, http://haggstrom.blogspot.se/2013/10/guest-post-by-david-sumpter-why.html.

Sutherland, W., Spiegelhalter, D. and Burgman, M. (2013) Twenty tips for interpreting scientific claims, *Nature* 503, 335–337.

Swift, J. (1729) *A Modest Proposal*, http://www.gutenberg.org/files/1080/1080-h/1080-h.htm.

Takahashi, K. and Yamanaka, S. (2006) Induction of pluripotent stem cells from mouse embryonic and adult fibroblast cultures by defined factors, *Cell* 126, 663–676.

Takahashi, K., Tanabe, K., Ohnuki, M., Narita, M., Ichisaka, T., Tomoda, K. and Yamanaka, S. (2007) Induction of pluripotent stem cells from adult human fibroblasts by defined factors, *Cell* 131, 861–872.

Taleb, N.N. (2007) *The Black Swan: The Impact of the Highly Improbable*, Random House, New York.

Taleb, N.N., Read, R., Douady, R., Norman, J. and Bar-Yam, Y. (2014) The precautionary principle (with application to the genetic modification of organisms), NYU School of Engineering working paper series, http://www.fooledbyrandomness.com/pp2.pdf.

Tännsjö, T. (2010) *Privatliv*, Fri Tanke, Stockholm.

Tegmark, M. (2014a) *Our Mathematical Universe: My Quest for the Ultimate Nature of Reality*, Knopf, New York.

Tegmark, M. (2014b) Friendly artificial intelligence: The physics challenge, *arXiv* 1409.0813, http://arxiv.org/abs/1409.0813.

The Harward Crimson (2005) Psychoanalysis Q-and-A with Steven Pinker, January 19. http://www.thecrimson.com/article.aspx?ref=505366.

Tilmes, S., Müller, R. and Salawitch, R. (2008) The sensitivity of polar ozone depletion to proposed geo-engineering schemes, *Science* 320, 1201–1204.

Tomasik, B. (2009) The importance of wild-animal suffering, http://foundational-research.org/publications/importance-of-wild-animal-suffering/.

Tomasik, B. (2013) Applied welfare biology and why wild-animal advocates should focus on not spreading nature, http://reducing-suffering.org/applied-welfare-biology-wild-animal-advocates-focus-spreading-nature/.

Toon, O.B., Turco, R.P., Robock, A., Bardeen, C., Oman, L. and Stenchikov, G.L. (2007) Atmospheric effects and societal consequences of regional scale nuclear conflicts and acts of individual nuclear terrorism, *Atmospheric Chemistry and Physics* 7, 1973–2002.

Trachtenberg, Z. (2012) Human engineering and the value of autonomy, *Ethics, Policy & Environment* 15, 244–247.

Tsien, J.Z. (2007) The memory code, *Scientific American* 297, 52–59.

Turing, A. (1936) On computable numbers with an application to the Entscheidungsproblem, *Proceedings of the London Mathematical Society* 42, 230–267.

Turing, A. (1950) Computing machinery and intelligence, *Mind* LIX(236), 433–460.

Turing, A. (1951) Intelligent machinery: A heretical theory, BBC http://philmat.oxfordjournals.org/content/4/3/256.

Turner, M.S. and Wilczek, F. (1982) Is our vacuum metastable?, *Nature* 298, 635–636.

Tversky, A. and Kahneman, D. (1983) Extensional versus intuitive reasoning: The conjunction fallacy in probability judgment, *Psychological Review* 90, 293–315.

Urmson, C. (2014) The latest chapter for the self-driving car: Mastering city street driving, *Google Official Blog*, April 28, http://googleblog.blogspot.se/2014/04/the-latest-chapter-for-self-driving-car.html.

Vakoch, D. (2014) The importance of active SETI, *Cato Unbound*, December 8. http://www.cato-unbound.org/2014/12/08/douglas-vakoch/importance-active-seti.

Veness, J., Ng, K.S., Hutter, M., Uther, W. and Silver, D. (2011) A Monte-Carlo AIXI approximation, *Journal of Artificial Intelligence Research* **40**, 95–142.

Vickers, J. (2014) The Problem of Induction, *The Stanford Encyclopedia of Philosophy* (ed. Zalta, E.), http://plato.stanford.edu/archives/sum2014/entries/induction-problem/.

Vinge, V. (1993) The coming technological singularity: How to survive in the post-human era, in *Vision-21: Interdisciplinary Science and Engineering in the Era of Cyberspace*, Nasa Lewis Research Center, Cleveland, OH, pp 11–22.

Vuolo, M. (2013) Let's resolve in the new year to stop using that expression about breaking eggs and making omelets, *Lexicon Valley*, December 30.

de Waal, F. (2006) *Primates and Philosophers: How Morality Evolved*, Princeton University Press, Princeton, NJ.

Wai, J. and Putallaz, M. (2011) The Flynn effect puzzle: A 30-year examination from the right tail of the ability distribution provides some missing pieces, *Intelligence* **39**, 443–455.

Wanjek, C. (2006) Milky Way churns out seven new stars per year, scientists say, *Goddard Space Flight Center*, http://www.nasa.gov/centers/goddard/news/topstory/2006/milkyway_seven.html.

Warwick, K. (2004) *I Cyborg*, University of Illinois Press, Champaign, IL.

Weart, S. (2003) *The Discovery of Global Warming*, Harvard University Press, Cambridge, MA.

Webb, S. (2002) *If the Universe Is Teeming with Aliens . . . Where is Everybody? Fifty Solutions to the Fermi Paradox and the Problem of Extraterrestrial Life*, Copernicus Books, New York.

Wehrwein, P. (2012) Repeat to fade, *Nature* **492**, S12–S13.

Weitzman, M. (1998) Recombinant growth, *Quarterly Journal of Economics* **CXIII**, 331–360.

Weitzman, M. (2001) Gamma discounting, *American Economic Review* **91**, 260–271.

Weitzman, M. (2007) A review of The Stern Review on the Economics of Climate Change, *Journal of Economic Literature* **XLV**, 703–724.

Weitzman, M. (2011) Fat-tailed uncertainty in the economics of catastrophic climate change, *Reviews of Environmental Economics and Policy* **5**, 275–292.

Weizenbaum, J. (1966) ELIZA – a computer program for the study of natural language communication between man and machine, *Communications of the ACM* **9**, 36–45.

Whitmire, D. and Wright, D. (1980) Nuclear waste spectrum as evidence of technological extraterrestrial civilizations, *Icarus* **42**, 149–156.

WHO (2014) *Global Health Observatory Data Repository*, http://apps.who.int/gho/data/node.main.688?lang=en.

Wijkman, A. and Rockström, J. (2011) *Den stora förnekelsen*, Medströms, Stockholm.

Wilczek, F. (2008) Big troubles, imagined and real, in Bostrom and Ćirković (2008), pp 346–362.

Witten, E. (1984) Cosmic separation of phases, *Physical Review D* **30**, 272–285.

Wolford, G., Miller, M.B. and Gazzaniga, M. (2000) The left hemisphere's role in hypothesis formation, *Journal of Neuroscience* **20**, RC64.

Wolpe, P.R. (2002) Treatment, enhancement, and the ethics of neurotherapeutics, *Brain and Cognition* **50**, 387–395.

Woo, H. (2013) Planets abound, *Caltech News*, February 1. http://www.caltech.edu/content/planets-abound.

Wood, G. (2009) Re-engineering the Earth, *The Atlantic*, July/August. http://www.theatlantic.com/magazine/archive/2009/07/re-engineering-the-earth/307552/.

Wright, J.T., Mullan, B., Sigurdsson, S. and Povich, M.S. (2014) The $\hat{G}$ infrared search for extraterrestrial civilizations with large energy supplies. I. Background and justification, *The Astrophysical Journal* **792**, 26.

Yeomans, D. and Chodas, P. (2013) Additional details on the large fireball event over Russia on Feb. 15, 2013, *NASA Near Earth Object Program*, March 1. http://neo.jpl.nasa.gov/news/ fireball_130301.html.

Yong, E. (2012) Mutant-flu paper published, *Nature* **485**, 13–14.

Yong, E. (2013) Will we ever . . . simulate the human brain?, *BBC*, February 8, http://www.bbc. com/future/story/20130207-will-we-ever-simulate-the-brain.

Yudkowsky, E. (1996) Staring into the Singularity, http://yudkowsky.net/obsolete/singularity. html.

Yudkowsky, E. (2004) Coherent extrapolated volition, Singularity Institute, https://intelligence. org/files/CEV.pdf.

Yudkowsky, E. (2007) Levels of organization in general intelligence, in *Artificial General Intelligence* (eds Goertzel, B. and Pennachin, C.), Springer, Berlin, pp 389–501.

Yudkowsky, E. (2008a) Cognitive biases potentially affecting judgement of global risks, in Bostrom and Ćirković (2008), pp 91–119.

Yudkowsky, E. (2008b) Artificial intelligence as a positive and negative factor in global risk, in Bostrom and Ćirković (2008), pp 308–345.

Yudkowsky, E. (2008c) The weak inside view, *Less Wrong*, November 18, http://lesswrong.com/ lw/vz/the_weak_inside_view/.

Yudkowsky, E. (2008d) Hard takeoff, *Less Wrong*, December 2, http://lesswrong.com/lw/wf/ hard_takeoff/.

Yudkowsky, E. (2008e) Engelbart: insufficiently recursive, *Less Wrong*, November 26, http:// lesswrong.com/lw/w8/engelbart_insufficiently_recursive/.

Yudkowsky, E. (2008f) Timeless identity, *Less Wrong*, June 3, http://lesswrong.com/lw/qx/ timeless_identity/.

Yudkowsky, E. (2008g) Against discount rates, *Less Wrong*, January 21, http://lesswrong.com/ lw/n2/against_discount_rates/.

Yudkowsky, E. (2009) *Three Worlds Collide*, http://robinhanson.typepad.com/files/ three-worlds-collide.pdf.

Yudkowsky, E. (2011) Complex value systems are required to realize valuable futures, in *Artificial General Intelligence: 4th International Conference, AGI 2011* (eds Schmidhüber, J., Thórisson, R. and Looks, M.), Springer, Berlin, 388–393.

Yudkowsky, E. (2013a) *Intelligence Explosion Microeconomics*, Machine Intelligence Research Institute, Berkeley, CA. http://intelligence.org/files/IEM.pdf.

Yudkowsky, E. (2013b) Five theses, two lemmas, and a couple of strategic implications, Machine Intelligence Research Institute, May 5, http://intelligence.org/2013/05/05/ five-theses-two-lemmas-and-a-couple-of-strategic-implications/.

Zaitsev, A. (2011) METI: Messaging to ExtraTerrestrial Intelligence, in *Searching for Extraterrestrial Intelligence: SETI Past, Present, and Future* (ed. Shuch, P.), Springer, New York, pp 399–428.

Ziliak, S. and McCloskey, D. (2008) *The Cult of Statistical Significance: How the Standard Error Costs Us Jobs, Justice and Lives*, University of Michigan Press, Ann Arbor, MI.

INDEX